Integrating Smart Grids and Renewable Technology

Volume I

Integrating Smart Grids and Renewable Technology
Volume I

Edited by **Benjamin Wayne**

NY RESEARCH PRESS
New York

Published by NY Research Press,
23 West, 55th Street, Suite 816,
New York, NY 10019, USA
www.nyresearchpress.com

Integrating Smart Grids and Renewable Technology: Volume I
Edited by Benjamin Wayne

International Standard Book Number: 978-1-63238-298-6 (Hardback)

Printed in China.

Contents

	Preface	**IX**
Chapter 1	**A Power Electronic Conditioner Using Ultracapacitors to Improve Wind Turbine Power Quality** Murali Bottu, Mariesa L. Crow, Andrew C. Elmore, Stan Atcitty	**1**
Chapter 2	**Wind Energy Conversion Systems Adapted to the Tunisian Sites** Fathi Ben Amar, Mustapha Elamouri	**8**
Chapter 3	**Efficient Power Conditioner for a Fuel Cell Stack-Ripple Current Reduction Using Multiphase Converter** Sampath Naveen Kumar, Natarajan Rajalakshmi, Kaveripatnam Samban Dhathathreyan	**20**
Chapter 4	**Experimental Research on Monomial Cooling Measure of Greenhouse in Summer** Jian Hu	**24**
Chapter 5	**Hybrid Prediction Method for Solar Power Using Different Computational Intelligence Algorithms** Md Rahat Hossain, Amanullah Maung Than Oo, A. B. M. Shawkat Ali	**29**
Chapter 6	**Design and Performance Analysis of an Innovative Single Basin Solar NanoStill** Moses Koilraj Gnanadason, Palanisamy Senthil Kumar, Vincent H. Wilson, Gajendiran Hariharan, Navaneethakrishnan Shenbaga Vinayagamoorthi	**41**
Chapter 7	**Modeling, Control & Fault Management of Microgrids** Mehdi Moradian, Faramarz Mahdavi Tabatabaei, Sajad Moradian	**52**
Chapter 8	**Parasitic Effects on the Performance of DC-DC SEPIC in Photovoltaic Maximum Power Point Tracking Applications** Nur Mohammad, Muhammad Quamruzzaman, Mohammad Rubaiyat Tanvir Hossain, Mohammad Rafiqul Alam	**66**
Chapter 9	**Towards Implementation of Smart Grid: An Updated Review on Electrical Energy Storage Systems** Md Multan Biswas, Md Shafiul Azim, Tonmoy Kumar Saha, Umama Zobayer, Monalisa Chowdhury Urmi	**75**

Chapter 10	**Demand and Response in Smart Grids for Modern Power System** Muhammad Qamar Raza, Muhammad Usman Haider, S. Muhammad Ali, Muhammad Zeeshan Rashid, Farooq Sharif	**86**
Chapter 11	**The Interaction between the Large-Scale EVs and the Power Grid** Yuan Huang, Junyong Liu, Xiaodong Shen, Tingting Dai	**90**
Chapter 12	**Modeling and Simulation of a Renewable Energy Based-Medical Herb Dryer** Abdel-Shafy A. Nafeh, Emad Ahmed Sweelem, Faten Hosny Fahmy	**97**
Chapter 13	**Temperature Control in a PV-WIND Medical Herb Dryer System** Abdel-Shafy A. Nafeh, Emad Ahmed Sweelem, Faten Hosny Fahmy	**106**
Chapter 14	**Uniformly Illuminated Efficient Daylighting System** Irfan Ullah, Seoyong Shin	**114**
Chapter 15	**Significance of Storage on Solar Photovoltaic System — A Residential Load Case Study in Australia** Mohammad T. Arif, Amanullah M. T. Oo, A. B. M. Shawkat Ali, G. M. Shafiullah	**120**
Chapter 16	**Very Short-Term Generating Power Forecasting for Wind Power Generators Based on Time Series Analysis** Atsushi Yona, Tomonobu Senjyu, Funabashi Toshihisa, Chul-Hwan Kim	**134**
Chapter 17	**Hybrid Power Systems Energy Controller Based on Neural Network and Fuzzy Logic** Emad M. Natsheh, Alhussein Albarbar	**140**
Chapter 18	**Design of a Monitoring System of Micro-Grid** Xiangyang Zhao, Shiyang Liu	**151**
Chapter 19	**Holonic Architecture of the Smart Grid** Ebisa Negeri, Nico Baken, Marjan Popov	**155**
Chapter 20	**A Low-Cost, Smart AC Charging System for Electric Vehicle** Bo Wang	**166**
Chapter 21	**Sizing and Design of the PV-WIND Energy Dryer for Medical Herbs** Emad Ahmed Sweelem, Abdel-Shafy A. Nafeh, Faten Hosny Fahmy	**170**
Chapter 22	**Feasibility Study for Self-Sustained Wastewater Treatment Plants—Using Biogas CHP Fuel Cell, Micro-Turbine, PV and Wind Turbine Systems** Ahmed Helal, Walid Ghoneim, Ahmed Halaby	**180**

Chapter 23 **Role of Energy Storage on Distribution Transformer Loading in Low Voltage Distribution Network** **189**
Mohammad Taufiqul Arif, Amanullah M. T. Oo, A. B. M. Shawkat Ali

Permissions

List of Contributors

Preface

The concept of a Smart Grid was introduced with the creation of the first alternating current power grid system in 1886. Today, it is a general term, for a system which utilizes analog or digital signals, to gather information about shortage of energy. Information about the behaviors of suppliers and consumers constitutes the data which is provided to the smart grid. This helps in improving the efficiency, sustainability and reliability of the production and distribution of energy.

The grid is called smart due to its ability to sense the precise requirement along the transmission lines. Smart Grid allows a two-way communication between the utility and its customers, through its controls, automation and latest technologies. It is similar to an electric superhighway which provides infrastructure for electricity. What makes a smart grid score over a conventional grid is the fact that conventional grid has difficulty accommodating variable sources of power like solar and wind energy. This is not the case with a Smart Grid.

The advantages of a Smart Grid are many. It helps in quicker restoration of electricity after power disturbances, covers renewable energy systems and will eventually result in lower power costs for consumers. The Smart Grid is a smart attempt to herald an era of efficiency, availability, and reliability, while ensuring environmental and economic well-being.

I would like to thank all the contributing researchers of this book, who helped in the successful assimilation and completion of the text.

Editor

A Power Electronic Conditioner Using Ultracapacitors to Improve Wind Turbine Power Quality

Murali Bottu[1], Mariesa L. Crow[1*], Andrew C. Elmore[2], Stan Atcitty[3]

[1]Electrical & Computer Engineering, Missouri University of Science & Technology, Rolla, USA; [2]Geological Engineering, Missouri University of Science & Technology, Rolla, USA; [3]Wind Energy Technologies, Sandia National Laboratories, Albuquerque, USA.

ABSTRACT

The large variability in wind output power can adversely impact local loads that are sensitive to poor power quality. To mitigate large swings in power, the wind turbine output power can be conditioned by using a small energy buffer. A power conditioner is developed to smooth the wind power output by utilizing the energy of an electrochemical capacitor, or ultracapacitor. The conditioner is based on a single phase voltage source inverter connected between the grid interconnection point and the ultracapacitor. The VSI shunt inverter injects or absorbs active power from the line to smooth the wind power output by utilizing the short term storage capabilities of the ultracapacitor. The ultracapacitor is connected to the DC link through a bidirectional DC-DC converter. The bidirectional DC-DC converter and VSI are constructed and field tested on a Skystream 3.7 wind turbine installed at the Missouri University of Science & Technology.

Keywords: Power Electronics; Electrochemical Capacitors; Ultracapacitors; Wind Power

1. Introduction

Due to the price volatility and carbon impact of fossil fuels, wind power generation is rapidly growing as an alternative energy source in many parts of the world. Due to the intermittency of wind speed, wind turbine output power can be highly variable. Power fluctuations from the wind turbine may cause severe power quality problems when connected to the grid. The large variability in wind turbine output power can adversely impact local loads that are sensitive to pulsating power, posing a challenge to the use of wind power extensively. The rapid growth of the wind power and its immense potential as a future energy source encourage us to find a way to smooth the intermittent wind power. Energy storage technologies can be used to improve the quality of the wind power [1,2]. In this paper, a power quality conditioner with the ultracapacitor is proposed to smooth the variable wind turbine output power. The short term storage capabilities of the ultracapacitor can be effectively used to smooth the wind power to minimize rapid power excursions that may damage sensitive local loads.

Ultracapacitors (also known as supercapacitors) have been proposed as a short term energy storage mechanism to provide improved active power output performance for wind turbines [3-7]. All of these articles address the application of ultracapacitors to *three-phase* wind turbine systems and none of them present *actual experimental results* for their designs. In [3], the authors present an approach to use ultracapacitors as part of the converter system for a Doubly Fed Induction Generator (DFIG) wind turbine to smooth the output during normal operation and to provide active power support to the DFIG during low-voltage conditions. They propose a fuzzy-logic rule based system to determine the charging modes of the UCAP system. In [6], the authors present an in-depth simulation analysis of the ultracapacitor charge and discharge capabilities to provide a procedure for sizing an ultracapacitor storage system. The authors in [5] use a state-space-based model of a permanent magnet synchronous generator coupled with blade-pitch control as the wind turbine model. The emphasis of this paper was to study the detailed behavior of the wind turbine as opposed to the characteristics of the ultracapacitor behavior.

The power conditioner mainly consists of power converters to shape the injected current at the point of common coupling [8]. The conditioner is based on a single phase shunt connected VSI connected between the grid interconnection point and the ultracapacitor. The shunt VSI injects or absorbs active power from the line to smooth the intermittent wind power by charging or discharging the ultracapacitor [9]. The ultracapacitor is

*Corresponding author.

connected to the DC link through a DC-DC converter. Traditionally, the VSI DC link voltage is maintained relatively constant by the shunt inverter control. In this application, we use a bidirectional DC-DC converter to maintain the DC link voltage. The bidirectional DC-DC converter acts in buck mode during discharge of the DC link and in boost mode during charging to maintain the voltage of the DC link to provide good controllability of the VSI.

Control of the injected active power via the shunt inverter is presented in this paper. The VSI controller calculates the compensating active power, which is then synthesized by using the bipolar pulse width modulation (PWM) switching sequence. The reference signal to the shunt inverter controller is obtained from a low pass filter, which has a large time constant. The fluctuating wind power is passed through the low pass filter to get the smoothed reference value. The conditioner ensures the smooth power is available at the grid interconnection point. The simulation results are presented to show the efficiency of the conditioner in smoothing the variable wind turbine output power.

2. The Wind Turbine

The power conditioner design and control will be validated on the Skystream3.7 wind turbine installed at the Missouri University of Science and Technology. The installed wind turbine is shown in **Figure 1**. The wind turbine is rated 2.4 kW at a wind speed of 29 mph with a cut-in wind speed of 8 mph. The three rotor blades have a 12 ft spinning diameter. The wind turbine alternator is a slotless permanent magnet brushless motor with passive yaw control. The Skystream3.7 wind turbine output is 120/240 VAC Split single phase, 60 Hz and is connected to the campus grid as shown in **Figure 2**.

The neutral and ground terminals of the wind turbine are tied together and connected to the neutral of the grid. The power quality conditioner is connected in shunt between the output of the wind turbine and the grid interconnection point.

Figure 1. Skystream3.7 wind turbine used for experimental validation.

3. The Power Conditioner

As shown in **Figure 3**, the power quality conditioner consists of a shunt inverter and a bidirectional DC-DC converter. The voltage source inverter acts as a shunt active filter compensating the active power of the wind turbine. The VSI is connected to the line through an RL filter which reduces the unwanted harmonics. The shape of the output current of the conditioner depends on the inductor value. The value of the resistor and the inductor determines the damping in the circuit. On the other side, the VSI is connected to the DC link capacitor. The DC-DC converter with the ultracapacitor is used to reduce the size of the DC link capacitor and to maintain the voltage of the DC link relatively constant as the ultracapacitor discharges and charges. The bidirectional DC-DC converter charges the ultracapacitor in buck mode by reducing the voltage of the DC link. In the other direction, it acts in boost mode, discharging the ultracapacitor to increase the voltage of the DC link. The power conditioner injects or absorbs active power from the line through the filter to smooth the variable wind turbine output power. The DC link acts as the voltage source for the VSI.

The primary objective of the conditioner is to inject a current I_{inv} at the point of common coupling (PCC) such

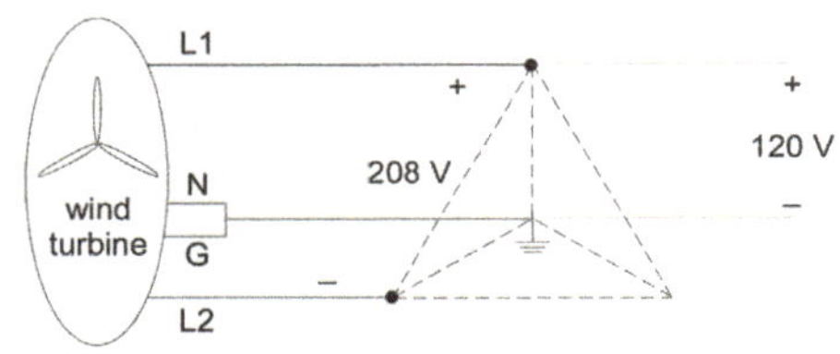

Figure 2. Skystream3.7 wind turbine connection to the grid.

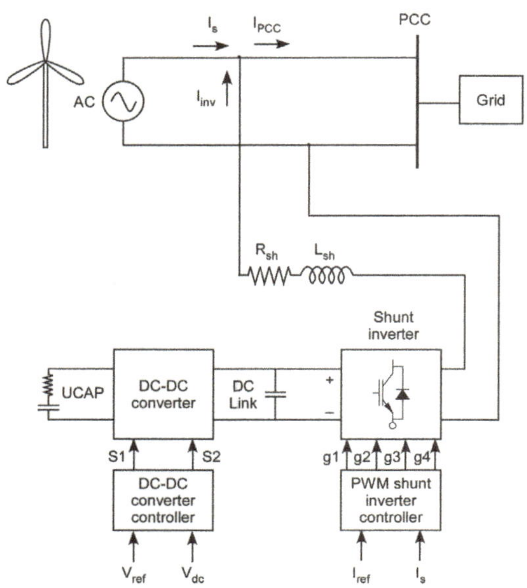

Figure 3. Power quality conditioner.

that the current supplied to the grid (I_{PCC}) is relatively smooth. The smoothed current is obtained by passing the (measured and scaled) wind turbine current through a low pass filter that is tuned to provide the appropriate high-frequency cutoff. The ultracapacitor is charged and discharged rapidly to supply the required current while holding the DC link constant. Note that the reference current is not a constant, but rather a slowly varying current. If the reference current were held constant, this would imply that the electrochemical capacitor would have infinite ability to charge and discharge. By allowing the reference current to slowly vary, the energy supplied to the grid will track the energy supplied by the wind turbine.

3.1. The DC-DC Converter

The three primary components of the proposed power conditioner are the shunt inverter to control the injected current, the ultracapacitor, and the DC-DC converter to regulate the DC link voltage and control the ultracapacitor injected current. The topology of the bidirectional DC-DC converter is shown in **Figure 4**. The bidirectional DC-DC converter acts a buck converter in one direction and as a boost converter in the other direction [10-13]. Power MOSFETS are used as the switching devices in the circuit. The operation of the converter is controlled by the DC link voltage and the voltage of the ultracapacitor. The main purpose of the bidirectional DC-DC converter is to maintain the voltage of the DC link relatively constant at a reference value. The 2200 μF DC link capacitor is used as an intermediate element between the DC-DC converter and the inverter.

The DC-DC converter operating modes can be divided into four modes:

- Mode 1: The DC-DC converter acts in buck mode when the DC link voltage V_{dc} is greater than the reference value V_{ref}. In this mode, the DC-DC converter controls the current to charge the ultracapacitor.
- Mode 2: The DC-DC converter acts in boost mode when the DC link voltage V_{dc} falls below the reference value. In this mode, the ultracapacitor discharges.

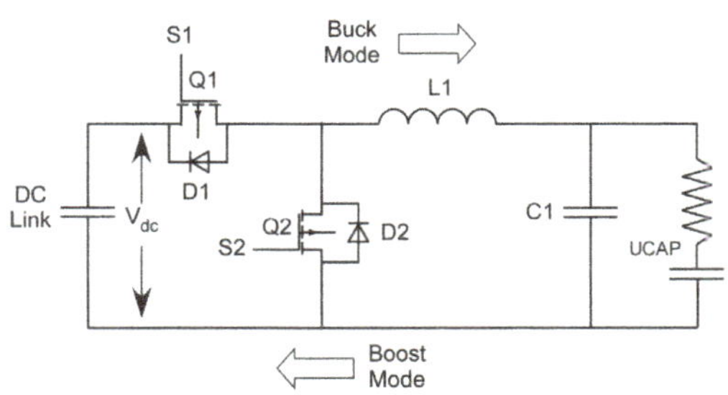

Figure 4. DC-DC converter.

- Mode 3: When the ultracapacitor is fully charged, the DC-DC converter shuts down to avoid damaging the ultracapacitor.
- Mode 4: When the ultracapacitor is fully discharged, the conditioner shuts down until the wind turbine produces sufficient current to resume charging of the ultracapacitor.

The bidirectional DC-DC converter shown in **Figure 5** was constructed from IXFK80N50P power MOSFETS as the switching devices in the circuit. The Microchip PIC24FJ128GA010 microcontroller along with the Explorer16 development board was used in implementing the control scheme and to obtain the gate signals for the MOSFETs in the bidirectional DC-DC converter. LEM LV 25-P voltage sensors are used to measure the voltages of the DC link and the ultracapacitor. The ACNW3190 5.0 Amp High Output Current Gate Drive optocoupler is used to drive the MOSFETs. A TP-75C DC power supply is used to supply power to the LEM sensors and gate drivers. The voltage measurement circuits and the gate driver circuits are built on the vector board and are soldered permanently. The high current inductor used in the circuit is manufactured by West Coast Magnetics.

3.2. The Ultracapacitor

Ultracapacitors are electrochemical double layer capacitors that have unique characteristics when compared to other energy storage devices. Ultracapacitors (UCAPS) have high energy density and large time constants as well. Although multiple time-scale models of UCAPs have been developed, a simple UCAP model such as the one in **Figure 3** containing only one RC branch was used in the simulation of the converter system. This model is composed of an equivalent series resistor (ESR) and a

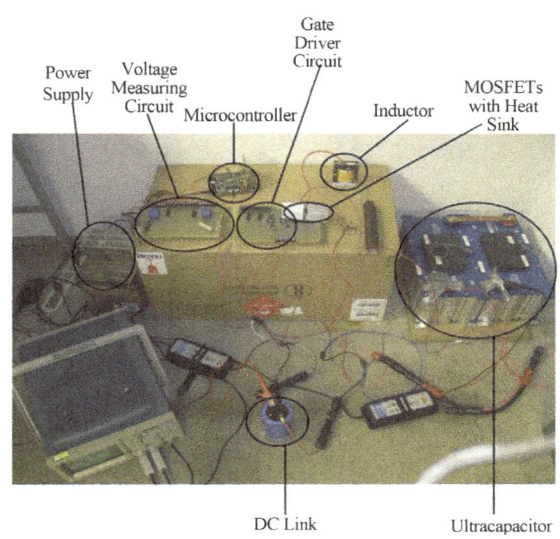

Figure 5. DC-DC buck-boost converter experimental hardware.

capacitor (C) [10]. The ESR represents the ohmic losses in the ultracapacitor. Higher order ultracapacitor models are essential for simulation studies in which the timescale of interest is on the order of microseconds. In the current application, the timescale of interest is on the order of minutes; therefore the single RC branch model is sufficient to capture the UCAP behavior of interest.

The benefits of using ultracapacitors are quite extensive. Ultracapacitors have low losses while charging and discharging. They have a very low ESR, allowing them to deliver and absorb very high currents and to be charged very quickly, making them well suited for energy buffer applications. Ultracapacitors are highly efficient components even at very high currents. The characteristics of the UCAP allow it to be charged and discharged at the same rates, something most batteries cannot tolerate. Ultracapacitors have a wide voltage window and can be deeply discharged. The energy storage mechanism of an UCAP is a highly reversible process. The process moves charge and ions only. It does not make or break chemical bonds like batteries; therefore it is capable of millions of cycles with minimal change in performance. It is therefore capable of many years of continuous duty with minimal change in performance. These advantages make ultracapacitors well suited for power quality conditioning applications.

The power conditioner was constructed using two series-connected Maxwell BMOD0165 ultracapacitor modules of 165F nominal capacitance and rated voltage 48.6 V. This ultracapacitor has a maximum continuous current of 77 A at 15°C [11].

3.3. The Shunt Inverter

A full-bridge IGBT based inverter topology is used in this application. The full-bridge inverter consists of four switching devices, which are connected to form the full-bridge inverter circuit shown in **Figure 6**. The gating signals for the IGBTs are obtained by the pulse width modulation controller. Anti-parallel diodes are connected across the power IGBTs to protect the devices and to provide the power flow in the reverse direction [12,13]. The voltage source inverter connected in shunt to the line acts as a current source, injecting or absorbing the compensating current from the line [14]. The shunt inverter is connected to the line through a series interference RL filter, which reduces the unwanted harmonics. The filter provides smoothing and isolation from high frequency components. On the downstream side, the full-bridge inverter is connected to the DC link. The injected current is in phase with the line voltage to produce a unity power factor.

The VSI operates in the following two modes:

- Mode 1: When the wind turbine power is greater than the reference value, the converter acts like a rectifier drawing active power from the line and charging the DC link capacitor.
- Mode 2: When the wind turbine power is less than the reference value, the converter acts like a VSI injecting active power into the line by discharging the DC link capacitor.

The variable wind power is passed through a low pass filter to get a smoothing reference signal for the inverter controller. The output of the low pass filter is given to the shunt inverter controller as the reference value [15]. The reference signal I_{ref} is obtained:

$$I_{ref} = \frac{I_S}{1 + sT_S} \quad (1)$$

where T_s is the time constant of the filter and I_s is the output of the wind turbine. The smoothing performance of the wind turbine output power depends on the time constant of a low pass filter. The time constant of the low pass filter is in the range of several seconds and is tuned to provide the desired smoothing. The power fluctuation is smoothed by drawing or injecting the difference of the reference power and the variable wind power.

The experimental setup and different components in the construction of the single phase shunt inverter are shown in **Figure 7**. A SEMIKRON SK 30 GBB 066 T IGBT module is used as the switching devices for the circuit. A 2.5 mH inductor is used as the output filter of

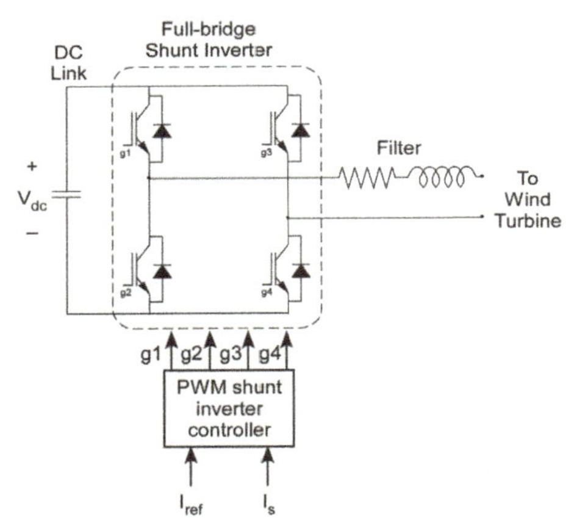

Figure 6. Shunt inverter circuit.

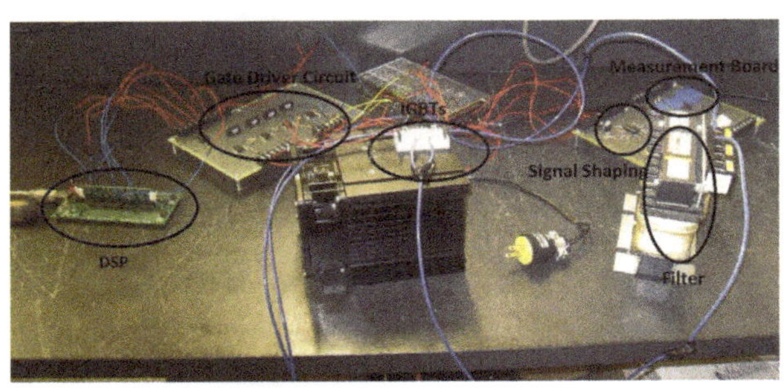

Figure 7. Shunt inverter hardware.

the shunt inverter. A TI TMDSDOCK28335 DSP is used in the closed loop control of the shunt inverter. The current and voltage sensors measures and sends signals to the DSP through the signal shaping circuit. DSP drives the IGBTS with the ACNW3190 drivers.

4. Experimental Results

Figure 8 shows the actual currents measured at the output of the wind turbine (unconditioned—thin) and at the PCC (conditioned—bold). The smoothing time constant was 60 seconds in this experiment. The average wind speeds were 15 mph and the average active power output is approximately 600 W. The corresponding voltage and current of the ultracapacitor are shown in **Figures 8-10**. The ultracapacitors were initially charged to a mid-range voltage of approximately 50 V to provide sufficient charge and discharge capabilities. Note that when the wind turbine current exceeds the desired (smoothed) current, the ultracapacitor will absorb the excess current (*i.e.* charge) and the voltage will increase. These results indicate that the conditioner is working as designed under normal operating conditions.

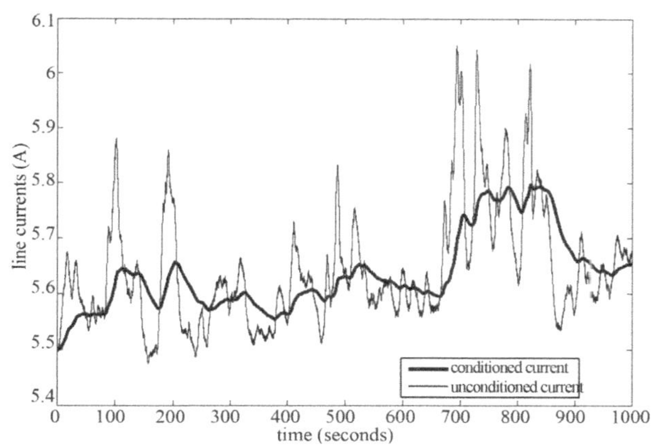

Figure 8. Line currents (measured).

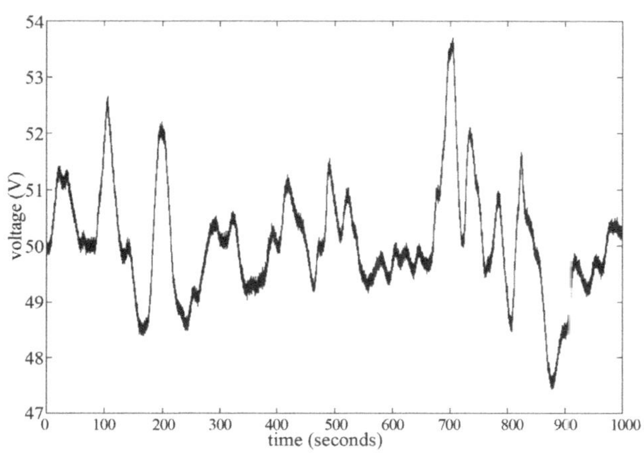

Figure 9. Ultracapacitor voltage (measured).

5. Further Considerations

The experimental results shown in **Figures 8-10** are somewhat limited in that they provide only a snapshop of the conditioner's behavior. To fully explore the conditioner's capabilities in a variety of situations, a simulation of the conditioner was developed and a variety of wind speeds and charge/discharge situations were tested.

5.1. Wind Speed Simulation

Wind variability is frequently modeled as having a Weibull probability distribution function (as shown in **Figure 11**) [16]:

$$f(x)=\begin{cases}\frac{k}{\lambda}\left(\left(\frac{x}{\lambda}\right)^{k-1}\exp\left(-\left(\frac{x}{\lambda}\right)^{k}\right)\right) & x\geq 0\\ 0 & x<0\end{cases} \quad (2)$$

where $\lambda>0$ is the scale factor and $k>0$ is the shape factor. In wind applications, k can range from 1 to 2.5, with regions with low wind having smaller k factors. In our system, we chose $k=1.5$ to best match measured wind speeds locally. The performance characteristic of the Skystream3.7 is shown in **Figure 12** [17].

5.2. Increasing Wind Speeds

The first trial considers the case in which the wind speed

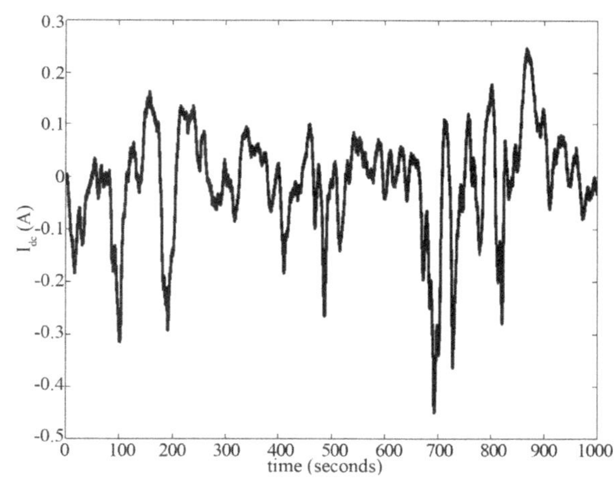

Figure 10. Ultracapacitor current (measured).

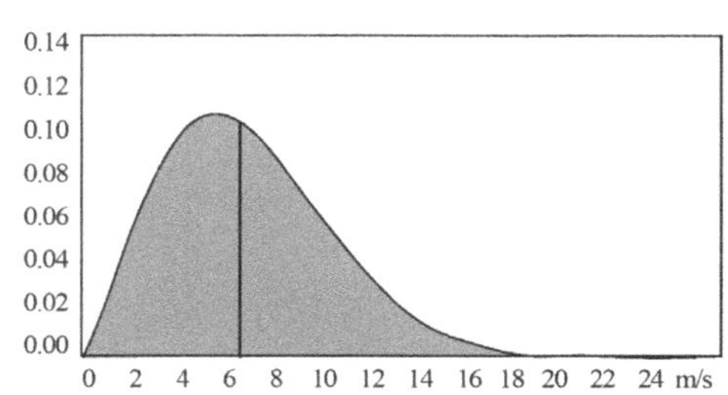

Figure 11. Windspeed with Weibull distribution.

slowly increases from zero to rated speeds over the time interval of two hours (7200 seconds). This case is intended to show the behavior of the ultracapacitor voltage and current in this situation. Note that the mean voltage change over the course of the two hours is nearly zero. This is to be expected because the mean of the filtered line current should be the same as the mean of the raw line current over long periods.

The currents and UCAP voltage is shown in **Figures 13** and **14**.

5.3. Decreasing Wind Speeds

The second case considers the situation in which the wind speed slowly decreases to zero over the time interval. Note that the mean voltage change over the course of the two hours is nearly zero. The currents and UCAP voltage is shown in **Figures 15** and **16**.

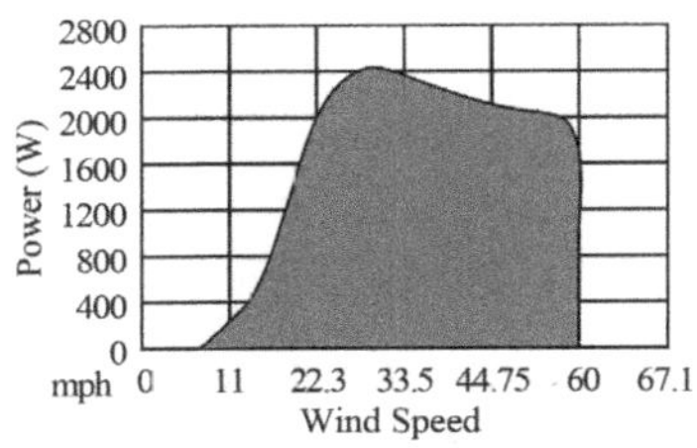

Figure 12. Skystream3.7 wind turbine performance.

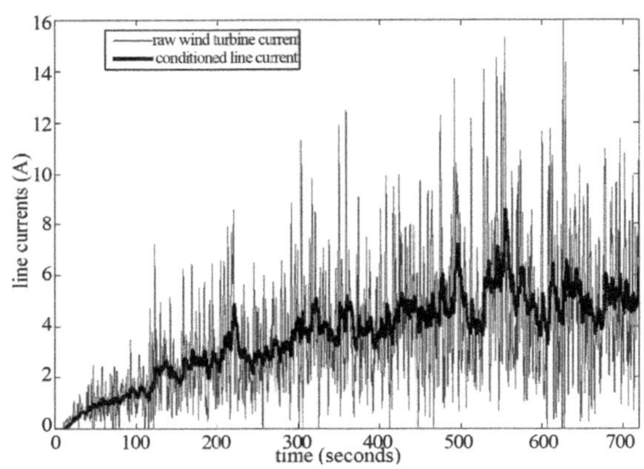

Figure 13. Line currents under increasing wind speeds.

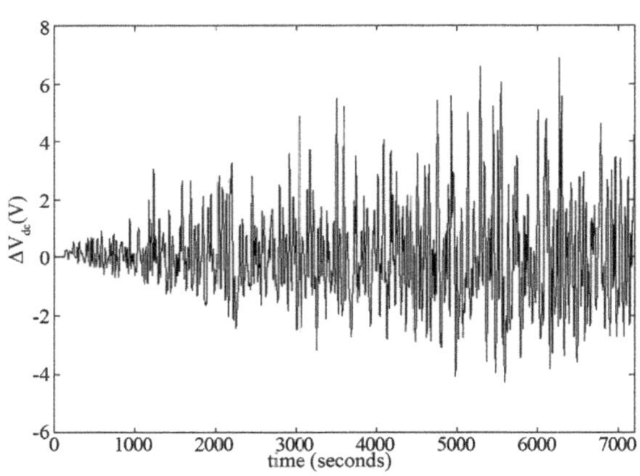

Figure 14. Change in ultracapacitor voltage with increasing wind speeds.

5.4. Discussion

The simulation results indicate that the proposed controller will work well under steady, increasing, and decreasing wind speeds. The selection of the filtered line current reference guarantees that the ultracapacitor will not charge or discharge beyond its operating range since the average of the filtered and raw line currents are the same over long periods of time. The injected or absorbed ultracapacitor current does not exceed the maximum allowable ultracapacitor at any time, thus the ultracapacitor is well-suited to this application.

6. Conclusion

This paper provided the schematic of an ultracapacitor-based smoothing circuit to condition wind turbine line currents. This controller would find application in

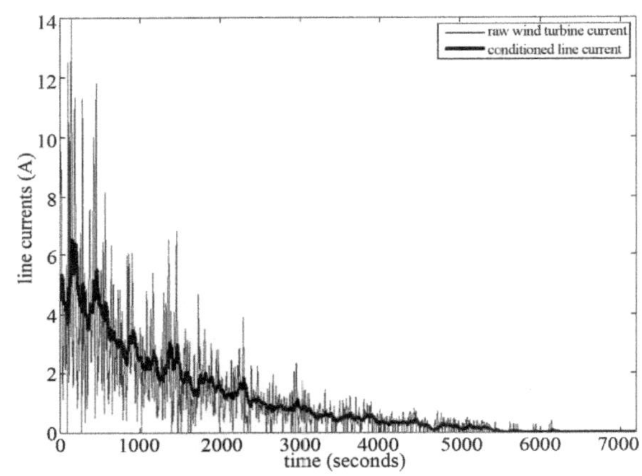

Figure 15. Line currents under decreasing wind speeds.

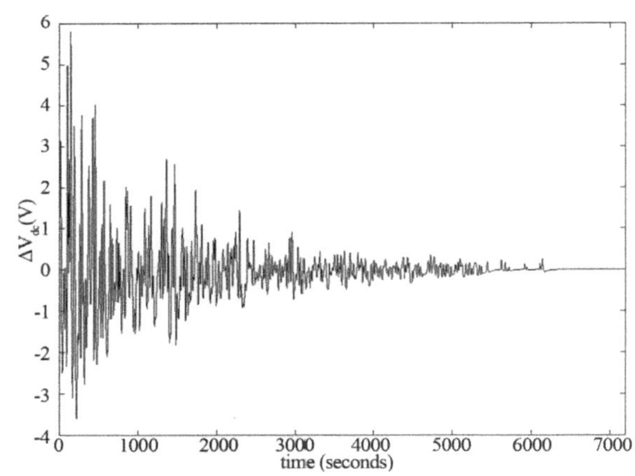

Figure 16. Change in ultracapacitor voltage with decreasing wind speeds.

situations in which rapidly varying currents may cause poor performance of loads such as pumps, induction motors, electronics, etc. in weak distribution systems. The ultracapacitor was sized to connect to the PCC without the need for a transformer. The DC-DC converter was designed to hold the dc-link capacitor of the voltage source converter near constant to provide good controllability. The proposed conditioner was validated in both experiment and simulation for a variety of conditions.

7. Acknowledgements

The authors gratefully acknowledge the support of the DOE Energy Storage Program through Sandia National Laboratories under BD-0071-D. Sandia is a multi-program laboratory operated by Sandia Corporation, a Lockheed Martin Company, for the United States Department of Energy's National Nuclear Security Administration, under contract DE-AC04-94AL85000.

REFERENCES

[1] P. F. Ribeiro, B. K. Johnson, M. L. Crow, A. Arsoy and Y. Liu, "Energy Storage Systems for Advanced Power Applications," *Proceedings of the IEEE*, Vol. 89, No. 12, 2001, pp. 1744-1756.

[2] M.-S. L. Lu, C.-L. Chang, W.-J. Lee and W. Li, "Combining the Wind Power Generation System With Energy Storage Equipment," *IEEE Transactions on Industry Applications*, Vol. 45, No. 6, 2009, pp. 2109-2115.

[3] C. Abbey and G. Joos, "Supercapacitor Energy Storage for Wind Energy Applications," *IEEE Transactions on Industry Applications*, Vol. 43, No. 3, 2007, pp. 769-776.

[4] W. Li, G. Joos and J. Belanger, "Real-Time Simulation of a Wind Turbine Generator Coupled with a Battery Supercapacitor Energy Storage System," *IEEE Transactions on Industrial Electronics*, Vol. 57, No. 4, 2010, pp. 1137-1145.

[5] G. Mandic and A. Nasiri, "Modeling and Simulation of a Wind Turbine System with Ultracapacitors for Short-Term Power Smoothing," *Proceedings of the* 2010 *International Symposium on Industrial Electronics*, Bari, 4-7 July 2010, pp. 2431-2436.

[6] E. Naswali, C. Alexander, H.-Y. Han, D. Vaniaux, A. Bistrika, A. von Jouanne, A. Yokochi and T. Brekken, "Supercapacitor Energy Storage for Wind Energy Integration," *Proceedings of the* 2011 *Energy Conversion Congress and Exposition*, Phoenix, 17-22 September 2011, pp. 298-305.

[7] L. Qu and W. Qiao, "Constant Power Control of DFIG Wind Turbines with Supercapacitor Energy Storage," *IEEE Transactions on Industry Applications*, Vol. 47, No. 1, 2011, pp. 359-367.

[8] C.-C. Hua and C.-C. Tu, "Design and Implementation of Power Converters for Wind Generator," *Proceedings of the IEEE Industrial Electronics and Applications Conference*, Sapporo, 21-24 June 2010, pp. 323-328.

[9] M. S. A. Dahidah, N. Mariun, S. Mahmod and N. Khan, "Single Phase Active Power Filter for Harmonic Mitigation in Distribution Power Lines," *National Power Engineering Conference*, Bangi, 15-16 December 2003, pp. 359-362.

[10] L. Shi and M. L. C. Crow, "Comparison of Ultracapacitor Electric Circuit Models," *Proceedings of the* 2008 *IEEE PES General Meeting*, Pittsburgh, 20-24 July 2008, pp. 1-6.

[11] "Maxwell Technologies." http://www.maxwell.com/products/ultracapacitors/docs/datasheet_48v_series_1009365.pdf

[12] R. W. Erickson and D. Maksimovic, "Fundamentals of Power Electronics," Kluwer Academic Publishers, New York, 2000.

[13] P. K. Krein, "Elements of Power Electronics," Oxford University Press, New York, 1998.

[14] H. Akagi, "New Trends in Active Filters for Power Conditioning," *IEEE Transactions on Industry Applications*, Vol. 32, No. 6, 1996, pp. 1312-1322.

[15] T. Kai and A. Tanaka, "A New Smooth Scheme for Power Fluctuation Using Inverter of Wind Power Generation with Doubly Fed Induction Generator," *Proceedings of the IEEE Electrical Machines and Systems International Conference*, Calgary, 17-20 October 2008, pp. 2390-2395.

[16] T. Burton, D. Sharpe, N. Jenkins and E. Bossanyi, "Wind Energy Handbook," Wiley, Chichester, 2001.

[17] S. Windpower, "Optimum Energy in High Wind Environments." www.windenergy.com

2

Wind Energy Conversion Systems Adapted to the Tunisian Sites

Fathi Ben Amar[1,2], Mustapha Elamouri[1]

[1]Department of Technology, Preparatory Institute for Engineering Studies of Sfax (IPEIS), Sfax, Tunisia; [2]Electric Networks and Machines Research Unit, National Institute of Applied Sciences and Technology (INSAT), Tunisia.

ABSTRACT

The choice of a wind turbine for a given site is a very significant operation. It is advised to make an in-depth study on the characteristics of the aerogenerator and the implantation site before any installation of a wind farm for choosing well the system best adapted to the energy needs. The objective of this study is to sensitize the users of the wind energy on the various problems involved in the aerogenerator installation on any site and to provide a decision-making aid to the judicious choice of the wind system best adapted. The idea is to estimate, for a selected implantation site, the energy production and the operating time of several wind systems. We can thus select the aerogenerator best adapted by making a compromise between a maximum recovery of energy and a high time of production. The influence height of the aerogenerator on energy profitability is also studied. We apply this step to the twenty six Tunisian synoptic sites by implantation scenarios of seven wind systems of various nominal speeds in function of the height.

Keywords: Tunisia; Wind Energy; Wind Turbine; Meteorological Method

1. Introduction

Face to the environmental problems caused by the greenhouse gas emissions during exploitation of the conventional resources, other alternative energy resources were and must continue to be developed. Among them, we can quote the inexhaustible renewable energy resources, which enable us to produce electricity more and more easily and properly.

In this study, we are mainly interested in the wind power which is regarded as a promising source in the improvement of the energy balance and the environmental preservation. In reality, the realization of a wind power station must take into account the wind resource. Thus, the study of the wind potential is a fundamental component of a wind project which determines the annual production and which allows the project manager to establish its business plan [1-8]. An error of a few percentages in the evaluation of the wind potential can have disastrous consequences on the future profitability of the project.

In this study, we propose a practical approach enabling to estimate energy provided by a given wind system on an implantation site chosen. This approach, taken with several machines and for various heights above ground level, results in a phase of optimal choice of the best adapted aerogenerator to the retained site while basing itself on energy efficiency, economic and technical of wind installation.

This approach is applied to the twenty six synoptic sites distributed on all the Tunisian territory. The implantation scenarios carried out on seven different wind systems and according to the height, enable us to select the best adapted aerogenerators to the Tunisian sites.

2. Analysis of Measured Wind Data

The wind is a clean, renewable and increasingly economical energy; but it is a variable phenomenon. However, the wind energy production depends on this variation. Consequently, we need to evaluate the wind characteristics of the site to be able to conceive a wind project. In the same way, this study offers to the investors to determine the estimated production of the future wind installation.

2.1. Geographical Situation and Relief of Tunisia

First, Tunisia is a Mediterranean country of a surface of 164,154 km^2. It is located between altitude 30˚N and 37˚N latitude and between 8˚E and 12˚E longitude [4,9-11]. Its littoral extends on 1300 km and comprises five large gulfs. Tunisia has common borders with Algeria to the West and with Libya to the South. The country can be divided into three zones (**Figure 1**):

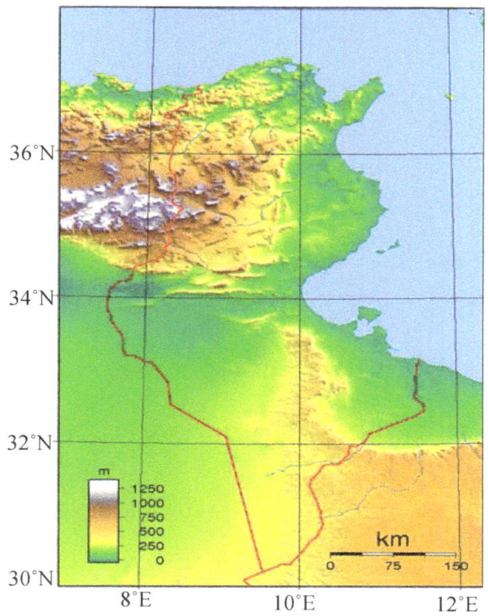

Figure 1. Geographical position and relief of Tunisia.

- Littoral Tunisia which extends to the East from Bizerte to the Sahel (Sfax) composed primarily of hills and plains.
- Interior Tunisia composed essentially of mountains and valleys.
- Tunisian desert, which represents 55% of the whole of the Tunisian territory.

Surrounded by the Mediterranean, Tunisia has a Mediterranean climate to the North of the country and desert to the South. The climate is generally soft. The annual temperature varies between 10° and 20° in winter and between 20° and 36° in summer. The temperature can reach 45° in summer for the areas of the Tunisian South.

2.2. Methodology Used

To reveal the good sites of wind in Tunisia, it is necessary to evaluate the wind characteristics (mean speed, power density and available energy) in the 26 meteorological sites distributed on all the Tunisian territory. The analysis of the wind is carried out using the meteorological method [9-11] using the measurement tables of the I.N.M. (Meteorology National Institute) indicating the cumulated frequency $F(V)$ of the wind speeds "superior or equal" at the classified speeds V. The occurrence frequency $f(V)$ is defined by the following relation

$$f(V) = F(V) - F(V+1) \quad (1)$$

The characteristic speeds of the wind (the mean V_m, the most energetic V_e and the most frequent V_f) are defined by the following expressions:

$$V_m = \sum_{i=1}^{n} V_i f(V_i) \quad (2)$$

$$V_f = V\left(f(V)_{\max}\right) \quad (3)$$

$$V_e = V\left(E_d(V)_{\max}\right) \quad (4)$$

The power density varies considerably according to the places and days'. It depends on the speed and on the wind frequency. The power density of the wind, per unit of surface swept by the turbine blades, is given by the following expression:

$$P_d = \frac{1}{2}\rho\sum_{i=1}^{n} V_i^3 f(V_i) \quad (5)$$

with ρ is the air density, equal to 1.225 Kg/m^3 for a temperature of 15°C and a pressure of 1013 mb.

Similarly, the available wind potential in the study site, per unit area and for a year, is expressed by:

$$E_d = 8.76\, P_d \quad (6)$$

2.3. Frequency and Energy Distribution of the Wind Speed

Based on hourly meteorological data provided by the I.N.M. in the form of tables to 16 directions and for a five year period (2004-2009), we trace the histograms of the frequency and the energy of classified wind speeds (all confused directions) for the twenty six Tunisian sites (**Figure 2**).

The measurements are taken with a standard height of 11 m above ground level and in an open area. Thus we can determine the areas most favorable for the exploitation of the wind in Tunisia.

The result of the analysis of **Figure 2** shows that the statistical distribution of the available wind energy is, because of the variation with the cube of the speed, shifted towards the high values of the wind speed: this explains why, for a given site, the most frequent speed contributes very slightly to total energy and is always lower at the most energetic speed; while the frequency of the most frequent speed is always superior than the frequency of the most energetic speed.

2.4. Analysis of Wind Characteristics

Table 1 gives the numerical results of the wind characteristics (the mean speed V_m, the most frequent speed V_f, the most energetic speed V_e, the power density P_d and the available energy E_d) for the twenty six synoptic sites.

At the height 11 m above ground level, the wind characteristics in Tunisia are modest. Indeed, the annual mean wind speed varies between a minimum of 2.09 m/s (in Kairouan) and a maximum of 5.45 m/s (in Thala). In

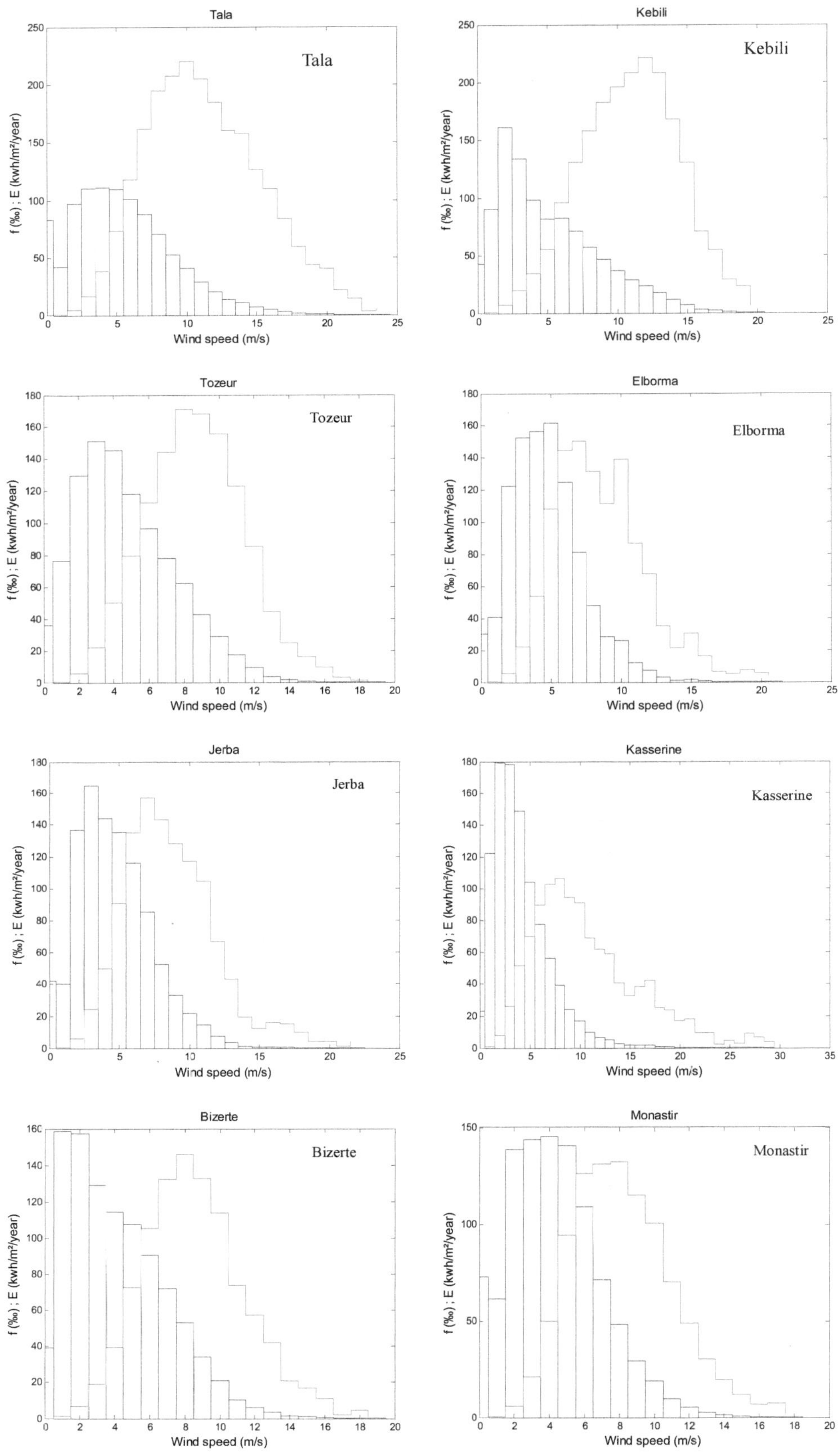
Tala
f (‰) ; E (kwh/m²/year)
Wind speed (m/s)
Kebili
f (‰) ; E (kwh/m²/year)
Wind speed (m/s)
Tozeur
f (‰) ; E (kwh/m²/year)
Wind speed (m/s)
Elborma
f (‰) ; E (kwh/m²/year)
Wind speed (m/s)
Jerba
f (‰) ; E (kwh/m²/year)
Wind speed (m/s)
Kasserine
f (‰) ; E (kwh/m²/year)
Wind speed (m/s)
Bizerte
f (‰) ; E (kwh/m²/year)
Wind speed (m/s)
Monastir
f (‰) ; E (kwh/m²/year)
Wind speed (m/s)

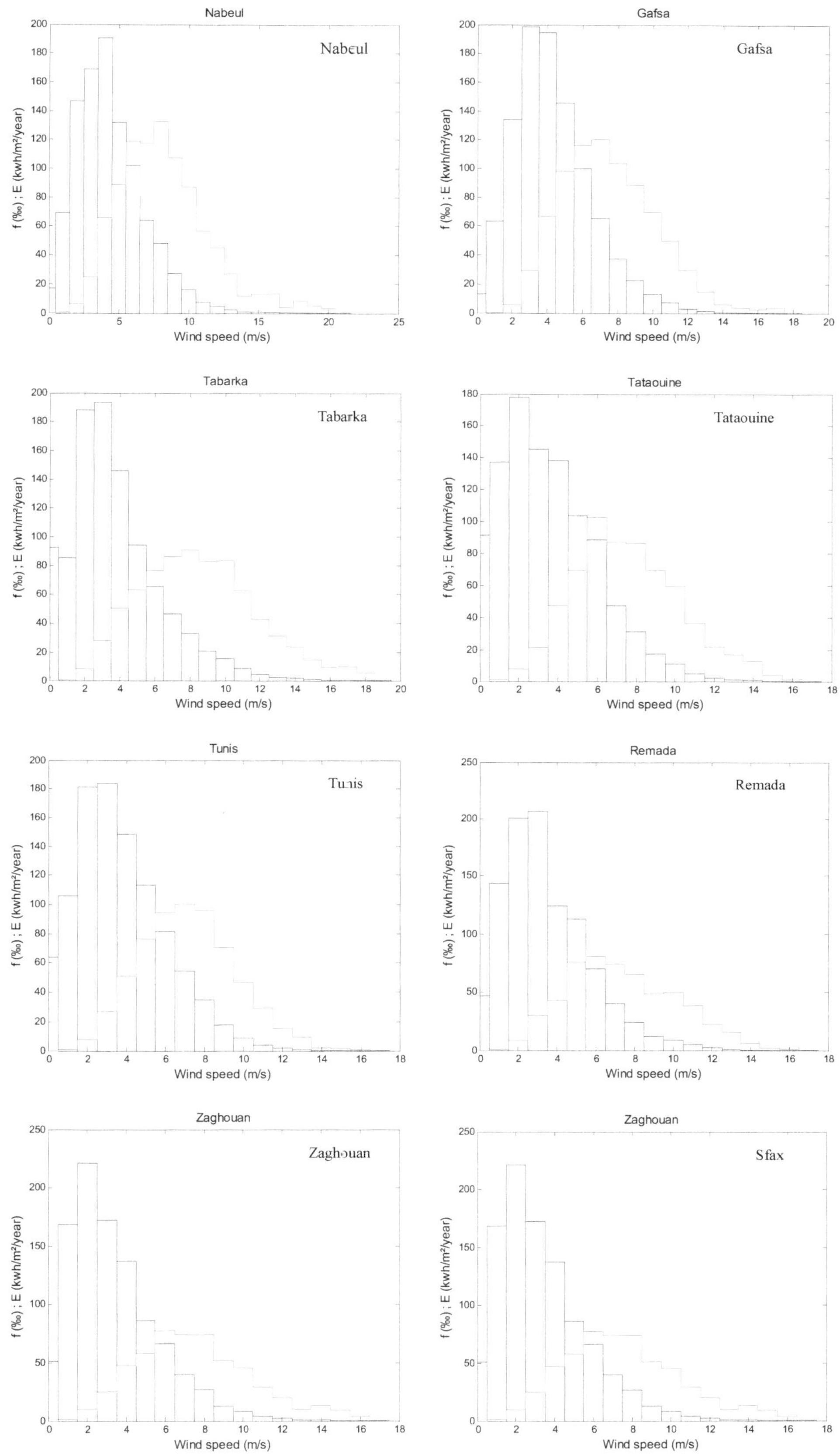
Nabeul
Nabeul
Gafsa
Gafsa
Tabarka
Tabarka
Tataouine
Tataouine
Tunis
Tunis
Remada
Remada
Zaghouan
Zaghouan
Zaghouan
Sfax
f (‰) ; E (kwh/m²/year)
Wind speed (m/s)

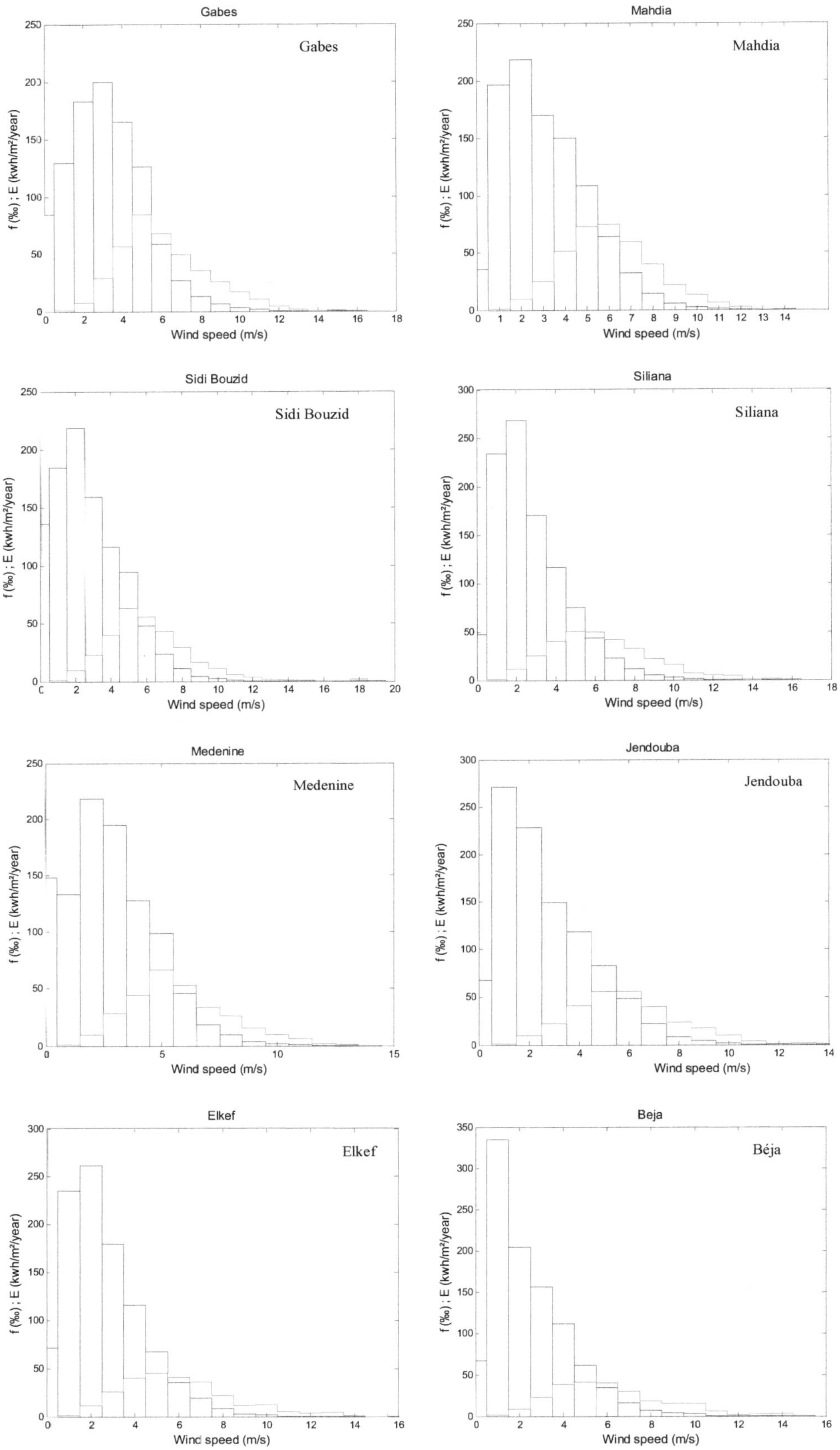
Gabes
Gabes
f (‰) ; E (kwh/m²/year)
Wind speed (m/s)
0
50
100
150
200
250
0 2 4 6 8 10 12 14 16 18
Mahdia
Mahdia
f (‰) ; E (kwh/m²/year)
Wind speed (m/s)
0
50
100
150
200
250
0 1 2 3 4 5 6 7 8 9 10 11 12 13 14
Sidi Bouzid
Sidi Bouzid
f (‰) ; E (kwh/m²/year)
Wind speed (m/s)
0
50
100
150
200
250
0 2 4 6 8 10 12 14 16 18 20
Siliana
Siliana
f (‰) ; E (kwh/m²/year)
Wind speed (m/s)
0
50
100
150
200
250
300
0 2 4 6 8 10 12 14 16 18
Medenine
Medenine
f (‰) ; E (kwh/m²/year)
Wind speed (m/s)
0
50
100
150
200
250
0 5 10 15
Jendouba
Jendouba
f (‰) ; E (kwh/m²/year)
Wind speed (m/s)
0
50
100
150
200
250
300
0 2 4 6 8 10 12 14
Elkef
Elkef
f (‰) ; E (kwh/m²/year)
Wind speed (m/s)
0
50
100
150
200
250
300
0 2 4 6 8 10 12 14 16
Beja
Béja
f (‰) ; E (kwh/m²/year)
Wind speed (m/s)
0
50
100
150
200
250
300
350
0 2 4 6 8 10 12 14 16

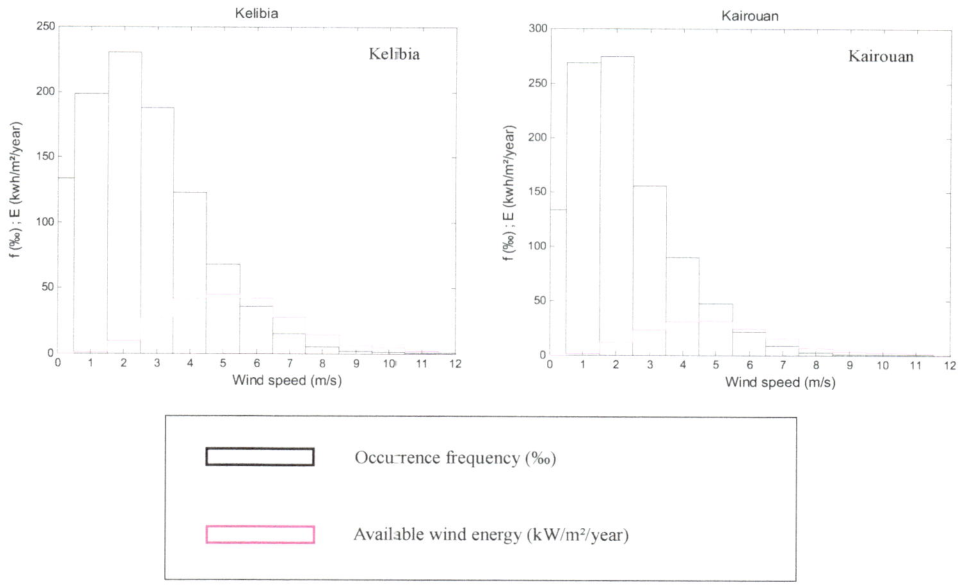

Figure 2. Frequency and energy distributions of the classified wind speeds for various sites to 11 m high in an open area.

general, an average is needed exceeding 4 m/s to justify the wind turbine installation. Similarly, the annual power density (respectively available wind energy), per unit of area, also varies between 17.29 W/m^2 (respectively, 151.43 kW/m^2/year) in Kairouan and 256.49 W/m^2 (respectively 2246.83 kW/m^2/year) in Thala.

The analysis of **Table 1** has enabled us to classify these sites in four groups [9]:

- A first group of excellent wind potential, noted (A) whose energy is superior to 2000 kWh/m^2/year. It is formed of two sites Thala and Kebili. It covers 20.49% of the available total energy in the twenty six sites. The annual mean wind speed exceeds 5 m/s. This group presents the energy zone and most promising for the exploitation of the wind power.
- A second group of good wind potential, noted (B), in which energy varies from 800 to 2000 kWh/m^2/year. It comprises eight sites (Elborma, Tozeur, Jerba, Monastir, Nabeul, Gafsa, Bizerte and Kasserine) and accounts for 45.5% of total energy. The annual mean wind speed is between 4 and 5 m/s.
- A third group of average wind potential, noted (C), interests eight sites (Tunis, Tabarka, Sfax, Tataouine, Remada, Zaghouan, Gabès and Mahdia) and covers only 23.1% of total energy. Its wind energy varies from 350 to 800 kWh/m^2/year. The annual mean wind speed is between 3 and 4 m/s.
- A fourth group of low wind potential, noted (D), concerns eight sites (Siliana, Kelibia, Jendouba, Sidi Bouzid, Medenine, Elkef, Béja and Kairouan) and covers only 10.91% of total energy. Its wind energy is lower than 350 kWh/m^2/year. The annual mean wind speed is lower than 3 m/s.

Also, it was possible to establish maps of the annual spatial distribution of mean speed and available energy of wind for the whole country (**Figures 3** and **4**). These cards, as was expected, have significant similarities and clearly indicate the regions most favorable for the exploitation of the wind energy in Tunisia.

Thus, the most interesting zones for the exploitation of the wind energy arise clearly (**Figures 1** and **2**): they lie in the North, North-East, West and South of Tunisia. On the other hand, the zones where the annual mean wind speed is lower than 4 m/s, like the regions located inside the country in the North-West and South-East, are much less favorable to the use of the wind energy. As an example, in the Jandouba region, to recover a total energy identical to that which one can discount in Bizerte, it will be necessary to use a wind turbine of diameter more important and/or located on a higher pylon. The cost of the installation will be consequently much larger.

We also note that a variation of 30% of the annual mean wind speed from one site to another, which is not exceptional, involves a variation of wind energy of a factor 2 (for instance, see the characteristics of the Kasserine and Kebili sites).

Practically, it is interesting to implant the wind turbine distant from any obstacle and sufficiently high to take the maximum of energy [12-14]. That is why we are interested in the study of the influence of the mast height of the wind systems on energy profitability.

2.5. Influence of Height on the Wind Resource

The Von Karman theory showed that, for a flow under constant flux and neutral atmospheric conditions, the profile of the wind speed is a logarithmic curve of the

Table 1. Wind characteristics for different sites at 11 m height in an open area.

Sites	Groups	V_f (m/s)	V_e (m/s)	V_m (m/s)	P_d (W/m²)	E_d (kWh/m²/year)
Thala	A	4	10	5.45	256.49	2246.83
Kebili		2	12	5.08	227.99	1997.21
Tozeur		3	8	4.66	138.75	1215.41
El Borma		5	7	4.73	131.50	1151.97
Jerba		3	7	4.64	131.24	1149.58
Kasserine		2	8	4.11	126.73	1110.12
Bizerte	B	1	8	4.12	113.55	994.72
Monastir		4	8	4.29	111.04	972.63
Nabeul		4	8	4.36	106.63	934.09
Gafsa		3	7	4.27	91.88	804.88
Tabarka		3	8	3.66	87.94	770.39
Tataouine		2	6	3.50	73.59	644.63
Tunis		3	7	3.65	71.56	626.83
Remada	C	3	6	3.43	64.17	562.14
Zaghouan		2	6	3.31	62.76	549.80
Sfax		2	6	3.62	60.23	527.64
Gabes		3	5	3.15	44.96	393.85
Mahdia		2	6	3.10	42.97	376.41
Sidi Bouzid		2	5	2.66	35.40	310.10
Siliana		2	5	2.74	35.22	308.54
Mednine		2	5	2.70	34.53	302.48
Jendouba	D	1	6	2.63	32.27	282.71
El Kef		2	5	2.58	29.93	262.18
Beja		1	5	2.42	28.00	245.17
Kelibia		2	5	2.44	25.38	222.32
Kairouan		2	5	2.09	17.29	151.43

Figure 3. Annual mean wind speed at 11 m height in an open area (in m/s).

Figure 4. Available power density from wind at 11 m height in an open area (in W/m²).

form [11-15]:

$$V(h) = \frac{U_*}{k}\text{Log}\left(\frac{h}{Z_0}\right) \quad \text{pour} \quad h \geq Z_0 \tag{7}$$

with: $V(h)$ is the mean flow speed to altitude h.

In the Tunisian meteorological stations, the wind speeds measurements are taken to 11 m height above ground level. These stations are often located at airports and having an open area roughness (Z_0 = 0.07 m). So, to transpose the data that is available (at 11 m) at the level that interests us (such as the propeller height), it is nec- essary to correct the wind measurements.

If we relate the measured speed (at 11 m) at height h, we obtain the following correction factor $C(h)$ (**Figure 5**):

$$C(h) = \frac{V(11m)}{V(h)} = \frac{\text{Log}\dfrac{11}{Z_0}}{\text{Log}\dfrac{h}{Z_0}} \tag{8}$$

The new values of the wind characteristics $(V_m, V_f, V_e, P_d \text{ and } E_d)$ calculated at the height h will be estimated directly starting from the wind characteristics to the height 11 m by the following relations:

Corrected mean speed:

$$V_m(h) = \sum_i V_i(h) \cdot f[V_i(h)] = \frac{V_m(11\text{ m})}{C(h)} \tag{9}$$

Corrected most frequent speed:

$$V_f(h) = V_i(h)\left(f\left[V_i(h)\right]\max i\right) = \frac{V_f(11\text{ m})}{C(h)} \tag{10}$$

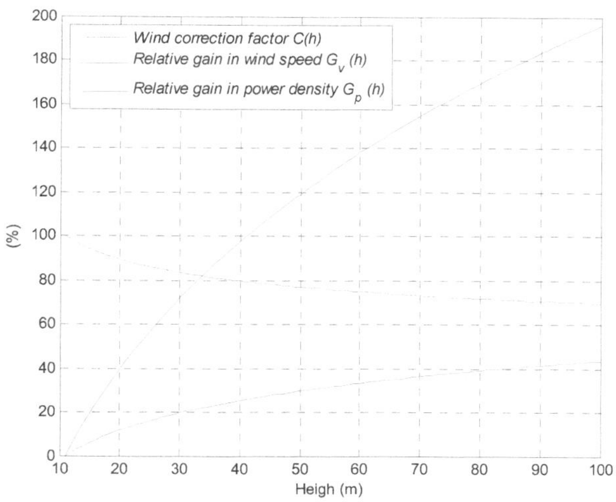

Figure 5. Influence of height on wind characteristics.

Corrected most energetic speed:

$$V_e(h) = V_i(h)\left(E_d\left[V_i(h)\right]\max i\right) = \frac{V_e(11\text{ m})}{C(h)} \quad (11)$$

Corrected power density:

$$P_d(h) = \frac{1}{2}\rho\sum_i\left[V_i(h)\right]^3 \cdot f\left[V_i(h)\right] = \frac{P_d(11\text{ m})}{\left[C(h)\right]^3} \quad (12)$$

Corrected available wind energy:

$$E_d(h) = 8.76\, P_d(h) = \frac{E_d(11\text{ m})}{\left[C(h)\right]^3} \quad (13)$$

The vertical extrapolation of the measured wind characteristics at 11 m height to a desired height h implies a relative gain in speed G_v and a relative gain in power G_p of expressions:

$$G_V(h) = \frac{V(h) - V(11\text{ m})}{V(11\text{ m})} = \frac{1 - C(h)}{C(h)} \quad (14)$$

$$G_P(h) = \frac{P_d(h) - P_d(11\text{ m})}{P_d(11\text{ m})} = \frac{1 - \left[C(h)\right]^3}{\left[C(h)\right]^3} \quad (15)$$

For example, extrapolation of data measured at 100 m height enables an increase of 43% on the characteristic speeds (mean V_m, energetic V_e and frequent V_f) of the wind and 190% on the power density (**Figure 5**).

3. Site-Wind Turbine Adaptation

3.1. Usable Wind Energy

The typical curve of a wind turbine is defined by **Figure 6**. Because of the operating mode of the wind systems, recoverable energy at the exit of a wind system represents only one portion of the available energy E_d in the wind [16-19].

The usable wind energy varies both with the values of the machine characteristic speeds (cut-in speed V_d, nominal speed V_n and cut-out speed V_c), *i.e.* with the type of wind turbine considered, and with the site via the wind frequency distribution $f(V)$. It is defined, per unit area and during one year, as follows:

$$E_u = 4.38\rho\left(\sum_{i=d}^{n} V_i^3 f(V_i) + V_n^3 \sum_{i=n}^{c} f(V_i)\right) \quad (16)$$

The extrapolation of this energy in function of height h is written:

$$\begin{aligned} E_u(h) &= 4.38\rho\left(\sum_{i=d}^{n}\left[V_i(h)\right]^3 \cdot f\left[V_i(h)\right] + V_n^3 \sum_{i=n}^{c} f\left[V_i(h)\right]\right) \\ &= 4.38\rho\left(\sum_{i=d}^{n}\left(\frac{V_i(11\text{ m})}{C(h)}\right)^3 \cdot f\left[V_i(11\text{ m})\right] \right. \\ &\quad \left. + V_n^3 \sum_{i=n}^{c} f\left[V_i(11\text{ m})\right]\right) \end{aligned} \quad (17)$$

The values of the characteristic speeds of a wind turbine depend on the type and the size of the wind machines.

They must be adapted in function of the wind characteristics of the establishment site. We study, for the 26 meteorological stations and the 7 types of wind turbines defined by V_n, the wind energy ratio E_u/E_d:

$$E_u/E_d = \frac{\sum_{i=d}^{n} V_i^3 f(V_i) + V_n^3 \sum_{i=n}^{c} f(V_i)}{\sum_{i=1}^{N} V_i^3 f(V_i)} \quad (18)$$

and the annual operating time T_f:

$$T_f = P(V_d \le V \le V_c) = F(V_d) - F(V_c) \quad (19)$$

Let us note that T_f is independent of the nominal speed V_n of the machine.

3.2. Application to the Tunisian Meteorological Sites

The objective of this paragraph is to provide a help to

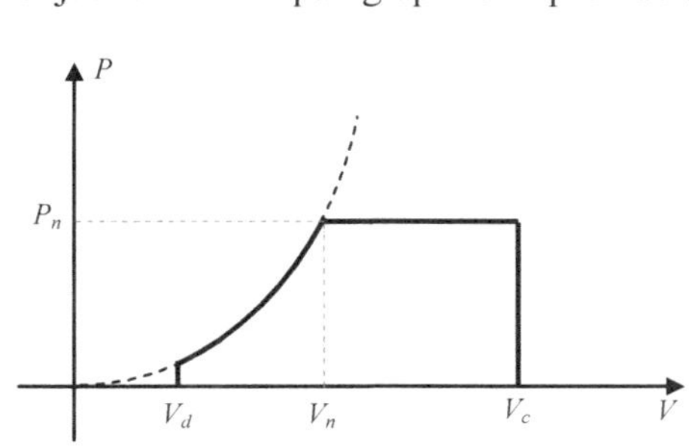

Figure 6. Ideal power curve of the wind turbine.

judiciously choose the best wind system adapted to a selected implantation site [6].

For this purpose, we have chosen 7 types of wind turbines defined by their nominal speeds V_n $(V_n = 11, \cdots, 17\ \text{m/s})$. The cut-in V_d and cut-out V_c speeds are constants and are respectively equal 3 and 25 m/s.

The selected height is of 60 m above ground level, which corresponds to a corrective factor C (60) = 75%, a relative gain in speed G_v = 33% and a relative gain in power density G_p = 137% (compared to the speed and the power density at 11 m height). **Table 2** gives the new values of the wind characteristics to the 60 m height.

For the 26 meteorological stations distributed on the whole of the Tunisian territory and for 7 chosen types of wind turbines, **Figures 7** and **8** give the annual wind energy ratio E_u/E_d and the annual operating time T_f $(V_d \le V \le V_c)$ at the 60 m height.

The analysis of **Figure 7** shows that:

Table 2. Extrapolation of the wind characteristics for different sites at 60 m height in an open area.

Sites	Groupes	V_f (m/s)	V_e (m/s)	V_m (m/s)	P_d (W/m^2)	E_d (kWh/m^2/year)
Thala	A	5.33	13.33	7.27	607.98	5325.82
Kebili		2.66	16.00	6.77	540.42	4734.13
Tozeur	B	4.00	10.66	6.21	328.89	2880.97
El Borma		6.66	9.33	6.31	311.70	2730.60
Jerba		4.00	9.33	6.19	311.09	2724.93
Kasserine		2.66	10.66	5.48	300.40	2631.40
Bizerte		1.33	10.66	5.49	269.16	2357.85
Monastir		5.33	10.66	5.72	263.21	2305.49
Nabeul		5.33	10.66	5.81	252.75	2214.14
Gafsa		4.00	9.33	5.69	217.79	1907.86
Tabarka	C	4.00	10.66	4.88	208.45	1826.11
Tataouine		2.66	8.00	4.66	174.44	1528.01
Tunis		4.00	9.33	4.86	169.62	1485.82
Remada		4.00	8.00	4.57	152.11	1332.48
Zaghouan		2.66	8.00	4.41	148.76	1303.23
Sfax		2.66	8.00	4.83	142.77	1250.70
Gabes		4.00	6.66	4.20	106.57	933.57
Mahdia		2.66	8.00	4.13	101.85	892.23
Sidi Bouzid	D	2.66	6.66	3.55	83.91	735.05
Siliana		2.66	6.66	3.65	83.48	731.35
Mednine		2.66	6.66	3.60	81.85	716.99
Jendouba		1.33	8.00	3.51	76.49	670.13
El Kef		2.66	6.66	3.44	70.95	621.46
Beja		1.33	6.66	3.23	66.37	581.14
Kelibia		2.66	6.66	3.25	60.16	526.98
Kairouan		2.66	6.66	2.79	40.98	358.95

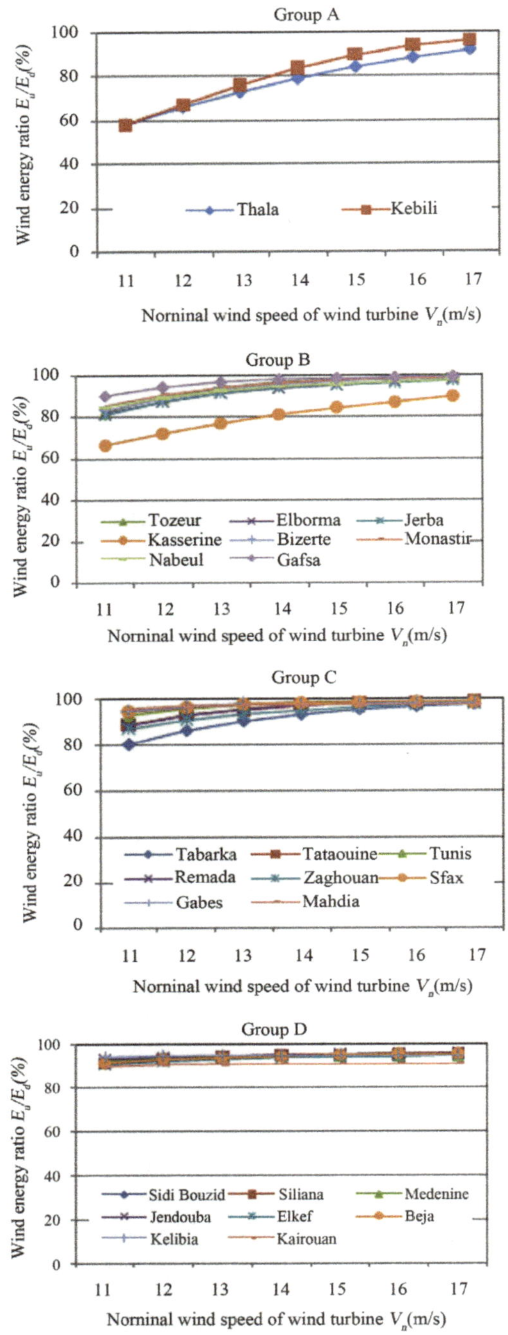

Figure 7. Annual percentages of wind energy ratio E_u/E_d at 60 m height for each sites group.

- For the most windy regions (sites of group A), the wind energy ratio increases with the nominal speed of the aerogenerator and varies between 60% and 95%.
- For the regions of good wind potential (sites of group B), the wind energy ratio is less sensitive to the machines of nominal speeds superior or equal to 15 m/s and varies between 96% and 98% (except for the Kasserine site).

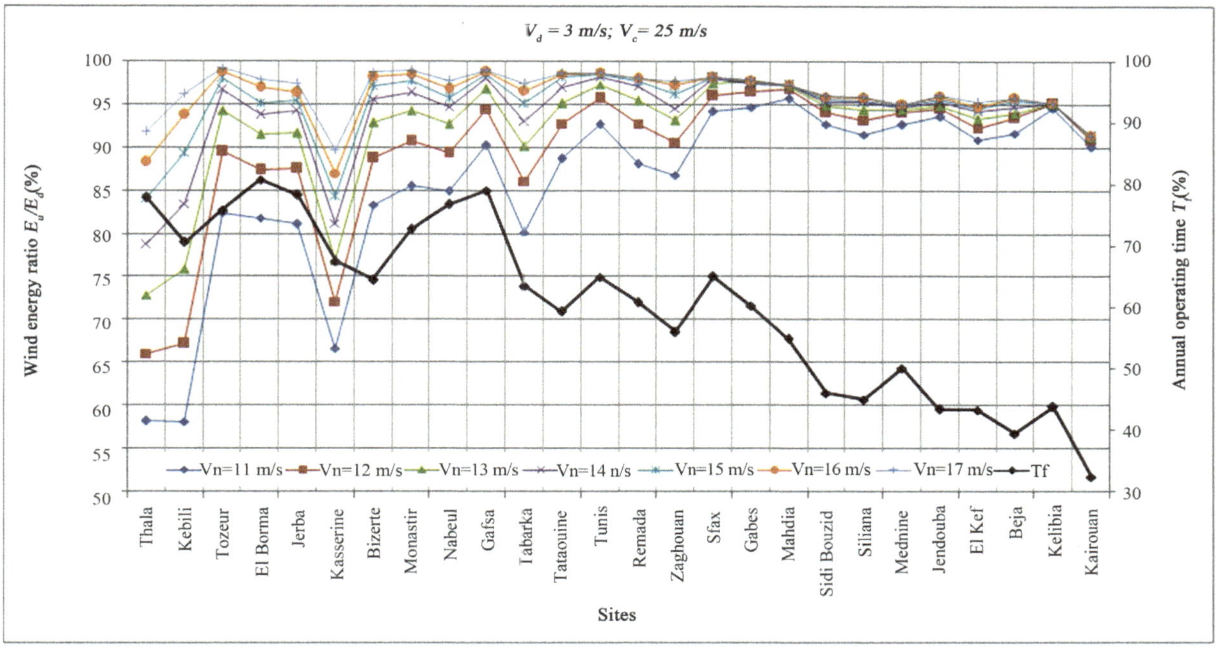

Figure 8. Annual percentages of operating time T_f and wind energy ratio E_u/E_d at 60 m height for seven types of wind turbine.

- For the regions of average wind potential (sites of group C), the variation of the wind energy ratio is weak and varies between 97% and 98% for machines of nominal speeds superior or equal to 14 m/s (except for the Tabarka site).
- For the slightly windy regions (sites of group D), the wind energy ratio is very high (superior to 90%) and varies little with the aerogenerator type.

The examination of **Figure 8** shows that the annual percentage of the operating time of a wind system is better for the windiest regions (groups A and B). Indeed, this percentage varies between 64.5% (in Bizerte) and 80.6% (in Elborma). For the less windy regions (groups C and D), this percentage degrades and reaches the minimal value of 32.5% (in Kairouan).

The simultaneous comparison of the annual percentages of the operating time and the wind energy ratio shows the need to choose the type of machine which adapts better to the implantation site. Indeed, it is necessary to realize a compromise between a maximum recovery of energy, a high time of production, and a good economic and technical profitability of the installation. A satisfactory compromise would be obtained while choosing (**Figure 8**):

- Machine of V_n = 17 m/s for the very windy sites (group A).
- Machine of V_n = 15 m/s for the windy sites (group B).
- Machine of V_n = 14 m/s for the fairly windy sites (group C).
- Machine of V_n = 11 m/s for the slightly windy sites (group D).

All these observations are valid for an open area roughness, at the 60 m height and machines of cut-in speed 3 m/s and cut-out speed 25 m/s. They will be reconsidered for different roughnesses, heights and speeds [12-14].

4. Conclusions

This study presents a practical approach of assistance for the wind systems exploitation adapted to the Tunisian sites. After the evaluation of the wind characteristics of the twenty six meteorological sites distributed on the Tunisian territory, we estimated the annual operating time and the wind energy which can provide a wind turbine given on the Tunisian sites. This estimate which was carried out for seven types of machines and a 60 m height above ground level gives a choice of the wind systems best adapted for the Tunisian meteorological sites, by taking into account energy, economic and technical efficiencies of the wind farm.

This estimate could be carried out for other types of machines and several heights above ground level and constitutes a decision-making aid tool to the appropriate choice of wind system.

The results of this work also enable us to avoid the bad wind turbine implantations in Tunisia that can have adverse consequences on the energy production.

REFERENCES

[1] A. Ouammi, V. Ghigliotti, M. Robba, A. Mimet and R.

Sacile, "A Decision Support System for the Optimal Exploitation of Wind Energy on Regional Scale," *Article in Press*, *Renewable Energy*, Vol. 37, No. 1, 2012, pp. 299-309.

[2] K. Xie and R. Billinton, "Energy and Reliability Benefits of Wind Energy Conversion Systems," *Renewable Energy*, Vol. 36, No. 7, 2011, pp. 1983-1988.

[3] C. Kongnam and S. Nuchprayoon, "A Particle Swarm Optimization for Wind Energy Control Problem," *Renewable Energy*, Vol. 35, No. 11, 2010, pp. 2431-2438.

[4] I. Jarraya, F. B. Amar, M. Elamouri and R. Dhifaoui, "Choix des Systèmes Eoliens Appropriés Pour les Sites Tunisiens," *International Renewable Energy Congress*, 5-7 November 2010, Sousse, pp. 162-168.

[5] F. Y. Ettoumi, A. H. Adane, M. L. Benzaoui and N. Bouzergui, "Comparative Simulation of Wind Park Design and Sitting in Algeria," *Renewable Energy*, Vol. 33, No. 10, 2008, pp. 2333-2338.

[6] K. A. Nigima and P. Parkerb, "Heuristic and Probabilistic Wind Power Availability Estimation Procedures: Improved Tools for Technology and Site Selection," *Renewable Energy*, Vol. 32, No. 4, 2007, pp. 638-648.

[7] A.-H. Marafia and H. A. Ashour, "Economics of Off-Shore/On-Shore Wind Energy Systems in Qatar," *Renewable Energy*, Vol. 28, No. 12, 2003, pp. 1953-1963.

[8] R. C. Bansal, T. S. Bhatti and D. P. Kothari, "On Some of the Design Aspects of Wind Energy Conversion Systems," *Energy Conversion and Management*, Vol. 43, No. 16, 2002, pp. 2175-2187.

[9] M. Elamouri and F. B. Amar, "Wind Energy Potential in Tunisia," *Renewable Energy*, Vol. 33, No. 10, 2008, pp. 758-768.

[10] F. B. Amar, M. Elamouri and R. Dhifaoui, "Energy Assessment of the First Wind Farm of Sidi Daoud, Tunisia," *Renewable Energy*, Vol. 33, No. 10, 2008, pp. 2311-2321.

[11] M. Elamouri, F. B. Amar and A. Trabelsi. "Vertical Characterization of the Wind Mode and Its Effect on the Wind Farm Profitability of Sidi Daoud, Tunisia," *Energy Conversion and Management*, Vol. 52, No. 2, 2011, pp. 1539-1549.

[12] S. Yahaya, "Dynamique de la Couche Limite de Surface Semi Aride: Approche des Caractéristiques Turbulentes par Anémométrie à Coupelles et Effet des Traitements du Surface du sol," Thesis, Paris University, Paris, 2004.

[13] B. Lange, S. Larsen, J. Hójstrup and R. Barthelmie, "Importance of Thermal Effects and Sea Surface Roughness for Offshore Wind Resource Assessment," *Journal of Wind Engineering and Industrial Aerodynamics*, Vol. 92, No. 11, 2004, pp. 59-88.

[14] B. Henderson-Sellers, "Plume Rise Modelling: The Effects of including a Wind Shear and a Variable Surface Roughness," *Ecological Modelling*, Vol. 37, No. 3-4, 1987, pp. 269-286.

[15] N. K. Merzouk, M. Merzouk and B. Benyoucef, "Pofil Vertical de la Vitesse du Vent Dans la Basse Couche Limite Atmosphérique," *Journées Internationales de Thermique*, 28-30 August 2007, Albi, pp. 1-5.

[16] T. J. Chang, Y. T. Wu, H. Hsu, C. Chu and C. Liao, "Assessment of Wind Characteristics and Wind Turbine Characteristics in Taiwan," *Renewable Energy*, Vol. 28, No. 6, 2003, pp. 851-871.

[17] M. H. Abderrazzaq, "Energy Production Assessment of Small Wind Farms," *Renewable Energy*, Vol. 29, No. 15, 2004, pp. 2261-2272.

[18] A. E. Kavak and S. Akpinar, "An Assessment on Seasonal Analysis of Wind Energy Characteristics and Wind Turbine Characteristics," *Energy Conversion and Management*, Vol. 46, No. 11-12, 2005, pp. 1848-1867.

[19] W. C. Cliff, "The Effect of Generalized Wind Characteristics on Annual Power Estimates from Wind Turbine Generators," Report PNL-2436, US Department of Energy, October 1977.

Nomenclature

$C(h)$	Correction factor of wind speed
G_v	Relative gain in wind speed
G_p	Relative gain in power density
E_u	Usable energy (kWh/m^2/year)
V_d	Cut in wind speed of wind turbine (m/s)
V_n	Nominal wind speed of wind turbine (m/s)
V_c	Cut out wind speed of wind turbine (m/s)
Tf	Availability rate of wind turbine
U_*	Friction wind speed (m/s)
ρ	Air density (1.225 kg/m^3)
K	Von Karman constant (k = 0.4)
Z_0	Roughness length of ground surface (m)
V	Classified wind speed (m/s)
$f(V)$	Occurrence frequency (%)
$F(V)$	Cumulated frequency (%)
n	Number of wind speed classes
P_d	Wind power density (W/m^2)
E_d	Available wind energy (kW/m^2/year)
V_m	Annual mean wind speed (m/s)
V_f	Most frequent wind speed (m/s)
V_e	Most energetic wind speed (m/s)
h	Altitude above ground level (m)

3

Efficient Power Conditioner for a Fuel Cell Stack-Ripple Current Reduction Using Multiphase Converter

Sampath Naveen Kumar, Natarajan Rajalakshmi*, Kaveripatnam Samban Dhathathreyan

Centre for Fuel Cell Technology, ARCI, IIT Madras Research Park, Chennai, India.

ABSTRACT

Integration of fuel cell stack with an inverter is complex in nature. A number of factors have to be taken into account in designing the inverter as well as during the integration. One of these factors is ripple current which could affect the life of the fuel cell stack if there is fuel and/or oxidant starvation. In this paper an inverter topology is investigated which significantly reduces or even nullifies the ripple content in the fuel cell system. The investigations have been carried out using indigenously developed 1 KW PEMFC stack and a 4 kW PEMFC stack with single and multi phase inverter. The results are presented here.

Keywords: Fuel Cell Stack; Ripple Current Reduction; Multiphase Inverter

1. Introduction

In recent times fuel cells of different types are under active consideration to meet the power needs of a variety of applications owing to a number of their positive characteristics. Fuel cells are electrochemical devices that convert chemical energy to electricity energy with high efficiency. In the case of stationary power applications, electricity can either be directly given to utility with the help of inverters or supply directly in decentralised/stand alone power applications. Fuel cells can also be integrated with DC-DC converters to cater to DC power requirements such as in telecom applications. In spite of the various beneficial features of fuel cell technology, certain challenges still remain, before they can be applied to real world applications. A fuel cell system includes besides the fuel cell stack, a reactants supply system, thermal management and power management system. There are many of challenges in the power conditioning unit (PCU) which vary depending on the application. Conventional inverters used in battery powered UPS systems are not suitable for fuel cell applications as the latter requires an inverter which can accept a large variable input DC voltage (sometimes 1:2) at high current. In addition, by its nature an inverter could impose its own time varying load on fuel cell which would lead to peak currents unsustainable by the fuel cells especially under low reactant conditions [1]. A typical fuel cell PCU employs a switch mode DC-DC and DC-AC converters or direct DC-AC inverter with transformer boost as shown in **Figures 1(a)** and **(b)**.

In order for low voltage DC fuel cell to generate 120/240 V AC at 50/60 Hz, the DC-AC inverter is designed considering the output voltage regulation of the fuel cell stack. The low voltage ac output of the inverter is stepped up using low frequency transformer. The inverter by its nature will draw current from the fuel cell stack twice that of the ac voltage frequency *i.e.* 100 Hz for 50 Hz ac output system as shown in **Figure 2**. The ripple current that is experienced by the stack could affect the fuel cell durability.

Chagrong Liu and Jih-Sheng Lai [2] have reported some methods to overcome low frequency ripple current. Woojin Choi *et al.*, [3] have carried out experiments using a commercially available fuel cell stack and the relationship between current and fuel cell stack performance such as power loss and fuel consumption has been investigated. Experimental results indicated that ripple current can contribute up to 10% reduction in the available

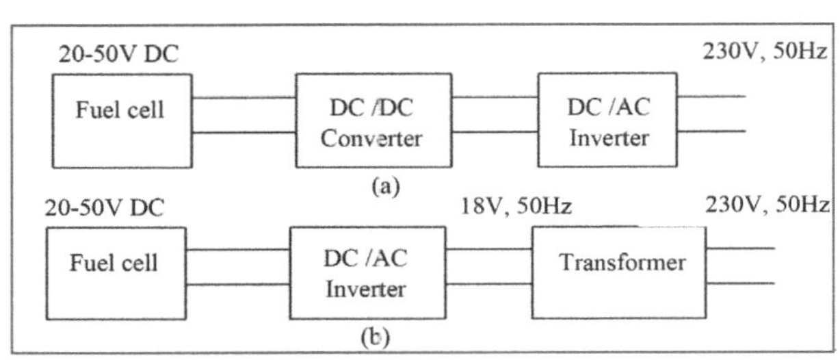

Figure 1. (a) Dc/Dc converter boost; (b) Transformer boost.

*Corresponding author.

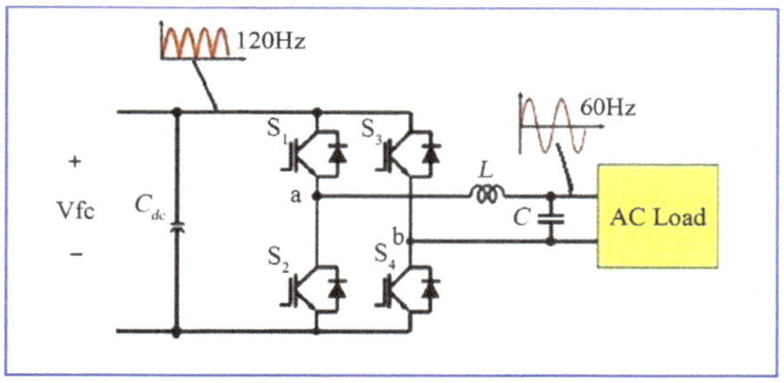

Figure 2. Low frequency ripple current generation by a single phase full bridge inverter [2].

output power. In the present paper we have attempted to reduce the ripple current of the inverter by multiphase interleaving by simulation and validated by experimental results for a fuel cell of 4 kW capacity.

2. Experimental

A transformer based inverter has been used, considering the following important design variables of the PCU like variation of fuel cell terminal voltage from no-load to full load and the maximum ripple current. A large variation in fuel cell terminal voltage from no-load to full load, results in larger volt-amp rating of the PCU. We have carried out various studies using indigenously developed 1 and 4 kW stacks with sine wave inverters, which delivers 220 V AC, 50 Hz and. The 1 kW PEM fuel cell stack consists of 52 cells, delivering a DC voltage was 30 V and the DC current 35 A. The 4 kW stack consisting of 40 cells, the nominal full load output voltage V_{fc} of the fuel cell stack is 24 V, and the average current I_{fc} is 165 A. Chauvin arnoux-C.A.8230 power analyzer has been used to measure the voltage, current and ripple content of the inverter power.

3. Results and Discussions

The voltage—current profile of the 1 kW fuel cell stack when integrated with the 1 kVA, 1 Φ inverter is shown in **Figure 3**. The maxima and minima of the current waveform for various average current along with ripple factor is presented in **Table 1**. From **Figure 3** it can be seen that for delivering a power of 850 W the average current of the fuel cell stack is 27 A. However, since the stack is loaded by inverter the peak current is 42 A. It can also be seen that the current ripple leads to ripple content in the fuel cell stack output voltage due to inherent impedance of fuel cell stack. The "*Ripple Factor*" which is defined as the ratio RMS value of the ripple content to the dc component is 40% in this case, in which the peak current exceeds the acceptable current limit of the stack, when the stack is operated at its full capacity Based on the above results, the current drawn by a single phase inverter has been simulated, using Pspice for a 4 kW inverter by considering a fuel cell stack delivering a stack output voltage of 25 V at 160 A. If the power were to be delivered by a 1 Φ inverter with typical ripple factor of 40% from the previous results of ripple factor, one can obtain the peak value of the ripple component for a sinusoidal wave, which is 90 A. Hence the maximum current drawn from the stack will be 250 A and the minima will be 70 A, for an average current of 160 A as shown in **Figure 4**.

In order to reduce the ripple current, two inverters are connected in parallel with an average current of 80 A (@

Table 1. Ripple factor at various stack current.

Average current (Idc) (A)	Current minima (A)	Current maxima (A)	Ripple factor (%)
4.3	1.1	7.5	52%
10.5	3.7	17.2	45%
16.5	5.9	27.0	44%
17.0	6.2	27.8	44%
23.2	8.4	38.0	45%
27.2	10.4	44.0	43%

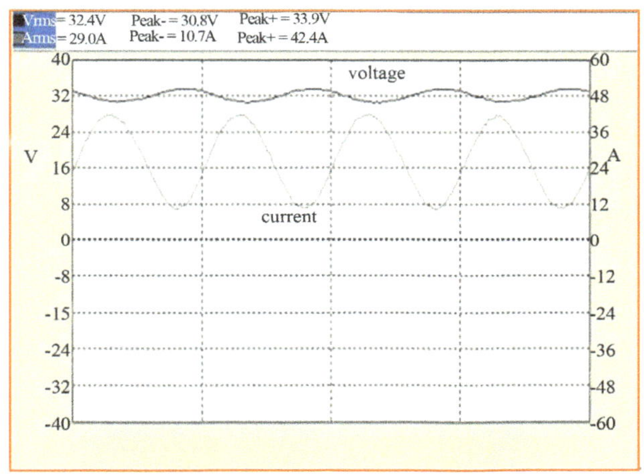

Figure 3. Voltage and current profile of fuel cell stack connected to inverter.

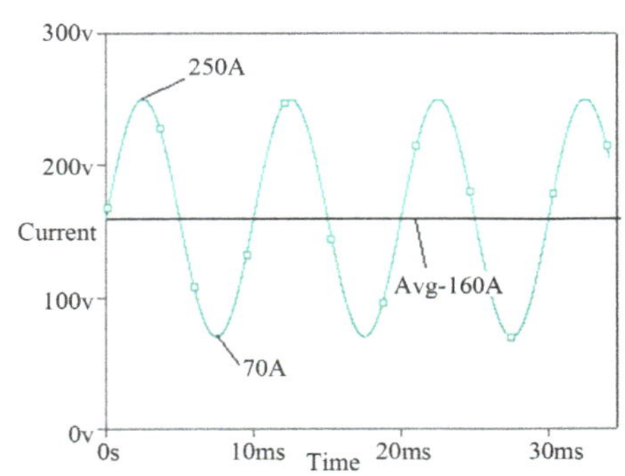

Figure 4. Ripple currents for a single phase inverter.

25 V) each, in order to deliver a total output power of 4 kW. If the two inverters are gated so that the AC output voltage is out of phase by 90° (*i.e.* 2 Φ converters) the current drawn from the stack by each inverter will be out of phase by 180° as shown in the **Figure 5**. Thus the resultant current drawn from the stack will be 160 A without ripple.

Experimental Results

The 2 Φ inverter is integrated with 4 kW fuel cell stack. The inverter is shown in **Figure 6**. The fuel cell stack is shown in **Figure 7**.

The fuel cell stack is loaded using the 2 Φ inverter, **Figure 8** shows the fuel cell voltage and the current drawn from the fuel cell stack, it can be seen that for average current of 142 A and stack voltage of 30 V the

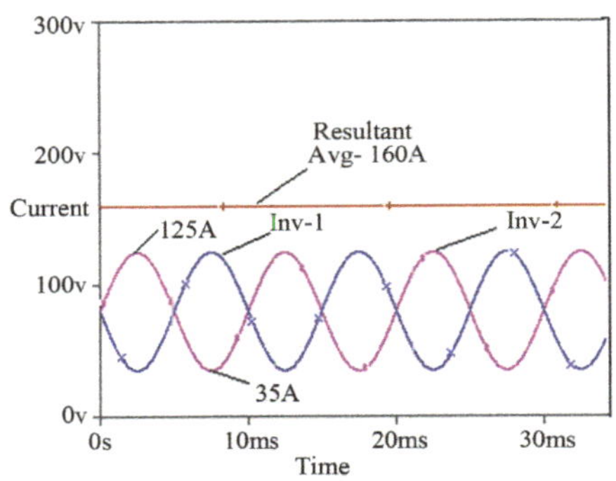

Figure 5. Stack current profile with 2 Φ converters.

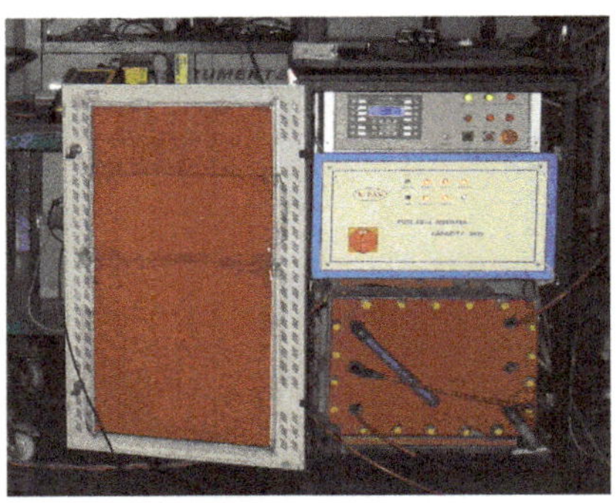

Figure 6. 2 Φ inverter.

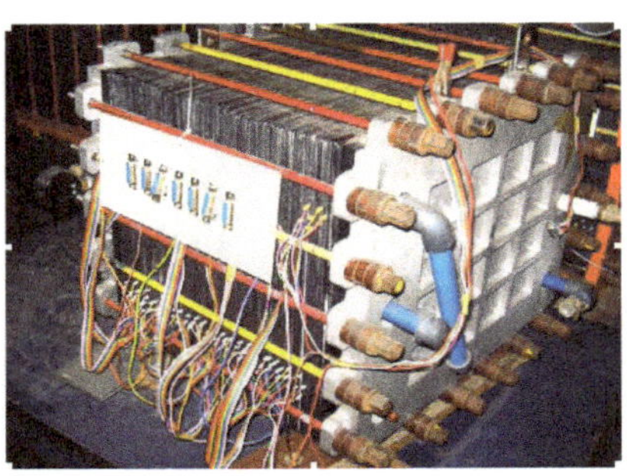

Figure 7. Fuel cell stack.

peak value of current is 150 A. The ripple factor is just 3%. **Figure 9** shows the same fuel cell stack connected to 1 Φ inverter loaded to 4.1 kW here we can see that the peak current drawn from the fuel cell is 194 A and the corresponding ripple factor is 32%.

Hence it can be seen that a 2 Φ inverter is highly beneficial in terms of ripple current reduction, thereby saving the fuel cell stack from damage.

4. Conclusion

This paper has investigated ripple generation in fuel cell stack connected to inverter, since it is claimed that ripple current and its corresponding peak current affects the stack performance, 2 Φ inverter is suggested which significantly mitigates ripple current and reduces the peak current imposed on the fuel cell. This also reduces the stack from damage by not drawing excess current than the rated current. This has been simulated and proven experimentally by connecting a fuel cell stack and a 2 Φ inverter and a 1 Φ inverter.

5. Acknowledgements

The authors would like to acknowledge Dr. G. Sundararajan, Director, ARCI for supporting this work. We also

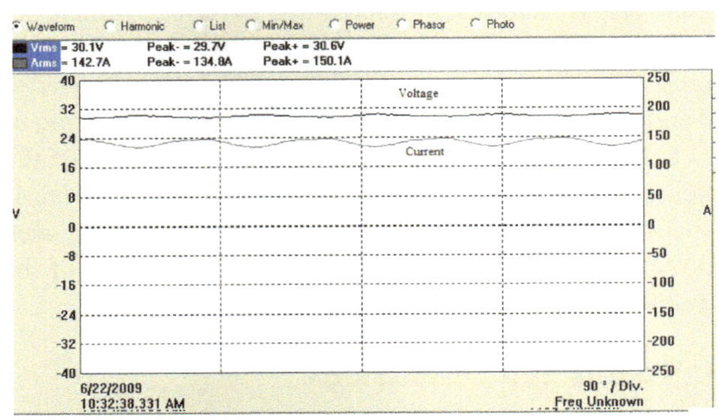

Figure 8. Voltage and current profile of fuel cell stack connected to 2 Φ inverter delivering 4.2 kW.

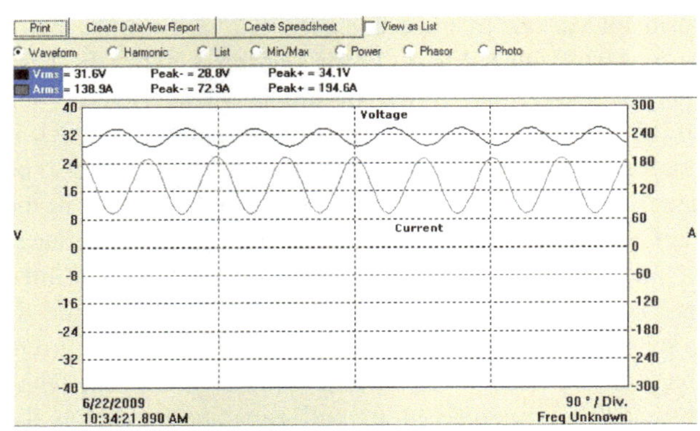

Figure 9. Voltage and current profile of fuel cell stack connected to 1 Φ inverter delivering 4.1 kW.

acknowledge the financial assistance from DST, New Delhi, Government of India.

REFERENCES

[1] R. S. Gemmen, "Analysis for the Effect of Inverter Ripple Current on Fuel Cell Operating Condition," *Journal of Fluids Engineering*, Vol. 125, No. 3, 2003, 10 p.

[2] C. R. Liu and J.-S. Lai, "Low Frequency Current Ripple Reduction Technique with Active Control in a Fuel Cell Power System with Inverter Load," *IEEE Transactions on Power Electronics*, Vol. 22, No. 4, 2007, pp. 1429-1436.

[3] W. Choi, P. N. Enjeti and J. W. Howze, "Development of an Equivalent Circuit Model of a Fuel Cell to Evaluate the Effects of Inverter Ripple Current," 19*th Annual IEEE Applied Power Electronics Conference and Exposition*, Vol. 1, 27 September 2004, pp. 355-361.

Experimental Research on Monomial Cooling Measure of Greenhouse in Summer

Jian Hu

College of Information and Engineering-Technology, Sichuan Agricultural University, Ya'an, China.

ABSTRACT

This paper describes several main monomial cooling methods of greenhouse in summer, including natural ventilation, mechanical ventilation, wet curtain cooling, external sun-shading etc., as well as analyzes the change law of the environment on temperature and humidity in greenhouse. The result shows that the four cooling methods stated above make the temperature gradients between internal and external respectively 1.1˚C, 0.4˚C, −2.6˚C and −1.2˚C, so it is obvious that the cooling effect of wet curtain cooling is the optimum; and the internal and external humidity difference is respectively: −13%, −6%, 8% and 1%.

Keywords: Glasshouse; Cooling Measure; Temperature; Humidity

1. Introduction

With the continuous improvement of science, technology and the modernization of agricultural production, agricultural facilities in China boom rapidly. Greenhouse, as the main type of agricultural facilities, has significant social and economic benefits, and it has been widely used in our country. But there are also a series of problems in the application process, especially for the cooling problem of greenhouse in summer that has always been the concern of agricultural scientific workers. July and August are the hottest months in Chongqing Municipality throughout the year. Mean monthly temperature is about 27˚C - 28˚C [1], and the existence of "greenhouse effect" makes the indoor temperature reach more than 50˚C. Crops are completely unable to survive, resulting in making greenhouse in an "idle" state over a third of time. The greenhouse in summer must be considered to take corresponding cooling measures to organize normal production, so as to improve the utilization rate. And it can provide the crops a good growth environment for achieving a high yield, a high quality, and a high efficiency of the intensive mode of production, so to meet the needs of people's life, and to improve the land utilization rate.

2. Experiment Conditions and Test Scheme

2.1. Experiment Conditions

This experiment is proceeded in water-saving irrigation efficiency demonstration laboratory of College of Engineering and Technology, Southwest University. The laboratory is four straight across Venlo type, which is a large high-grade glasshouse with one cross and three ridges. As **Figure 1** shows, the greenhouse runs north to south, with a total area of 9.6 m/span × 4 span × 36 m = 1382.4 m^2. It is divided into four areas: the first area is vegetable area, the second is flower area, the third area is for fruit nursery, and the last is for nursery garden. Each area is 9.6 m × 36 m = 345.6 m^2, and there is a central control room in front of flower area. All the top and the board of the greenhouse are covered with 4 mm domestic high quality float glass. Light transmittance is greater than 90%; the coefficient of heat transfer K = 6.6 W/(m^2·˚C); the greenhouse is fixed by special aluminum alloy; it is sealed by rubber strip.

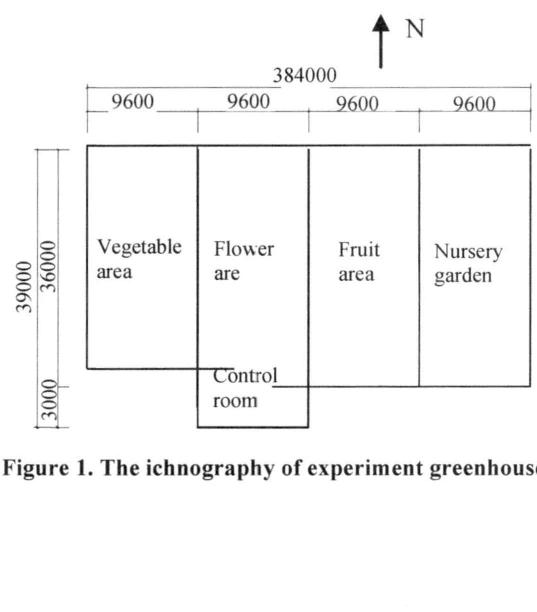

Figure 1. The ichnography of experiment greenhouse.

2.2. Test Instruments and Equipments

1) Greenhouse with natural ventilation system, mechanical ventilation system, wet curtain cooling system and external shading system [2].

2) 16 numbers of thermocouple sensors, one placed outdoor, the other fifteen placed indoor. They are located in three different planes. Thermal performance test system of TRM-2 Solar energy water heater (developed by Jinzhou Sunshine Technology Development Company) is used to record temperature automatically. The recording tome is setted for every 30 minutes.

3) Three wet and dry bulb thermometers are setted, 1 placed outdoor, the other 2 placed indoor, which are used to determine the dry bulb temperature, wet bulb temperature and relative humidity outdoors and indoors.

4) Computer is mainly used for the derivation of data. And some related softwares are also used for data processing.

2.3. Test Arrangement

This test is conducted in the second working place, the area—flower area, as is shown in **Figure 2**. Greenhouse is divided into two symmetrical work areas by corridor, and six observation positions are set up on the right side of the horizontal plane area for analyzing the temperature distribution rule in the horizontal plane during the cooling process indoors. And in the position 3, a temperature sensor is placed underground with a distance of 20 cm from the ground, which is used to measure the geothermal change rule in the whole test process. An observation point 7 is placed outdoors, which is used for monitoring the change of outdoor temperature. In addition, in the greenhouse there placed two dry wet bulbs, and out of the greenhouse there placed dry wet bulb, which are used to measure the relative humidity change rules outdoors and indoors.

Figure 3 shows the map of temperature vertical distribution of each point. Three planes are set up in vertical direction. The third plane is 500 mm distance from the ground. The first plane is 500 mm distance from the gutter. They are used for measuring the variation of temperature in vertical direction. All the temperature data are the automatic record of thermal performance test system of TRM-2 type solar energy water heater by every 30 minutes. The relative humidity is recorded by worker every 30 minutes.

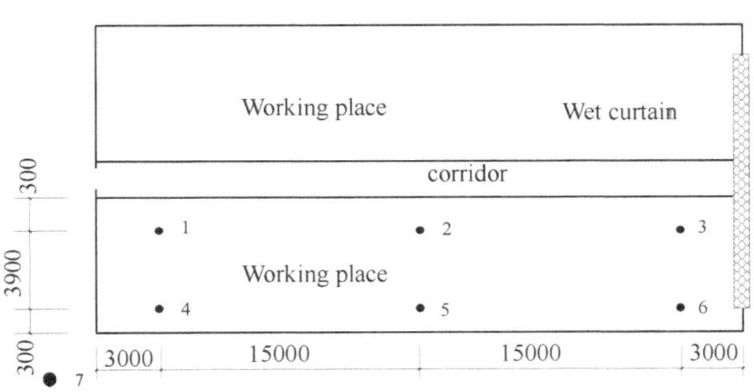

Figure 2. Horizontal distribute chart of every observation.

3. Test Results and Analysis

3.1. Natural Ventilation Cooling Effect

Before the test, the greenhouse is closed. At 9:00 a.m., the indoor temperature reaches 34˚C, and the outdoor temperature reaches 31.6˚C. Open the axial flow shutters, wet curtain insect nets and top windows for natural ventilating, as shown in **Figure 4**. After the ventilation, the temperature of every plane in the greenhouse rises slowly, but the rise speed of temperature is slower than outdoor. After 10:00 a.m., the increase trend of temperature is basically same as outdoors. To about 13:00 p.m., temperatures of each plane all reach the maximum. They are about 7.9˚C, 5.4˚C, 5.3˚C higher than the temperature outdoors. With the cooling down outdoors, the temperatures of each plane indoors slow declining as well. By the end of the test, the indoor minimum temperature is 37.4˚C, which is beyond the temperature range of normal

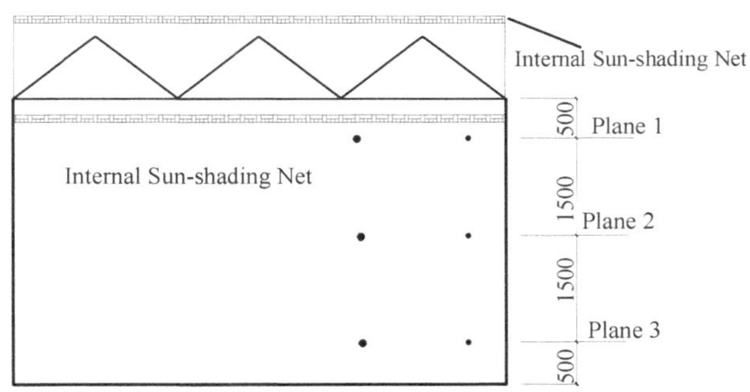

Figure 3. The vertical distribution of every measuring point in flower area.

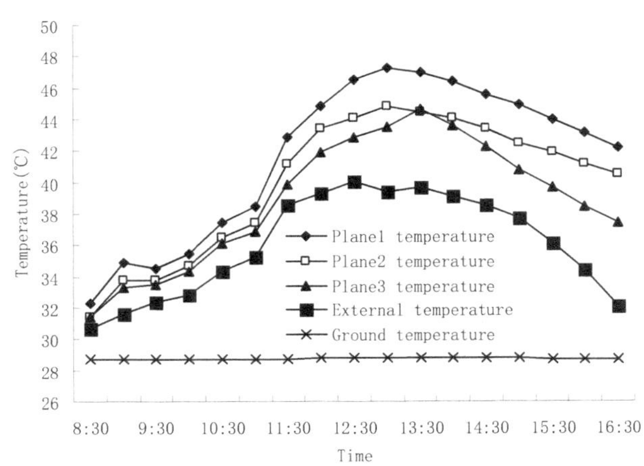

Figure 4. The change of temperature in natural ventilation cooling process.

growth of crops. During the whole process of natural ventilation, the humidity outdoors remains a relatively high level. While the relative humidity indoors changes obviously, which presents a trend of gradual declining, and the lowest point 33% appears at 14:30 p.m. After that it increases with the rising of humidity outdoors, as is shown in **Figure 5**.

The analysis above shows that it is impossible to use natural ventilation to reduce the temperature in greenhouse to achieve the demand of crop growth in the summer under such a climate situation of hot and humid in Chongqing area. It must adopt other ways to achieve the purpose of cooling.

3.2. Mechanical Ventilation and Cooling Effect

Making axial flow as power machine to extract the air indoors outside to form negative pressure, it can realize forcing gas exchange between indoors and outdoors [3]. Although this method consumes certain energy, the ventilation effect is reliable, and it's easy to achieve regulating and controlling according to the needs. While, the range of temperature controlling is limited, and it can only control to equal the internal and external temperatures, as is shown in **Figure 6**. Through the positive action of the mechanical ventilating, the air exchange degree between indoors and outdoors is strengthened. Before 12:00 a.m. and after 16:00 p.m., the temperature of plane 2 and 3 are basically the same as the temperature outside. From 12:00 to 16:00 p.m., due to the enhancement of sun radiation, the temperature of the two planes above are higher than the air temperature outside about 2°C. Plane 1, near the roof of the greenhouse, as its air exchange effect is weak, temperature here is higher than the temperature outdoors during the whole test process. And its highest temperature is up to 42.9°C, which is 5.8°C higher than the air temperature outside. After the

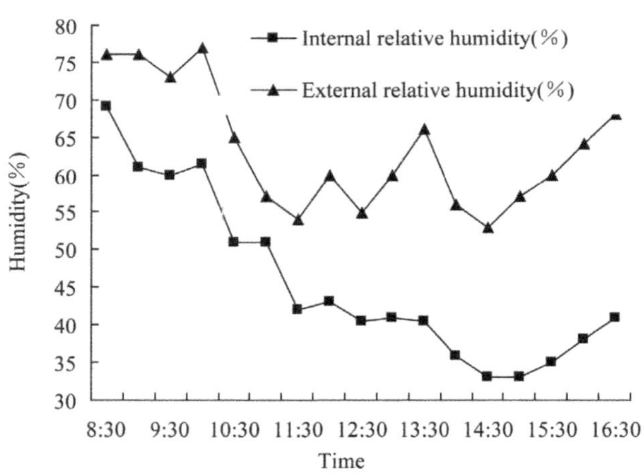

Figure 5. The change of internal and external humidity and temperature in natural ventilation cooling process.

use of mechanical ventilation, the indoor relative humidity reduces with the decrease of humidity outdoors gradually. By the time of 10:00 a.m., there appears an intersection (54%), and the humidity indoors and outdoors is equal, which is consistent with the actual fact. After words, with the increase of temperature indoors, the relative humidity presents a reducing trend, and the minimum declines to 44%, as is shown in **Figure 7**.

Through the analysis above, it is easy to see that there are a certain cooling effect and a dehumidifying effect through using mechanical ventilation to maintain temperature at about 35°C when the temperature outdoors is not more than 35°C. For the high temperature and humidity climate characteristic in Chongqing, it can be regarded as a kind of ideal cooling mode in a certain condition.

3.3. Wet Curtain Cooling Effect

Wet curtain cooling is the main cooling way of modern greenhouse [4]. It has been a mature technology, and the cooling effect is obvious, which has been widely used, as

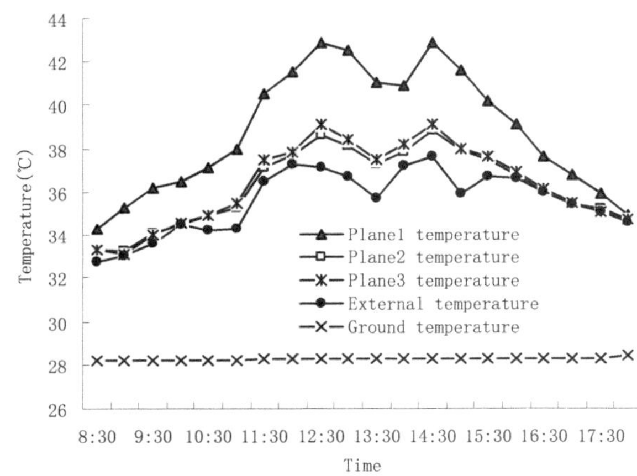

Figure 6. The change of temperature under mechanical ventilation cooling.

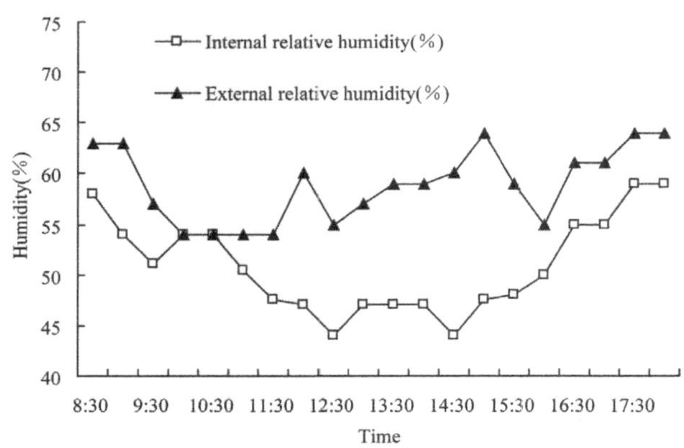

Figure 7. The change of internal and external temperature and humidity under mechanical ventilation cooling.

is shown in **Figures 8** and **9**. Open the wet curtain cooling system, the temperature indoors appears obviously decline. After 30 minutes, the temperature indoors drops to 31.7˚C, and at the same time, the indoor relative humidity increases from 60.5% to 75% after cooling. And the relative humidity increases by 24%. After opening the wet curtain cooling system, the temperatures of plane 2 and plane 3 are both significantly dropped below the temperature outdoors. After 9:30, the difference in temperature indoors and outdoors remains stable, and the highest temperature on the two planes is 35˚C. The cooling effect is very obvious. The temperature of plane 1 also declines, but it is higher than the outdoor temperature all the time, and the highest temperature can reach 44.6˚C.

Through the data analysis above, it shows that wet curtain cooling system has a good cooling effect, which can basically maintain temperature indoors within the scope of the crop normal growth. While, when the temperature reduces, the indoor humidity increases quickly.

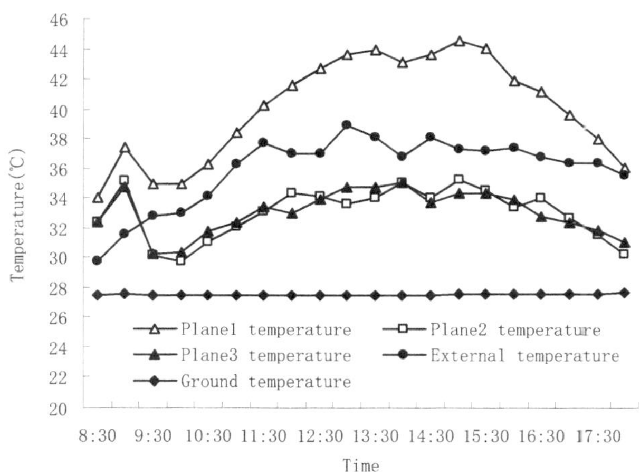

Figure 8. The change of temperature under wet curtain cooling.

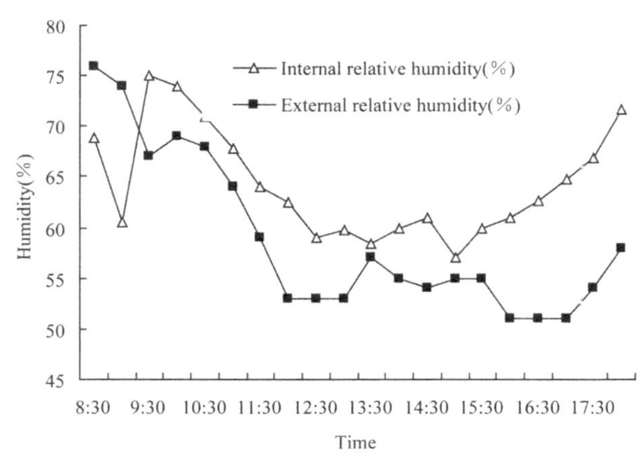

Figure 9. The change of temperature and humidity under wet pad curtain cooling.

If this condition continues for a too long time, it may cause plant diseases and insect pests. Besides, the high humidity will make open-strength degree of leaves stomata decrease, consequently inhibiting crop transpiration, making bad affects on the crops for the absorption and transport of moisture, nutrients. It can be considered to use mechanical ventilation with it in interval using, so as to achieve the purpose of dehumidification and eliminating the high humidity environment indoor.

3.4. The Cooling Effect of External Sun-Shading

According to the difference of placement, there are external sun-shading and internal sun-shading these two types of sun-shading [5]. The internal sun-shading will make a lot of heat gathered between the sun-shading net and the roof, preventing the heat indoors from dispersing and cooling in summer. So the external sun-shading is considered only to cool the temperature. There are two kinds of testing methods: one is under the condition of closing the greenhouse, unfolding the external sun-shading to collect temperature and humidity data indoors and outdoors data. Another one is to open the top window to ventilate to make the indoor and outdoor temperature and humidity in a relatively steady state, and then, unfold the external sun-shading and collect the temperature and humidity data indoors and outdoors. According to the actual situation, the first method has no real significance. Because in an airtight greenhouse, the air indoors can not circulate from the outside. Besides, the temperature indoors is very high, and crops can't survive. It is impossible that actual greenhouse production appears this kind of circumstance. So the second method is adopted, and the test result is shown in **Figures 10** and **11**.

Open the top window in the greenhouse to ventilate before the test. At the time of 10:00 a.m., when the temperature and humidity indoors and outdoors have been on the condition of relative stability, the temperature indoors

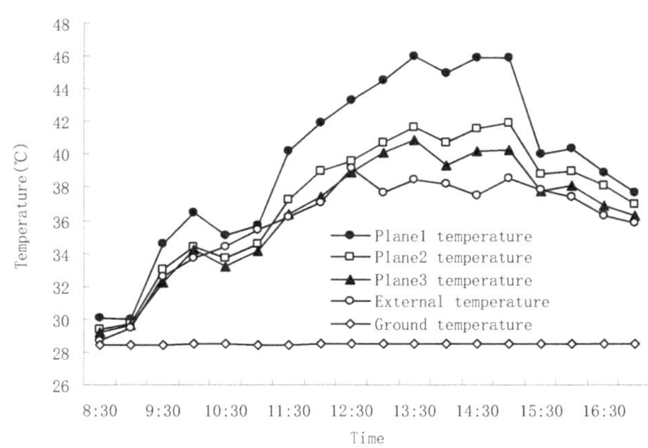

Figure 10. The change of temperature under outside sun-shading.

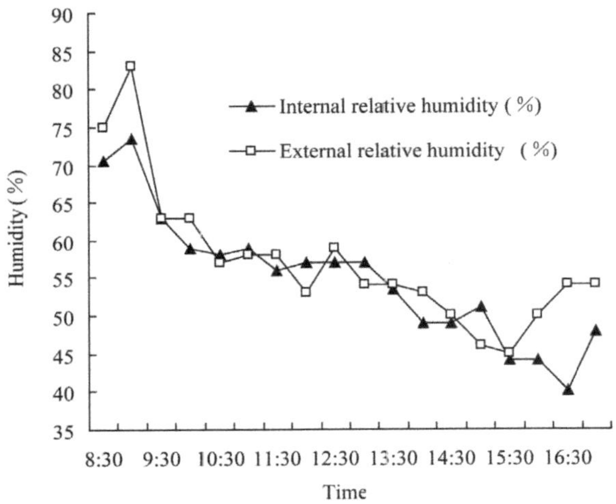

Figure 11. The temperature and humidity change indoors and outdoors under external sun-shading.

is about 35˚C and the temperature outdoor is about 33.7˚C, opening the external sun-shading and closing the top window. After the opening of the external sun-shading at 10:00 a.m., the temperature indoors of all planes drop. At the time of 10:30 a.m., the temperature of plane 1 drops from initial 36.5˚C to 35.1˚C, the temperature of plane 2 drops from initial 34.4˚C to 33.7˚C, and the temperature of plane 3 drops from initial 34.2˚C to 33.2˚C, while the temperature of outdoor is about 34.4˚C at that moment. The temperatures of both plane 2 and plane 3 are lower than that of the outdoors. The temperatures of the two planes and the outdoor don't appear intersection until 11:30. After that, as the sun radiation enhancing, temperature indoor presents a rise tendency. Cause the effect of the external sun-shading, the humidity indoors and outdoors maintain 57% basically. In the subsequent time, both of them present alternate change trend, but the difference is small.

4. Conclusions

Tests are respectively conducted with four cooling measures of natural ventilation, mechanical ventilation, wet curtain and external sun-shading. There are some relaxation effects on the high temperature environment indoors. To further reduce the temperature in greenhouse in summer, it must be combined with cooling methods.

1) Wet curtain has the best cooling effect. The change of the temperature and humidity of the internal planes is not very influenced by the change of the eternal. Temperature in horizontal direction distributes unevenly, and it fluctuates greatly. If the air outdoors has a large relative humidity, the cooling effect will be affected a lot, and this cooling mode will make humidity indoors increase greatly, which is up to 24%, and which restrains crop transpiration. It can be considered to use mechanical ventilation by interval, so as to achieve the purpose of dehumidification and eliminating the high humidity environment indoors.

2) After opening the external sun-shading, the temperature on working planes in greenhouse can be reduced to below outdoors. And the temperatures of the working planes in horizontal distribution are well distributed, which can continue to noon. In addition, the influence to the humidity indoors is not very obvious.

3) Through using mechanical ventilation, it can strengthen the air exchange between indoors and outdoors. When the temperature outside is under 35˚C, the temperature indoors can be controlled at 35˚C or so, which can basically satisfy the needs of crop growth. At the same time, it also has a dehumidifying effect, which can be used together with other cooling system to achieve the purpose of dehumidification.

4) It is difficult to achieve the effect of cooling temperature through the natural ventilation. The methods above can be used when there exists a feeble solar radiation intensity. Only in cloudy days, it can strengthen air exchange between indoors and outdoors by natural ventilation to reduce indoor humidity.

REFERENCES

[1] C. W. Ma, "Agricultural Biological Environment Engineering," China Agriculture Press, Beijing, 2005.

[2] J. Q. Wang, "Application of Some Cooling Measures for Greenhouse Cooling in Summer," *Transactions of the CSAE*, Vol. 22, No. 9, 2006, pp. 257-260.

[3] C. Y. Ran, "Temperature-Decreasing Research and Use for Glass-Greenhouse in Summer," *Journal of Jilin Architectural and Civil Engineering Institute*, Vol. 19, No. 4, 2002, pp. 31-34.

[4] B. G. Wang and X. W. Miao, "A Study on Comparison of Cooling Efficiency for the East-China Type Multi-Span Plastic Greenhouse in Summer," *Journal of Agricultural Mechanization Research*, Vol. 26, No. 5, 2004, pp. 187-193.

[5] J. Hu, "Cooling Technology of Modern Greenhouse in Summer," *Journal of Agricultural Mechanization Research*, Vol. 29, No. 6, 2007, pp. 18-21.

Hybrid Prediction Method for Solar Power Using Different Computational Intelligence Algorithms

Md Rahat Hossain, Amanullah Maung Than Oo, A. B. M. Shawkat Ali

Power Engineering Research Group (PERG), Central Queensland University, Rockhampton, Australia.

ABSTRACT

Computational Intelligence (CI) holds the key to the development of smart grid to overcome the challenges of planning and optimization through accurate prediction of Renewable Energy Sources (RES). This paper presents an architectural framework for the construction of hybrid intelligent predictor for solar power. This research investigates the applicability of heterogeneous regression algorithms for 6 hour ahead solar power availability forecasting using historical data from Rockhampton, Australia. Real life solar radiation data is collected across six years with hourly resolution from 2005 to 2010. We observe that the hybrid prediction method is suitable for a reliable smart grid energy management. Prediction reliability of the proposed hybrid prediction method is carried out in terms of prediction error performance based on statistical and graphical methods. The experimental results show that the proposed hybrid method achieved acceptable prediction accuracy. This potential hybrid model is applicable as a local predictor for any proposed hybrid method in real life application for 6 hours in advance prediction to ensure constant solar power supply in the smart grid operation.

Keywords: Computational Intelligence; Heterogeneous Regressions Algorithms; Performance Evaluation; Hybrid Method; Mean Absolute Scaled Error (MASE).

1. Introduction

Large scale penetration of solar power in the electricity grid provides numerous challenges to the grid operator, mainly due to the intermittency of sun. Since the power produced by a photovoltaic (PV) depends decisively on the unpredictability of the sun, unexpected variations of a PV output may increase operating costs for the electricity system as well as set potential threats to the reliability of electricity supply [1]. One of the main concerns of a grid operator is to predict changes of the solar power production, in order to schedule the reserve capacity and to administer the grid operations [2-6]. However, the prediction accuracy level of the existing methods for solar power prediction is not up to the mark; therefore, accurate solar power forecasting methods become very significant. Next to transmission system operators (TSOs), the prediction methods are required by various end-users as energy traders and energy service providers (ESPs), independent power producers (IPPs), etc. The accurate prediction methods are essential to provide inputs for different functions such as economic scheduling, energy trading and security assessment.

Many researches focus on providing a forecasting method in order to predict solar power production with expected accuracy. Depending on their input, these methods can be classified as physical or statistical approaches or jointly approach. The physical models use physical considerations, as meteorological (numerical weather predictions) and topological information, and technical characteristics of the PV (power curve, photo conversion efficiency and correction factor for photo conversion efficiency). Their intention is to get the most likely approximation of the local solar radiation and then use Model Output Statistics (MOS) to diminish the remaining error. Statistical models utilize descriptive variables and online measurements, typically employing recursive techniques, like recursive least squares or artificial neural networks. Moreover, physical models have to be used and statistical models may use Numerical Weather Prediction (NWP) models [7-11]. According to the literature review, an hourly irradiance producer [12] as well as based on a whole day [13] was invented by Gordon and Reddy. Knight *et al.* [14] demonstrated methods to generate irradiance data based on hour is known as GEN which involves the contribution of the month based standard irradiance data to produce hour based irradiance figures [14]. The hypothesis that simulations run with those produced data for a distinct one year time would provide fairly parallel outcomes to those gained from simulations determined by numerous years of calculated data was proved by Knight *et al.* [14]. Synthetically produced hourly varying solar radiation data and the temperature was employed as input data for the simulation program by Morgan [15].

To facilitate successful exploration for the best model, reliable techniques are required for combining different regression algorithms, creating ensembles, model testing etc. It is called meta-learning or ensemble learning research [16]. Since every inductive learning algorithm makes use of some biases, it performs well in some domains where its biases are suitable while it behaves poorly in other domains. One algorithm cannot be superior, in terms of generalization performance to another one among all the domains [17-18]. One approach to conquer this dilemma is to make a decision when a method may be appropriate for a given problem by selecting the best regression model according to cross validation [19]. A second, more accepted approach is to combine the capabilities of two or more regression methods [20]. To date, comparatively few researches have addressed ensembles for regression [20-22]. The success of the techniques that combine regression models comes from their ability to diminish the bias error as well as the variance error [23]. The bias is an assessment of how closely the model's average prediction, measured over all possible training sets of fixed size. Variance is a measure of how the models' predictions will differ from the average prediction of over all possible training sets of fixed size. Most of ensemble or hybrid methods described so far use models of one single class, e.g. neural networks [24] or regression trees [21] to predict renewable energy. Ensemble learning consists of two problems; ensemble generation: how the base models are generated and ensemble integration: how the base models' predictions incorporated to get better performance. Ensemble generation can be differentiated as homogeneous if each base learning model employs the identical learning algorithm or heterogeneous if the base models can be developed from a variety of learning algorithms. There has been much empirical work on ensemble learning for regression in the perspective of neural networks, however there has been fewer research carried out in terms of using heterogeneous ensemble techniques to improve the performance.

In this paper, a novel hybrid method for solar power prediction is proposed that can be used to estimate PV solar power with improved prediction accuracy. Here, the term "hybridization" is anchored in the top most three selected heterogeneous regression algorithm based local predictors and a global predictor. The proposed method focuses on one of the two decisive problems of ensemble learning namely heterogeneous ensemble generation for solar power prediction. **Figure 1** outlines the sequential structure of the proposed hybrid method of solar power prediction. For the time being this particular paper deals with the first two steps; presenting ensemble generation strategy and preliminary prediction performance of the local predictors based on the selected top most three regression algorithms.

Overview of the proposed hybrid prediction method is presented in Section 2. The data used in the experiment is described in Section 3. The experiment design for the ensemble generation, estimating and comparing the strengths of preliminary selected regression algorithms with 10 folds cross-validation and training and testing error estimator method are depicted in Section 4. Six hours ahead individual predictions performed by the initially selected regression algorithms and the accuracy of the prediction performances are validated with both the scale dependent and scale free error measurement metrics in Section 5. The detail theory about regression algorithms are presented in Section 6. Independent samples t-tests are carried out to evaluate the individual mean prediction performance of the preliminary selected regression algorithms as well as the pair wise comparisons for mean predicted values of those algorithms in Sections 7 and 8 respectively. Finally, the paper concludes with the concluding remarks and recommendations.

2. Description of the Hybrid Method for Solar Power Prediction

First of all, the ensemble generation is performed from a pool of regression algorithms. For this purpose the three top most regression algorithms to act as the local predictors are selected based on experimental results. Next, feature selection is carried out. The feature selection aspect is fairly significant for the reason that with the same training data it may happen that individual regression algorithm can perform better with different feature

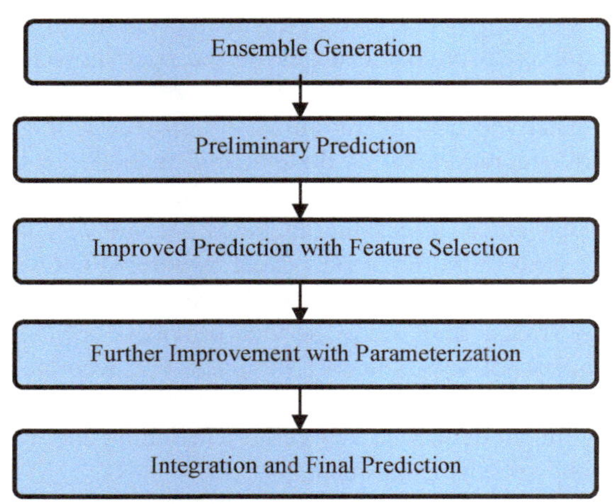

Figure 1. Outline of the proposed hybrid method.

sub sets [25]. The aim is to reduce the error of individual local predictors. The preliminary prediction using the selected three regression algorithms with feature selection is executed then. The structure of the proposed hybrid method for solar power prediction is shown in **Figure 2** where RA_i = Regression algorithm for local or base predictor.

In brief, the working mechanism of the proposed hybrid prediction method for solar power prediction can be stated in the following way. Raw set of data comprises stage one of data. Regression algorithm based local predictors operate in this stage. Stage two data are the preliminary predictions from the local predictors.

A further learning procedure takes place by means of

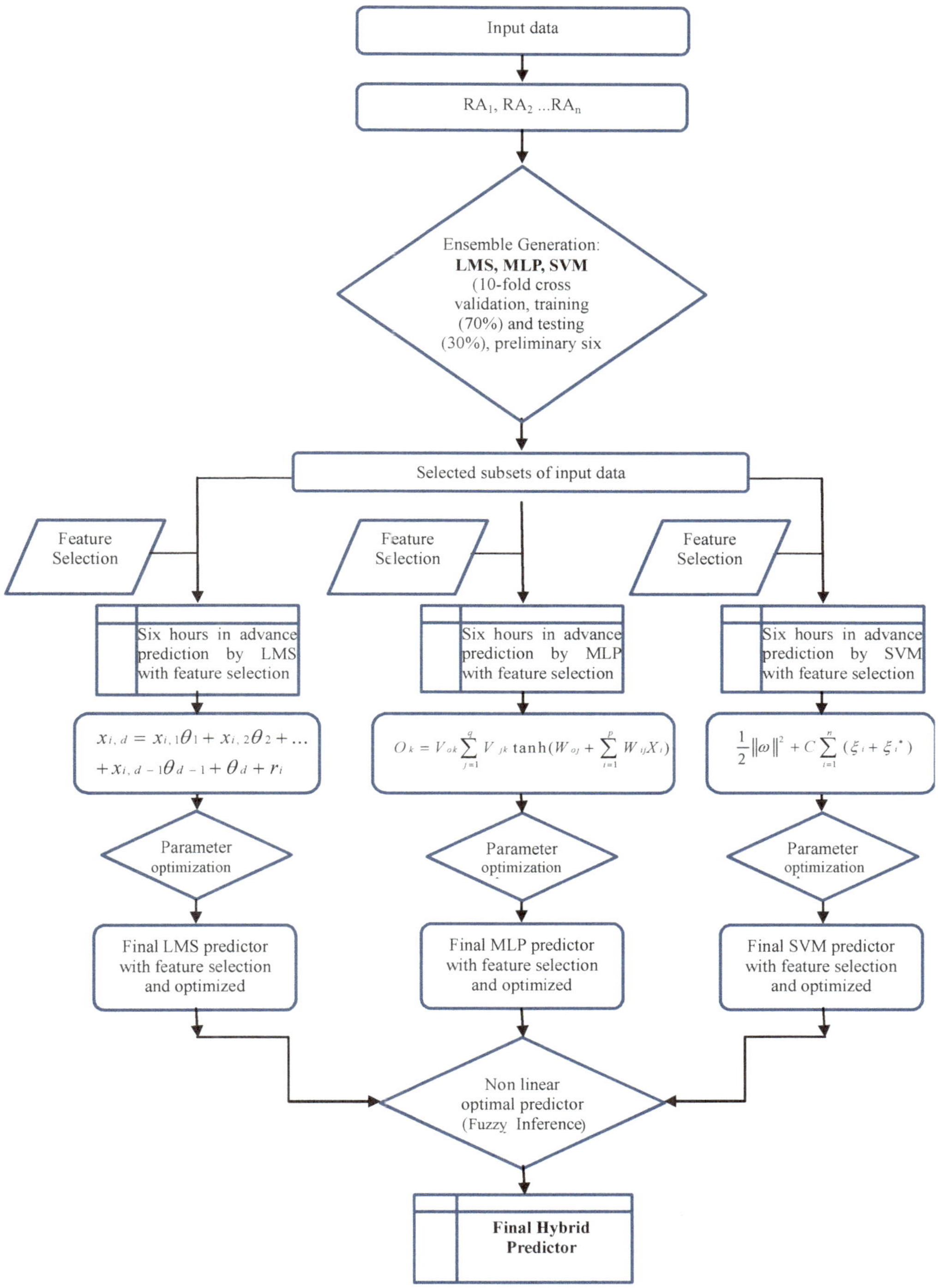

Figure 2. Structure of the hybrid method for solar power prediction.

placing level two data as input to produce the ultimate prediction. For this stage a regression algorithm is employed for finding out the way to integrate the outcomes from the foundation regression modules. One of the key conditions to successfully develop this hybrid prediction method is to collect recent, reliable, accurate and long term historical weather data of the particular location selected for the experiments. The following section continues with the description and analysis of the raw data used for this research.

3. Data Collection

Historical solar data are a key element in solar power prediction systems. Rockhampton, a sub tropical city in Australia was chosen for the experiment of the proposed heterogeneous regression algorithms based hybrid method. The selected station is "Rockhampton Aero", having latitude of −23.38 and longitude of 150.48. The recent data were collected from Commonwealth Scientific and Industrial Research Organization (CSIRO), Australia. In **Figure 3** Renewable Energy Certificates (RECs) zones within Australia is shown where Rockhampton is identified within the most important zone [26]. Data was also collected from the Australian Bureau of Meteorology (BOM), the National Aeronautics and Space Administration (NASA), the National Oceanic and Atmospheric Administration (NOAA). Free data is available from National Renewable Energy Laboratory (NREL) and NASA. These are excellent for multi-year averages but perform poorly for hourly and daily measurements. After rigorous study the data provided by CSIRO was finally selected as the raw data is available in hourly resolution which is a significant aspect of the data set. Hourly raw data were gathered for a period of 2005 to 2010. **Table 1** represents the attributes of the used data set as well as statistical properties of those attributes. The next sections will illustrate the procedure of base models generation *i.e.* ensemble generation.

4. Experiment Design

Ten popular regression algorithms namely Linear Regression (LR), Radial Basis Function (RBF), Support Vector Machine (SVM), Multilayer Perceptron (MLP), Pace Regression (PR), Simple Linear Regression (SLR), Least Median Square (LMS), Additive Regression (AR), Locally Weighted Learning (LWL) and IBK Regression have been used to find out ensemble generation. A unified platform is used with WEKA release 3.7.3 for all of the experiments. The WEKA is a Java based data mining tool [27] which is an efficient data pre-processing tool that encompasses a comprehensive set of learning algorithms with graphical user interface as well as command prompt. The accuracy of the model is justified by cross-validation method and training-testing method.

4.1. *K* Fold Cross-Validation Error Estimator

Cross-validation, at times called rotation estimation is a method for assessing how the results of a statistical analysis will generalize to an independent data set. In k-cv a data set S_n is uniformly at random partitioned into k folds of similar size $P=\{P_1,\cdots,P_k\}$. For the sake of clarity and without loss of generality; it is supposed that n is multiple of k. Let $T_i = S_n/P_i$ be the complement data set of P_i. Then, the algorithm $A(\cdot)$ induces a classifier from, $T_i\Psi_i = A(T_i)$ and estimates its prediction error with P_i. The k-cv prediction error estimator of $\Psi = A(S_n)$ is defined as follows [28]:

$$\xi_k(S_n,P) = \frac{1}{n}\sum_{i=1}^{k}\sum_{(x,c)\in P_i} 1(c,\psi_i(x)) \tag{1}$$

where $1(i,j)=1$ if $i \neq j$ and zero otherwise. So, the k-cv error estimator is the average of the errors made by the classifiers Ψ_i in their respective divisions P_i. The approximated error may be considered a random variable

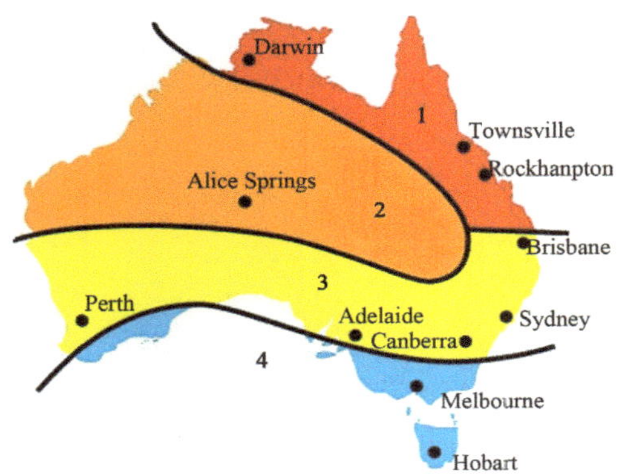

Figure 3. Renewable Energy Certificates (RECs) zones within Australia.

Table 1. Attributes of the raw data set and the corresponding statistics.

	Min.	Max.	Mean	Std Dev
Air Temperature (˚C)	−5.8	40.1	20.47	6.99
Wind Speed (Km/h)	0	27.1	6.99	4.78
Wind Direction (˚C)	0	359	158.91	103.66
Relative Humidity (%)	0	100	55.11	24.26
Rainfall (mm)	0	30.4	0.07	0.69
VWSP Wind Speed (Km/h)	0	24.83	5.77	4.38
VWDIR Wind Direction (˚)	0	360	169.91	109.84
Max. Wind Gust (Km/h)	0	106	20.45	11.33
Evaporation (mm)	−1.36	1.36	0.31	0.28
Abs. Barometer (hPa)	921	1020	966.59	12.09
Solar Radiation (W/m^2)	1	1660	300.75	325.17

which relies on the training set S_n and the division P. In **Table 2** the results of applying 10 folds cross-validation method on initially selected regression algorithms are demonstrated. The above results clearly show that in terms of the mean absolute error (MAE) the most accurate one is the MLP regression algorithm. Next to the MLP, SVM is in the second best position and LMS regression algorithm is in the third best position.

4.2. Training and Testing Error Estimator

In this error estimator method, a common rule of thumb is to use 70% of the data set for training and 30% for testing. In **Table 3** the results of applying training and testing error estimator method on initially selected regression algorithms are illustrated.

Based on the experimental outcome the MLP is the number one choice, SVM is in the second best position and LMS regression algorithm is in the third best position. Both of the 10 folds cross-validation and training and testing method suggests the top most three algorithms for this task are MLP, SVM and LMS in descending order. Next, six hours ahead individual predictions performed by the initially selected regression algorithms and the accuracy of the prediction performances are validated with both the scale dependent and scale free error measurement metrics.

Table 2. Results of applying 10 folds cross validation method on the data set.

		CC	RMSE	MAE	RRSE	RAE
10-Fold Cross Validation	LR	0.89	150.14	66.35	46.30	27.44
	RBF	0.13	321.58	240.09	99.16	99.30
	SVM	0.88	164.18	**46.58**	50.63	**19.26**
	MLP	**0.99**	**14.813**	**9.74**	**4.57**	**4.03**
	PR	0.89	150.14	66.31	46.30	27.42
	SLR	0.87	158.56	83.15	48.89	34.39
	LMS	0.88	165.15	**47.94**	50.92	**19.83**
	AR	**0.94**	**108.94**	80.99	**33.59**	33.50
	LWL	0.81	190.26	146.09	58.67	60.42
	IBK	**0.93**	**124.41**	90.86	**38.36**	37.58

Table 3. Results of applying training and testing method on the data set.

		CC	RMSE	MAE	RRSE	RAE
Training (70%) and Testing (30%)	LR	0.88	150.84	66.91	47.21	27.75
	RBF	0.12	317.38	239.64	99.34	99.40
	SVM	0.87	165.56	**47.19**	51.82	**19.58**
	MLP	**0.99**	**6.90**	**11.73**	**5.29**	**4.87**
	PR	0.88	150.88	66.65	47.22	27.65
	SLR	0.87	159.22	82.74	49.83	34.32
	LMS	0.87	164.69	**48.69**	51.55	**20.19**
	AR	**0.93**	**114.90**	83.97	**35.96**	34.83
	LWL	0.80	190.39	146.48	59.59	60.76
	IBK	**0.92**	**128.82**	94.23	**40.32**	39.09

5. Using Preliminary Short Term Prediction with Base Regression Algorithms

The six hours ahead solar radiation prediction with the potential regression algorithms were performed to compare the errors of the individual prediction to select three decisive regression algorithms for the ensemble generation. In **Table 4**, the summary of six hours in advance prediction error for different regression algorithms with both the scale dependent and scale free prediction accuracy validation metrics are presented.

Prediction Accuracy Validation Metrics

In general, there are four types of prediction-error metrics [29]. Those are: scale dependent metrics as for example MAE; metrics related to percentage of error such as mean absolute percentage error (MAPE); metrics regarding relative error which computes the ratio of error between a selected and a novel approach; and finally scale free error measurements that is not dependent to scale of the data and can be used to compare forecast systems on a distinct series and also to compare forecast accuracy between series such as MASE.

From the individual prediction results the regression algorithms are ranked. According to [30], MAE is strongly suggested for error measurement. Hence, the ranking is done based on the MAE of those regression algorithms' predictions.

Accuracy of the preliminary experimental results of distinct base predictor was also analyzed according to MASE [29]. MASE is scale free, less sensitive to outlier; and less variable to small samples. MASE is suitable for uncertain demand series as it never produces infinite or

Table 4. Six hours ahead prediction errors of different regression algorithms.

Six Hours Ahead Prediction Error Summary							
	CC	MAE	MAPE	RMSE	RMSPE	MASE	RANK
LMS	0.96	77.19	17.65	107.94	29.19	0.63	1
MLP	0.99	91.02	20.17	119.73	31.62	0.74	2
SVM	0.96	126.88	21.72	135.15	24.01	1.03	3
LR	0.96	148.41	24.07	155.82	25.12	1.21	4
LWL	−0.15	213.33	46.86	271.22	72.39	1.73	5
IBK	0.88	275.00	47.90	290.18	53.66	2.24	6
AR	0.93	298.45	48.21	306.03	49.05	2.43	7

undefined results. It indicates that the forecast with the smallest MASE would be counted the most accurate among all other alternatives [29]. Equation (2) states the formula to calculate MASE.

$$MASE = \frac{MAE}{(1/n-1)\sum_{i=2}^{n}|a_i - a_{i-1}|} \quad (2)$$

where, $$MAE = \frac{1}{n}\sum_{i=1}^{n}|p_i - a_i| \quad (3)$$

In Equations (2) and (3), n = number of instances, a_i = actual values and p_i = predicted outputs. In **Figure 4** the prediction performance of the preliminary selected regression algorithms in terms of various error validation metrics used are portrayed.

Based on MAE and MASE the top most three regression algorithms for ensemble generation are LMS, MLP and SVM. In **Figures 5-7** the comparison between the actual and predicted values of these three regression algorithms are graphically presented respectively.

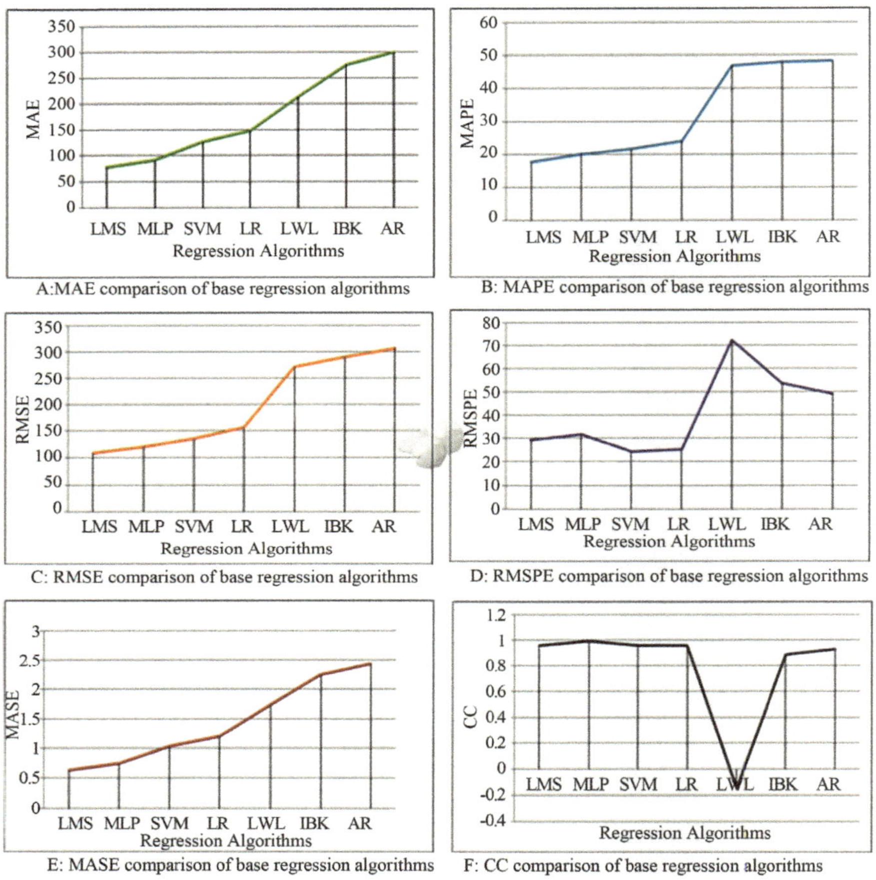

Figure 4. Prediction performance of the regression algorithms with A: MAE, B: MAPE, C: RMSE, D: RMSPE, E: MASE and F: CC error validation metrics.

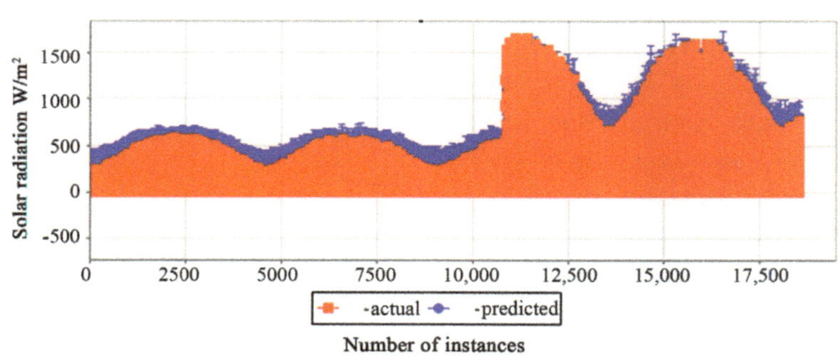

Figure 5. Prediction performance of the LMS regression algorithm.

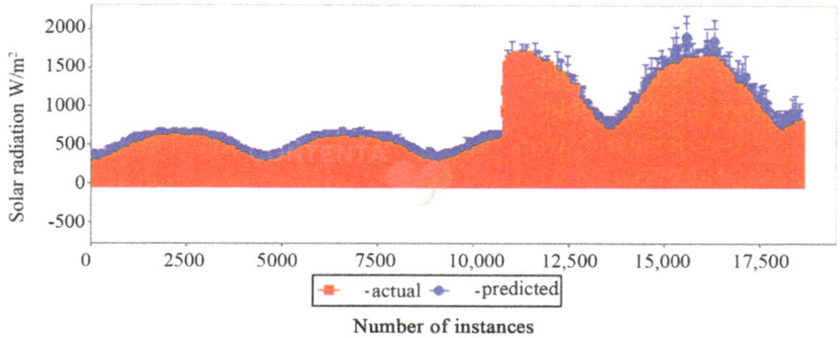

Figure 6. Prediction performance of the MLP regression algorithm.

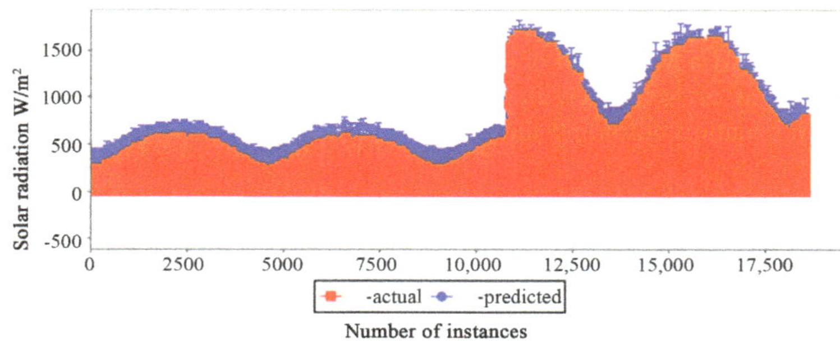

Figure 7. Prediction performance of the SVM regression algorithm.

6. Selected Regression Algorithms for Ensemble Generation

6.1. Support Vector Machine (SVM)

The Support Vector (SV) algorithm is a nonlinear generalization of the "Generalized Portrait" algorithm developed in Russia in the sixties [31]. One of the artificial intelligence based renewable energy forecasting methods described in the literature [32,33], which involves the so-called Support Vector Machines. Such models are less rigid in terms of architecture than Artificial Neural Networks (ANNs) and may allow one to save even more modeling efforts when designing prediction models. While other statistical models are estimated following the empirical risk minimization principle, *i.e.* the minimization of loss function over the learning set and the checking of the generalization ability with some criteria, the SVM theory is based on the structural risk minimization principle, which consists on directly minimizing an upper bound on the generalization error, and thus on future points [34]. There is today a large interest in applying SVMs for several purposes including renewable energy forecasting [35-39].

6.2. Least Median Square (LMS)

Strong and fast regression methods are necessary that are able to cope with the various challenges involved in the successful implementation of smart grid. LMS is one of the potential factors which can be used to serve the purpose of the successful implementation of smart grid [40].

The least median square is basically consists of adjusting the parameters of a model function to best fit a data set. Rousseeuw's LMS linear regression estimator [41] is amongst the best known and most extensively used robust estimators. The LMS estimator is defined formally as follows. Consider a set $P=\{p_1,p_2,\cdots,p_n\}$ of n points in R, where $p_i=(x_{i,1},x_{i,2},\cdots,x_{i,d-1},x_{i,d})$. A parameter vector $\theta=(\theta_1,\theta_2,\cdots,\theta_d)$ that best fits the data by the linear model is likely to be computed.

$$x_{i,d}=x_{i,1}\theta_1+x_{i,2}\theta_2+\cdots+x_{i,d-1}\theta_{d-1}+\theta_d+r_i \tag{4}$$

for all i where $r_1,r_2,\cdots,r_n$ are the (unknown) errors, or residuals.

6.3. Multilayer Perception (MLP)

MLP which is one of the types of ANNs have been used in diverse applications as well as in renewable energy problems. ANNs have been used by various authors in the field of solar energy; for modeling and design of a solar steam generating plant, for the estimation of a parabolic trough collector intercept factor and local concentration ratio and for the modeling and performance prediction of solar water heating systems. They have also been used for the estimation of heating loads of buildings, for the prediction of air flow in a naturally ventilated test room and for the prediction of the energy consumption of a passive solar building [42].

In all those models multiple hidden layer architecture has been used. Errors reported in these models are well within acceptable limits, which clearly suggest that ANNs can be used for modeling in other fields of renewable energy production and use. Research works using ANNs in the field of renewable energy as well as in other energy systems are found in the literature review. This includes the use of ANNs in solar radiation and wind speed prediction, photovoltaic systems, building services systems and load forecasting and prediction [42].

In general MLP can have numerous hidden layers (**Figure 8**), however according to Hornik; a neural network with single hidden layer is capable to estimate a function of any complexity [43]. If a MLP with one hidden layer is considered, tanh as an activation function and a linear output unit, the equation unfolding the network structure can be expressed as:

$$O_k = V_{ok} \sum_{j=1}^{q} V_{jk} \tanh\left(Woj + \sum_{i=1}^{p} W_{ij} X_i \right) \quad (5)$$

where O_k is the output of the k_{th} output unit, V_{jk} and W_{oj} are the network weights, p is the number of network inputs, and q is the number of hidden units. Throughout the training progression, weights are altered in such a way that the difference between the obtained outputs O_k and the desired outputs d_k is minimized, which is typically done by minimizing the following error function:

$$E = \sum_{k=1}^{r} \sum_{e=1}^{n} \left(d_{e,k} - O_{e,k} \right)^2 \quad (6)$$

where r is the number of network outputs and n is the number of training examples. The minimization of the error function is usually done by gradient descent methods, which have been comprehensively investigated in the field of optimization theory [44].

7. Statistical Test

T-test is employed to verify the algorithm accuracy performances in the experiments. In this paper the independent samples T-Tests are carried out in order to justify whether any significant difference exists between the actual and predicted values achieved by the selected three regression algorithms. These tests are done in order ensure the potentiality of those selected heterogeneous regression algorithms for the suggested hybrid method. The t-test is executed with the SPSS package-PASW Statistics 18.

7.1. T-Test to Compare the Mean Actual and Predicted Values of LMS

For the least median square (LMS) regression algorithm, an equal variances t test failed to reveal a statistically reliable difference between the mean number of the actual and predicted values of solar radiation with actual $(M = 639.167, s = 177.405)$ and predicted data $(M = 716.357,\ s = 104.778)$, $t(10) = 0.918$, two tailed significance value $p = 0.380$, significance level $\alpha = 0.05$. The decision rule is given by: if $p \le \alpha$, then reject the obtained result. In this instance, 0.380 is not less than or equal to 0.05 and fail to reject the obtained result. That implies that the test failed to observe any significant difference between the actual and predicted values performed by LMS on average. In **Table 5** the group statistics with Mean (M), Standard Deviation (S) and Standard Error Mean for the actual and predicted values of LMS is represented. In **Table 6** the independent samples t-test results for the actual and predicted values of LMS is illustrated.

In the same way t-test to compare the mean actual and predicted values of MLP and t-test to compare the

Table 5. Group statistics of LMS.

Group Statistics					
Group		N	Mean	Std. Deviation	Std. Error Mean
Data	Actual	6	639.17	177.41	72.43
	Predicted	6	716.36	104.78	42.78

Table 6. Independent samples t-test results for LMS.

Independent Samples Test										
		Levene's Test for Equality of Variances		T-Test for Equality of Means						
		F	Sig.	t	df	Sig. (2-Tailed)	Mean Difference	Std. Error Difference	95% Confidence Interval of the Difference	
									Lower	Upper
Data	Equal Variances Assumed	1.17	0.31	−0.92	10	0.38	−77.19	84.11	−264.61	110.23
	Equal Variances Not Assumed			−0.92	8.11	0.39	−77.19	84.11	−270.70	116.32

actual and predicted values of SVM are performed and the corresponding results are discussed in the results and discussions section of this paper.

Observing all the above mentioned results, it is failed to observe any significant difference between the actual and predicted values performed by LMS regression algorithm on average. Therefore, the test failed to reject the obtained result for LMS regression algorithm.

7.2. T-Test to Compare the Mean Predicted Values between LMS and MLP

For the LMS and MLP regression algorithm, an equal variances t test failed to reveal a statistically reliable difference between the mean number of predicted values of solar radiation with LMS ($M = 716.357$, $s = 104.778$) and MLP ($M = 550.216, s = 262.128$), $t(10) = 1.442$, two tailed significance value $p = 0.180$, significance level $\alpha = 0.05$. The decision rule is given by: if $p \leq \alpha$, then reject the obtained result. In this instance, 0.180 is not less than or equal to 0.05 and fail to reject the obtained result. That implies that the test failed to observe any significant difference between the predicted values performed by the pair LMS: MLP on average. In **Table 7** the group statistics with M, s and Standard Error Mean for the predicted values of LMS: MLP is represented.

In **Table 8** the independent samples t-test results for the predicted values of LMS: MLP is illustrated. In the same way t-test to compare the mean predicted values between LMS and SVM and t-test to compare the predicted values between MLP and SVM are performed and the corresponding results are discussed in the results and discussions section of this paper. All the above mentioned results in **Table 8** failed to observe any significant difference between the predicted values performed by the pair LMS: MLP on average. Therefore, the test failed to reject the obtained results for the pair LMS: MLP regression algorithm.

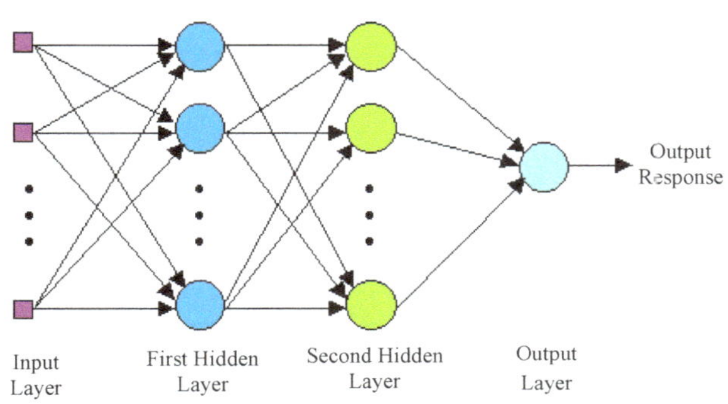

Figure 8. Architecture of a MLP with two hidden layers.

Table 7. Group statistics of the pair LMS and MLP.

Group Statistics					
Group		N	Mean	Std. Deviation	Std. Error Mean
Data	Actual	6	716.36	104.78	42.78
	Predicted	6	550.22	262.13	107.01

8. Results and Discussions

The ensemble generation through empirically selected heterogeneous regression algorithms are presented in this paper. Those potential regression algorithms were applied as local predictors of the proposed hybrid method for six hour in advance prediction of solar power. Several performance criteria found in the solar power prediction method literature as: the training time, the modeling time and the prediction error. As the training process was in offline mode, the first two criteria were not considered to be relevant for this paper. In this context, the prediction performance was evaluated only in term of prediction error, defined as the difference between the actual and the forecasted values and based on statistical and graphical approaches. T-tests were performed as statistical error test criteria. For the individual performance of LMS, MLP and SVM regression algorithm, an equal variances t-test failed to reveal a statistically reliable difference between the mean number of actual and predicted values of solar radiation with actual and predicted data. In **Table 5** the group statistics with M, s and Standard Error Mean for LMS was presented. In **Table 6** the independent samples t-test results for LMS was illustrated. To compare the mean predicted values between the LMS: MLP, LMS: SVM and MLP: SVM regression algorithm, an equal variances t test failed to reveal a statistically reliable difference between the mean number of predicted values of solar radiation with LMS and MLP; with MLP and SVM. But for the instance of the pair LMS: SVM trivial difference between the predicted values is observed. Therefore, the obtained result is rejected. But this inconsistency can be eliminated by the proper selection of the kernel function and the optimal parameters of SVM for this purpose. In **Table 7** the group statistics with M, s and Standard Error Mean for the predicted values of LMS: MLP was represented. In **Table 8** the independent samples t-test results for the mean predicted values of LMS: MLP was illustrated. All the results found from the t-tests clearly indicate that the heterogeneous regression algorithms selected for the suggested hybrid method are more accurate and prospective. In **Figure 5**, the prediction performance of the preliminary selected regression algorithms in terms of various error validation metrics used are presented graphically. While the 2-D error prediction form (**Figures 5-7**) for the three proposed regression algorithm based predictors were presented as graphical error performance. In **Table 4**, referring to the MAE and MASE criteria, it is observed that LMS, MLP and SVM have the lowest prediction error. Therefore, these three

Table 8. Independent samples t-test results for the pair LMS and MLP.

Independent Samples Test										
		Levene's Test for Equality of Variances		t-test for Equality of Means						
		F	Sig.	t	df	Sig. (2-Tailed)	Mean Difference	Std. Error Difference	95% Confidence Interval of the Difference: Lower	95% Confidence Interval of the Difference: Upper
Data	Equal Variances Assumed	1.17	0.31	−0.92	10	0.38	−77.19	84.11	−264.61	110.23
	Equal Variances Not Assumed			−0.92	8.11	0.39	−77.19	84.11	−270.70	116.32

regression algorithms are empirically proved to be potential for the proposed hybrid prediction method.

9. Conclusion

In this paper, based on heterogeneous regression algorithms a novel hybrid method for solar power prediction that can be used to estimate PV solar power with improved prediction accuracy is presented. The hybridization aspect is anchored in the top most selected heterogeneous regression algorithm based local predictors as well as a global predictor. One of the two decisive problems of ensemble learning namely heterogeneous ensemble generation for solar power prediction is mainly focused in this research. There are scopes to further improve the prediction accuracy of the selected individual regression algorithms namely LMS, MLP and SVM. As a consequence, the next step will be the efficient utilization of the feature selection aspect on the used data set to reduce the generalized error of those regression algorithms. Further tuning can be achieved through the parameter restructuring of those regression algorithms to make them as accurate and diverse as possible as well as to formulate the proposed hybrid method more effective. Finally, another regression algorithm or learning algorithm will be found out empirically to nonlinearly combine or integrate the individual predictions supplied from the improved local predictors. Further application of the proposed ensemble will include distributed intelligent management system for the cost optimization of a smart grid.

REFERENCES

[1] B. Parsons, M. Milligan, B. Zavadil, D. Brooks, B. Kirby, K. Dragoon and J. Caldwell, "Grid Impacts of Wind Power: A Summary of Recent Studies in the United States," National Renewable Energy Laboratory, Madrid, 2003.

[2] R. Doherty and M. O'Malley, "Quantifying Reserve Demands Due to Increasing Wind Power Penetration," 2003 *IEEE Bologna Power Tech Conference Proceedings*, Bologna, Vol. 2, 23-26 June 2003, pp. 23-26.

[3] R. Doherty and M. O'Malley, "A New Approach to Quantify Reserve Demand in Systems with Significant Installed Wind Capacity," *IEEE Transactions on Power Systems*, Vol. 20, No. 2, 2005, pp. 587-595.

[4] N. Hatziargyriou, A. Tsikalakis, A. Dimeas, D. Georgiadis, A. Gigantidou, J. Stefanakis and E. Thalassinakis, "Security and Economic Impacts of High Wind Power Penetration in Island Systems," *Proceedings of Cigre Session*, Paris, 30 August-3 September 2004, pp. 1-9.

[5] N. Hatziargyriou, G. Contaxis, M. Matos, J. A. P. Lopes, G. Kariniotakis, D. Mayer, J. Halliday, G. Dutton, P. Dokopoulos, A. Bakirtzis, J. Stefanakis, A. Gigantidou, P. O'Donnell, D. McCoy, M. J. Fernandes, J. M. S. Cotrim and A. P. Figueira, "Energy Management and Control of Island Power Systems with Increased Penetration from Renewable Sources," *Power Engineering Society Winter Meeting*, Vol. 1, No. 27-31, 2002, pp. 335-339.

[6] E. D. Castronuovo and J. A. P. Lopes, "On the Optimization of the Daily Operation of a Wind-Hydro Power Plant," *IEEE Transactions on Power Systems*, Vol. 19, No. 3, 2004, pp. 1599-1606.

[7] M. C. Alexiadis, P. S. Dokopoulos, and H. S. Sahsamanoglou, "Wind Speed and Power Forecasting Based on Spatial Correlation Models," *IEEE Transactions on Energy Conversion*, Vol. 14, No. 3, 1999, pp. 836-842.

[8] T. G. Barbounis, J. B. Theocharis, M. C. Alexiadis and P. S. Dokopoulos, "Long-Term Wind Speed and Power Forecasting Using Local Recurrent Neural Network Models," *IEEE Transactions on Energy Conversion*, Vol. 21, No. 1, 2006, pp. 273-284.

[9] L. Landberg, G. Giebel, H. A. Nielsen, T. Nielsen and H. Madsen, "Short-Term Prediction—An Overview," *Wind Energy*, Vol. 6, No. 3, 2003, pp. 273-280.

[10] G. Giebel, L. Landberg, G. Kariniotakis and R. Brownsword, "State-of-the-Art on Methods and Software Tools for Short-Term Prediction of Wind Energy Production," *EWEC*, Madrid, 16-19 June 2003, pp. 27-36.

[11] G. Giebel, R. Brownsword and G. Kariniotakis, "The State of the Art in Short-Term Prediction of Wind Power: A Literature Overview," 2003. http://130.226.56.153/rispubl/vea/veapdf/ANEMOS_giebel.pdf

[12] J. M. Gordon and T. A. Reddy, "Time Series Analysis of Hourly Global Horizontal Solar Radiation," *Solar Energy*, Vol. 41, No. 5, 1988, pp. 423-429.

[13] J. M. Gordon and T. A. Reddy, "Time Series Analysis of Daily Horizontal Solar Radiation," *Solar Energy*, Vol. 41, No. 3, 1988, pp. 215-226.

[14] K. M. Knight, S. A. Klein and J. A. Duffie, "A Methodology for the Synthesis of Hourly Weather Data," *Solar Energy*, Vol. 46, No. 2, 1991, pp. 109-120.

[15] T. R. Morgan, "The Performance and Optimization of Autonomous Renewable Energy Systems," Ph.D. Thesis, University of Wales, Cardiff, 1996.

[16] N. Jankowski, K. Grąbczewski and W. Duch, "Hybrid Systems, Ensembles and Meta-Learning Algorithms," 2008. http://www.is.umk.pl/WCCI-HSEMLA/

[17] S. B. Kotsiantis and P. E. Pintelas, "Predicting Students' Marks in Hellenic Open University," *Proceedings of 5th IEEE International Conference on Advanced Learning Technologies*, Kaohsiung, 5-8 July 2005, pp. 664-668.

[18] S. Kotsiantis, G. Tsekouras, C. Raptis and P. Pintelas, "Modelling the Organoleptic Properties of Matured Wine Distillates," *MLDM'05 Proceedings of the 4th International Conference on Machine Learning and Data Mining in Pattern Recognition*, Vol. 3587, 2005, pp. 667-673.

[19] A. Sharkey, N. Sharkey, U. Gerecke and G. Chandroth, "The Test and Select Approach to Ensemble Combination," Springer-Verlag, Cagliari, 2000.

[20] N. L. Hjort and G. Claeskens, "Frequentist Model Average Estimators," *Journal of the American Statistical Association*, Vol. 98, No. 464, 2003, pp. 879-899.

[21] L. Breiman, "Bagging Predictors," *Machine Learning*, Vol. 24, No. 2, 1996, pp. 123-140.

[22] R. Zemel and T. Pitassi, "A Gradient-Based Boosting Algorithm for Regression Problems," MIT Press, Cambridge, 2001.

[23] G. Brown, J. Wyatt and P. Tino, "Managing Diversity in Regression Ensembles," *Journal of Machine Learning Research*, Vol. 6, No. 1, 2005, pp. 1621-1650.

[24] U. Naftaly, N. Intrator and D. Horn, "Optimal Ensemble Averaging of Neural Networks," *Network*, Vol. 8, No. 3, 1997, pp. 283-296.

[25] P. Langley, "Selection of Relevant Features in Machine Learning," AAAI Press, New Orleans, 1994.

[26] J. Harburn, "RECs/STCs-What Are They and How Are They Calculated?" http://www.solarchoice.net.au/blog/recs-what-are-they-and-how-are-they-calculated/

[27] R. Remco, Bouckaert, E. Frank, M. Hall, R. Kirkby, P. Reutemann, A. Seewald and D. Scuse, "WEKA Manual for Version 3-7-3," The University of Waikato, New Zealand, 2010.

[28] S. Shevade, S. Keerthi, C. Bhattacharyya and K. Murthy, "Improvements to the SMO Algorithm for SVM Regression," *IEEE Transactions on Neural Networks*, Vol. 11, No. 5, 2000, pp. 1188-1183.

[29] R. J. Hyndman and A. B. Koehler, "Another Look at Measures of Forecast Accuracy," *International Journal of Forecasting*, Vol. 22, No. 4, 2005, pp. 679-688.

[30] C. J. Willmott and K. Matsuura, "Advantages of the Mean Absolute Error (MAE) over the Root Mean Square Error (RMSE) in Assessing Average Model Performance," *Climate Research*, Vol. 30, No. 1, 2005, pp. 79-82.

[31] V. Vapnik and A. Lerner, "Pattern Recognition Using Generalized Portrait Method," *Automation and Remote Control*, Vol. 24, No. 6, 1963, pp. 774-780.

[32] K. Larson and T. Gneiting, "Advanced Short-Range Wind Energy Forecasting Technologies-Challenges, Solutions and Validation," *CD-Proceedings of the Global Wind Power Conference*, Chicago, 28-31 March 2004, pp. 67-73.

[33] D. Monn, D. Christenson and R. Chevallaz-Perrier, "Support Vector Machines Technology Coupled with Physics-Based Modeling for Wind Facility Power Production Forecasting," *CD-Proceedings of the Global Wind Power Conference*, Chicago, 28-31 March 2004, pp. 97-102.

[34] G. Kariniotakis, "Estimation of the Uncertainty in Wind Power Forecasting," Ph.D. Thesis, Department of Energy, Ecole des Mines de Paris, Paris, 2006.

[35] M. Mohandes, T. Halawani and R. A. Hussain, "Support Vector Machines for Wind Speed Prediction," *Renewable Energy*, Vol. 29, No. 6, 2004, pp. 939-947.

[36] J. Zeng and W. Qiao, "Support Vector Machine-Based Short-Term Wind Power Forecasting," *Proceedings of the IEEE PES Power System Conference and Exposition*, Arizona, 20-23 March 2011, pp. 1-8.

[37] J. Shi, W. Lee, Y. Liu, Y. Yang and P. Wang, "Forecasting Power Output of Photovoltaic System Based on Weather Classification and Support Vector Machine," *IEEE Industry Applications Society Annual Meeting* (*IAS*), Orlando, 9-13 October 2011, pp. 1-6.

[38] N. Sharma, P. Sharma, D. Irwin and P. Shenoy, "Predicting Solar Generation from Weather Forecasts Using Machine Learning," *Proceedings of the 2nd IEEE International Conference on Smart Grid Communications*, Brussels, 17-20 October 2011, pp. 32-37.

[39] K. R. Müller, A. Smola, G. Rätsch, B. Schölkopf, J. Kohlmorgen and V. Vapnik, "Using Support Vector Machines for Time Series Prediction," In: B. Schölkopf, C. Burges and A. Smola, Eds., *Advances in Kernel Methods—Support Vector Learning*, MIT Press, Cambridge, 1998.

[40] O. Kramer, B. Satzger and J. Lässig, "Power Prediction in Smart Grids with Evolutionary Local Kernel Regression," International Computer Science Institute, Berkeley. http://www.icsi.berkeley.edu/pubs/algorithms/powerprediction10.pdf

[41] P. J. Rousseeuw, "Least Median of Squares Regression," *Journal of American Statistics*, Vol. 79, No. 388, 1984, pp. 871-880.

[42] S. A. Kalogirou, "Artificial Neural Networks in Renewable Energy Systems Applications: A Review," *Renew-*

able and Sustainable Energy Reviews, Vol. 5, No. 4, 2001, pp. 373-401.

[43] K. M. Hornik, M. Stinchcombe and H. White, "Multilayer Feed Forward Networks Are Universal Approximator," *Neural Networks*, Vol. 2, No. 2, 1989, pp. 359-366.

[44] R. Fletcher, "Practical Methods of Optimization," 2nd Edition, Wiley, Chichester, 1990.

Design and Performance Analysis of an Innovative Single Basin Solar NanoStill

Moses Koilraj Gnanadason[1], Palanisamy Senthil Kumar[2], Vincent H. Wilson[3], Gajendiran Hariharan[4], Navaneethakrishnan Shenbaga Vinayagamoorthi[5]

[1]Mechanical Engineering, PRIST University, Thanjavur, India; [2]Mechanical Engineering Department, KSR College of Engineering, Tiruchengode, India; [3]Engineering & Technology, PRIST University, Thanjavur, India; [4]Faculty of Mechanical Engineering, VOC Anna University College of Engineering, Tuticorin, India; [5]Department of Mechanical Engineering, Anna University of Technology, Tirunelveli, India.

ABSTRACT

The provision of fresh water is becoming an increasingly important issue in many areas of the world. Clean water is a basic human necessity, and without water life will be impossible. The rapid international developments, the industrial growth, agriculture and population explosion all over the world have resulted in a large escalation of demand for fresh water. The solar still is the most economical way to accomplish this objective. The sun's energy heats water to the point of evaporation. When water evaporates, water vapour rises leaving the impurities like salts, heavy metals and condensate on the underside of the glass cover. Solar distillation has low yield, but safe and pure supplies of water in remote areas. The attempts are made to increase the productivity of solar still by using nanofluids and also by black paint coating inside the still basin. Heat transfer enhancement in solar still is one of the key issues of energy saving and compact designs. The essential initiative is to seek the solid particles having thermal conductivity of several hundred times higher than those of conventional fluids. Recently, as an innovative material, nanosized particles have been used in suspension in conventional solar still water. The fluids with nanosized solid particles suspended in them are called "nanofluids". The suspended metallic or nonmetallic nanoparticles change the transport properties, heat transfer characteristics and evaporative properties of the base fluid, The aim of this paper is to analyze and compare the enhanced performance of a single basin solar still using nanofluids with the conventional water. They greatly improve the rate of evaporation and hence the increase in efficiency.

Keywords: Solar Still; Nanofluid; Nanoparticles; Productivity

1. Introduction

Water is a nature's gift and it plays a key role in the development of an economy and in turn for the welfare of a nation. Non-availability of drinking water is one of the major problem faced by both the under developed and developing countries all over the world [1]. Around 97% of the water in the world is in the ocean, approximately 2% of the water in the world is at present stored as ice in polar region, and 1% is fresh water available for the need of the plants, animals and human life [2]. Today, majority of the health issues are owing to the non availability of clean drinking water. In the recent decades, most parts of the world receive insufficient rainfall resulting in increase in the water salinity. The pollution of water resources is increasing drastically due to a number of factors including growth in the population, industrialization, urbanization, etc. These activities adversely affected the water quality in rural areas and agriculture. In developing countries, lack of safe and unreliable drinking water constitutes a major problem. Worldwide drought and desertification are expected to increase the drinking water shortage to become one of the biggest problems facing the world [3]. As population grows; there is less water per capita. At the current trend of growth, it is predicted that the global population will reach 8 billion by 2025 and the per captia water available will go down. Along with depletion and pollution of existing water supplies, the growing world population leads to the assumption that two thirds of the population will lack sufficient fresh water by the year 2025 [4]. Globally, 200 million hours are spent each day, mostly by females, to collect water from distant, often polluted sources. In the world, 3.575 million people die each year from water related diseases. Majority of the rural people are still unaware of the consequences of drinking untreated water [5].

Desalination is the oldest technology used by people

for water purification in the world. Solar energy is available in abundant in most of the rural areas and hence solar distillation is the best solution for rural areas and has many advantages of using freely available solar energy [6]. A solar still operates similar to the natural hydrologic cycle of evaporation and condensation. It is a simple technology and more economical than the other available methods [7].

The use of solar energy is more economical than the use of fossil fuel in remote areas having low population densities, low rain fall and abundant available solar energy. The attempts are also made to increase the productivity of water by painting black coating inside the basin and by providing insulation to the still basin. But the novel approach is to introduce the nanofluids in solar still with conventional water. The poor heat transfer properties of these conventional fluids compared to most solids are the primary obstacle to the high compactness and effectiveness of the system. The essential initiative is to seek the solid particles having thermal conductivity of several hundred times higher than those of conventional fluids [8]. The use of additives is a technique applied to enhance the heat transfer performance of water in the still basin. An innovative idea is to suspend ultrafine solid particles in the fluid for improving the thermal conductivity of the fluid. The fluids with solid-sized nanoparticles suspended in them are called nanofluids [9]. The suspended metallic or nonmetallic nanoparticles change the transport properties, heat transfer characteristics and evaporative rate of the base fluid. The carbon nanotube (CNT)-based nanofluids are expected to exhibit superior heat transfer properties compared with conventional water in the solar still and other type of nanofluids and hence the increase in the productivity and efficiency of the solar still.

2. Conventional Solar Still

As the available fresh water is fixed on earth and its demand is increasing day by day due to increasing population and the rapid increase of industry, hence there is an essential and earnest need to get fresh water from the saline/brackish water present on or inside the earth. This process of getting fresh water from saline/brackish water can be done easily and economically by desalination. The changing climate is one of the major challenges the entire world is facing today. Gradual rise in global average temperatures, increase in sea level and melting of glaciers and ice sheets have underlined the immediate need to address the issue. All these problems could be solved only through efficient and effective utilization of renewable energy resources such as solar, wind, biomass, tidal, and geothermal energy, etc. Owing to the diffuse nature of solar energy, the main problems with the use of solar thermal energy in large-scale desalination plants are the relatively low productivity rate, the low thermal efficiency and the considerable land area required. Apart from the cost implication, there are environmental concerns with regard to the burning of fossil fuels [10]. Solar energy can directly or indirectly be harnessed for desalination. The solar stills are simple and have no moving parts.

Working of solar still is based on the simple scientific principle of Evaporation and condensation. There are several types of solar stills the simplest of which is the single basin still. But the yield of this is low and falls in the range of 3 - 4 litres per day per square metre [11]. Different still designs have been used in different regions globally, where high quality drinking water supplies are scarce and the solar option is viable. The operation of Solar still is very simple and no special skill is required for its operation and maintenance. Solar stills use natural evaporation and condensation, which is the rainwater process. A solar still is a low-tech way of distilling water, powered by the heat from the sun. In the conventional solar still, saline water is stored in the basin of still, where it is evaporated by means of the sunlight through clear glass. The pure water vapour condenses inside the glass surface and the pure water is collected in the beaker as shown in **Figure 1**. The various factors affecting the productivity of solar still are solar intensity, wind velocity, ambient temperature, water glass temperature difference, and free surface area of water, absorber plate area, temperature of inlet water, glass angle and depth of water. The solar intensity, wind velocity, ambient temperature cannot be controlled as they are metrological parameters whereas the remaining parameters, free surface area of water, absorber plate area, temperature of inlet water, glass angle and depth of water can be varied to enhance the productivity of the solar stills. By considering the various factors affecting the productivity of the solar still, various modifications are being made to enhance the productivity of the solar still [12].

3. Experimental Setup

3.1. Solar Still Made up of Aluminium Sheet

A Single basin solar still made up of Aluminium Sheet is

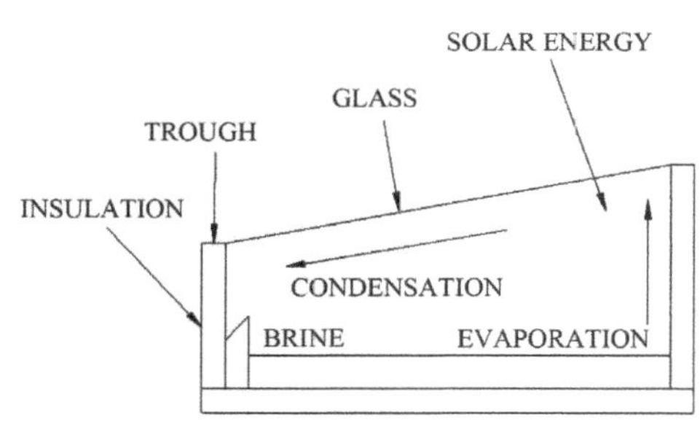

Figure 1. Solar still.

fabricated. It consists of a basin made up of Aluminum of 1 × 1 meter with maximum height of 50 mm and thickness 2 mm thick as shown in **Figure 2**. Aluminum has higher thermal conductivity compared to Galvanized Iron so that the rate of heat transfer to water in the still is more. The colour of aluminum sheet is silvery here it is painted with black colour so it will be better to absorb the maximum amount of solar radiation falling on them and convert it in to heat. The black colour paint needed for painting the basin is about 100 ml. The aluminum also has a smooth surface to make it easier to paint the black paint.

When the aluminum sheet is used without black colour painting, it absorbs less solar radiation. The basin is made water proof using, M-seal. The top of the basin is covered with transparent 5 mm window glass inclined by nearly 10° angles with the horizontal. There are certain specifications needed for the used glass cover in the still. They are (a) Minimum amount of reflection for solar radiation energy (b) High thermal resistance for heat loss from the basin to the ambient. The spacing between the glass cover and the basin water surface is 120 mm to 200 mm. The slope of the glass cover does not affect the rate at which the distillate runs down its inner surface to the collection trough. The glass cover that is no more than from the water surface will allow the still to operate efficiently. Consequently, a glass-to-water distance increases, heat loss due to convection become greater, causing the still efficiency to drop.

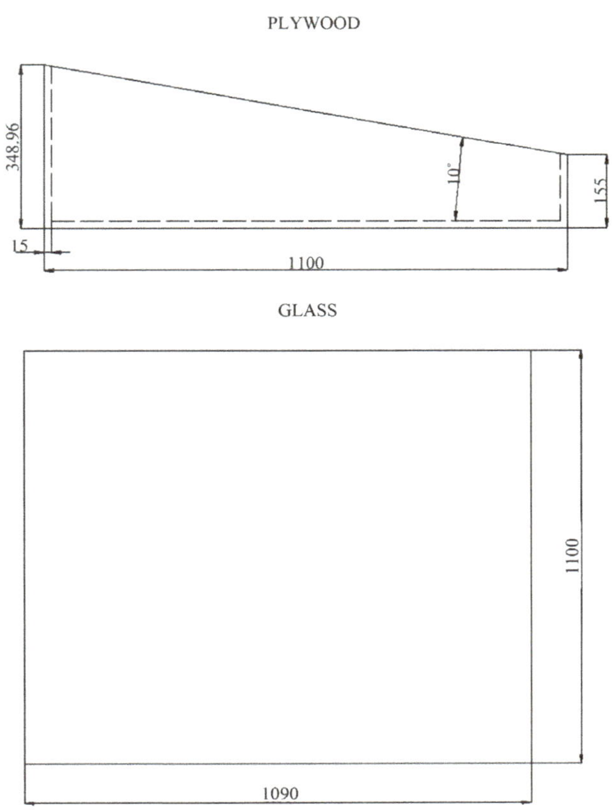

Figure 2. Schematic diagram of our solar still.

3.2. Experimental Setup

Figure 3 shows a single basin solar still is fabricated. This solar still consists of a basin made up of Aluminum of 1 × 1 meter with maximum height of 50 mm and thickness 2 mm thick. The cover is sealed tightly using silicon sealant to reduce the vapor leakage. Two pipes of 10 mm diameter are fitted to the basin; one for filling the brackish water in to the basin and other for flushing the brackish water out from the basin of the solar still. A condensate channel runs along the lower edges of the glass cover which collects the distillate and carries it outside the still. The entire assembly is placed on a stand structure made up of M.S. angles. The outlet is connected to a storage container through a pipe. Provision is made to change water in the still.

The still is filled with the brackish water in a thin layer as shown in **Figure 4**. The experiment is carried out keeping water depth of 1cm. During the experiment every day the solar radiation, atmosphere temperature and day time wind speed were also measured. The feed water is changed and the distilled water collected is measured at 7:00 a.m. every morning. The hourly productivity of fresh water is collected through a graduated flask. Day by day the salts deposited are removed manually. Each and every hour the potable water output is measured correspondingly the prevailing conditions are noted down. When the water is maintained at 1 cm level the output gain is more if it is more than 2 cm. A cross section pipe of diameter 6 cm of length 1 meter is used to carry down the pure condenses water from the solar stills to the collector which is provided outside the solar still. Eventually the output of the solar still is increased by hour by hour in the mid period of 11:00 a.m. to 3:00 p.m.

3.3. Terminology

Length of the base = 1 m

Figure 3. Still basin painted with black colour.

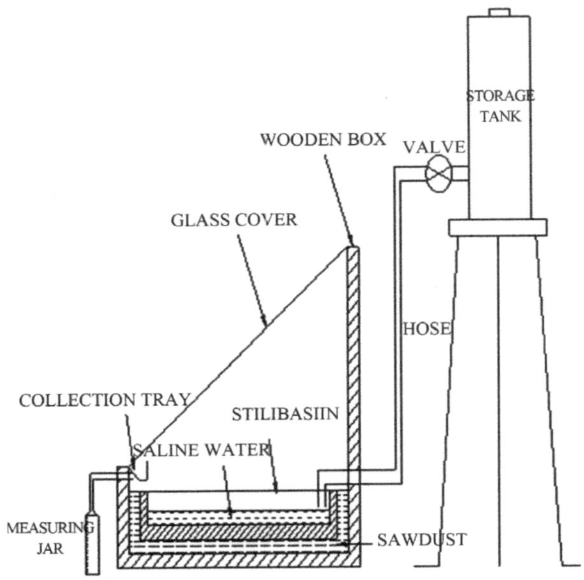

Figure 4. Skeleton of Solar Still.

Breadth of the base = 1 m

Thickness of the base = 1 mm

Volume = LBH = 1 × 1 × 0.001 = 0.001 m^3

Density of salt water = 1078 kg/m^3

Density of aluminum = 7833 kg/m^3

Mass of aluminum = Density∗volume = 7833 × 0.001 = 7.833 kg/m^3

Surface area of the base = 1 m∗1 m = 1 m^2

Volume of salt water = l∗b∗h= 1∗1∗0.005 = 0.005 m^3

Mass of salt water = Density∗volume = 1078 × 0.005 = 5.39 Kg

In solar still the basin is painted with black colour gives a daily production of 2220 ml of water/day. The average solar radiation is 906 W/m^2. The yield of distilled water is depending on the several factor wind velocity, solar radiation, and ambient temperature.

The measuring devices used in the system are as follows:

1) Solar meter (Sun meter) is used to measure the solar radiation. This device measures the instantaneous intensity of radiation in (W/m^2), Range 0 - 1999 $Watt/m^2$.

2) A hygrometer is an instrument used for measuring the moisture content in the environmental instrument.

3) A digital anemometer is used to measure wind speed.

4) Five thermocouples (type-k) coupled to digital Thermometer with a range from 0˚C to 99.9˚C with ±1˚C accuracy are used to measure the temperatures of the various components of the still system.

3.4. Measurements

The solar still made up of aluminum, inside bottom black paint coated is operated at ambient conditions from 6:00 a.m. to 7:00 p.m. during the months of April and May 2012. The measurements of the temperatures, solar radiation intensity, and the production of distilled water are taken hourly to study the effect of each parameter on the still productivity. In this study various operating conditions have been examined such as; different water depth, insulation thickness, ambient temperature and salt concentration without using nanofluids inside the till. The variables such as $T_{g\ in}$, $T_{g\ out}$, T_a, T_w, T_p and productivity are measured hourly. The total productivity and solar Intensity for each day are also measured. Also, different experimental tests are carried out at different ambient conditions. From about 2:00 p.m., water temperature decreases due to the losses from the solar still which becomes larger than the absorbed solar radiation. It can be noted that the basin temperature gets closer to the water temperature because of the continuous contact between them which leads to heat equilibrium.

As the glass temperature is much lower than the vapour temperature, it causes condensation of vapour on the glass. In the early hours of the morning 8:00 - 9:00 a.m., the glass temperature is higher than the water and vapour temperatures causing small productivity due to the small energy absorbed by the water at these times. Increase in the solar intensity in the early morning until it reaches the maximum at around 12:00 to 2:00 p.m., and then decreases in the late afternoon. The solar intensity has an important effect on the solar still productivity. As the solar intensity increases, the productivity increases due to the increase in heat gain for water vaporization inside the still.

The productivity rate varies as time passes from the early morning until late afternoon. In the morning, the temperature of water is low; therefore it needs high energy to change its phase from saturated liquid to saturated vapour phase. The results show that temperature and required heat are inversely proportional. In the early afternoon the temperature of water reaches the maximum so it needs less heat to vaporize, and vice versa in the late afternoon.

Experimental Readings

After the fabrication of the solar still successfully completed the experimental readings of the solar still is taken place. The various factors are considered along with the output. The water level in the solar basin is maintained to a level of 1 cm. The hour by hour reading is tabulated in the below tabular column. In **Table 1** the various parameter are predicted such as wind velocity, wet bulb temperature, dry bulb temperature, anemometer, and pyrometer. From the above tabular column we can able to understand that the output of the solar still is increased when there is a sufficient sequence parameter prevails. The experiment is carried out to determine the output of the solar still when its basin is painted with black colour.

Table 1. Time vs solar radiation.

Time duration	DBT	WBT	Wind velocity	Solar radiation	Water collection
Hour	°C	°C	m/s	W/m^2	ml/m^2
06 a.m. - 07 a.m.	27	24	0.5	1250	20
07 a.m. - 08 a.m.	28	25	0.9	1250	50
08 a.m. - 09 a.m.	30	26	1	1250	80
09 a.m. - 10 a.m.	32	27	1	1250	100
10 a.m. - 11 a.m.	34	27	0.7	1275	230
11 a.m. - 12 a.m.	35	28	4.5	340	350
12 p.m - 13 p.m	36	27	0.5	1143	350
13 p.m. - 14 p.m.	37	27	2.1	1060	280
14 p.m. - 15 p.m.	36.5	28	1.8	338	360
15 p.m. - 16 p.m.	36	27	2.5	750	300
16 p.m. - 17 p.m.	33	25	2.3	1100	220
17 p.m. - 18 p.m.	31	24	2.5	904	120
18 p.m. - 19 p.m.	28	23	2.0	700	30

4. Modified Solar Still

4.1. Nanofluids

Suspended CuO and Al_2O_3 (18.6 and 23.6 nm,) with two different base fluids: water and ethylene glycol (EG) and gives four combinations of nanofluids: CuO in water, CuO in EG, Al_2O_3 in water and Al_2O_3 in EG. Their experimental results showed that nanofluids have substantially higher thermal conductivities than the same liquids without nanoparticles. The CuO/EG mixture showed enhancement of more than 20% at 4 volume% of nanoparticles. In the low volume fraction range (<0.05 in test), the thermal conductivity ratios increase almost linearly with volume fraction. Although the size of Al_2O_3 particle is smaller than that of CuO, CuO-nanofluids exhibited better thermal conductivity values than Al_2O_3 nanofluids [13]. The thermal conductivity of four kinds of nanofluids such as MWCNTs in water, CuO in water, SiO_2 in water, and CuO in ethylene glycol are compared and found that the thermal conductivity of MWCNT nanofluid is increased up to 11.3% at 1 volume%, which s relatively higher than that of the other groups of nanofluids [14].

Using the MWCNT (dia 20 - 60 nm) with water as base fluid with temperature of 30°C shows that the thermal conductivity varies linearly below 30°C but above 30°C it is independent [15]. When 0.1% of MWNT is used it is found that thermal conductivity increased for 0.6% of volume fraction but both CMWNT and CDWNT gives 34% increased in thermal conductivity for 0.6% volume fraction [16].

Using the same nano particle with same base fluid but the size is reduced to 28.6 nm and 36 nm shows that the thermal conductivity depends on volume fraction, diameter, and bulb temperature. It is concluded that when the size of the nano particle decreases the thermal conductivity get increased based on the temperature range [17].

Multiwalled carbon nanotubes (MWCNTs) have been a topic of tremendous scientific interest in recent years due to their excellent thermal and electrical properties. They consist of several annular layers of rolled up grapheme sheets held together by interlayer van der Waal's forces. The typical diameter of the outermost layer (d_o) varies between a few nanometers (nm) and hundreds of nanometers, while the length (L_{CNT}) can be as high as 100 μm. Due to the high thermal conductivity (k_{CNT} = 3000 W/mK [18] and aspect ratio $(AR = L_{CNT}/d_o)$ of MWCNTs, adding them to a liquid improves the effective thermal conductivity of the suspension medium (nanofluid) significantly, compared to that of the original liquid. Depending on the properties of the base liquid, CNT geometry, and volume fraction, a wide range of enhancement has been reported in the literature. With as small as 1% volumetric fraction of MWCNT loading, the thermal conductivity of water is enhanced by ~40% [19].

4.2. Preparation of Carbon Nanotubes

The multiwalled carbon nanotubes are prepared by using the chemical vapor deposition methods using methane as an energy source and iron particles are used as a substrate. The as-produced CNT soot contains a lot of impurities. The main impurities in the soot are graphite (wrapped up) sheets, amorphous carbon, metal catalyst, and the smaller fullerenes. These impurities will interfere with most of the desired properties of the CNTs. The most common method used to purify the CNT is acid treatment [20]. First of all, the surface of the metal must be exposed to sonication. The CNTs remain in suspended form. When using a treatment in nitric acid (HNO_3), the acid only has an effect on the metal catalyst. It has no effect on the CNTs and other carbon particles if a treatment in hydrochloric acid (HCl) is used; the acid has also a little effect on the CNTs and other carbon particles. The diameter and the length are measured by transmission electron microscopy (TEM) and the structures of the CNTs are analyzed using scanning electron microscopy (SEM).

Mass production of carbon nanotubes (CNTs) by a cost effective process is still a challenge for further re- search and application of CNTs. Our group has focussed on the deposition of CNTs on a continuously-fed carbon substrate via arc discharge at atmospheric pressure. This process produces MWNTs using carbon from the substrate. The method differs in other respects from the conventional batch arc discharge method by using lower currents (<20 A) and larger inter-electrode gaps. To help define the local conditions of nanotube growth, the substrate surface temperature (Ts) was measured by optical pyrometry. Here, we report the influence of inter-electrode gap, substrate velocity and arc current on this temperature. It is found that carbon nanotube growth is favourable over a certain temperature range and retention time in the arc.

4.3. Preparation of Nanofluid

Figure 5 clearly shows that the multiwalled carbon nanotubes are entangled and not ready to be dispersed into fluids. Generally, carbon nanotubes are in hydrophobic nature, prone to agglomerate together, and settled quickly. To maintain stable and even suspension, two different methods are adopted for producing stable CNT nanofluids [20]. One is to use a surfactant, and sodium dodecyl sulfate (SDS) and is adopted as a surfactant in this study. At first, SDS is dissolved in DW at the rate of 1.0 wt% and then the mixture of CNTs and SDS solution is sonicated to make well-dispersed and homogenous suspendsions.

The other method is to attach hydrophilic functional group on to the surfaces of CNTs. Nitric/sulfuric acid mixture was used to modify the surfaces of CNTs. In a typical treatment of the present work, SDS surfactant is used to prepare the stable nanofluids. 2.0 wt% of SDS give homogeneous dispersion of carbon nanotubes in the nanofluids [21]. UV-vis spectrophotometric measurements are used to quantitatively characterize colloidal stability of the dispersions. Few existing nanofluid thermal properties studies have used this type of particle. Heat transfer behavior also depends on other properties (specific heat, density, viscosity). Macroscopic theory based on Maxwell's predictions for dielectric behavior of composites, Predicts increase in conductivity in nanofluids is approximately independent of particle size and particle conductivity liquids and materials that can be vaporized at low to moderate Temperature. Small particle size, but little control over size.

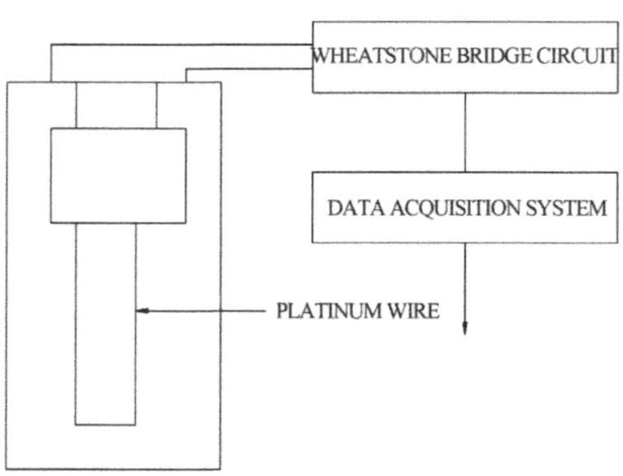

Figure 5. Thermal conductivity measurement.

4.4. Stability Evaluation

Usually, the stability of the oxide particles in suspension is determined by measuring the sediment volume versus the sediment time. However, this method is not suitable for the CNT dispersion. So the quality of the stability is characterized by using UV-vis spectrophotometer. It works on the principle of Beer-Lamberts law (*i.e.*) absorption of the solution is directly proportional to the solution concentration). In aqueous solution, the absorption of CNTs appeared at 283 nm. With increasing sediment time, the absorbance of CNTs in the supernatant aqueous solution is diminished.

4.5. Measuring the Thermal Conductivity of the Nanofluids

The thermal conductivity of the base fluid and nanofluid is measured by using transient hot-wire method. In this study, the transient hot-wire method for measuring electrically conducting fluid has been applied because the particles used in this experiment are electrically conductive. It is a well-known method and generally used to measure the thermal conductivity of nanofluids. Teflon-coated platinum wire, which diameter is 76 nm and the thickness of Teflon insulation layer is 17 nm, is used for a hot wire in the measurement system. Initially, the platinum wire immersed in media is kept at equilibrium with surroundings. When a uniform voltage is supplied to the circuit, the electric resistance of the platinum wire rises with the temperature of the wire and the voltage output is measured by an A/D-converting system at a sampling rate of ten times per second. The relation between the electric resistance and the temperature of platinum wire is well known. The measured data of temperature rise are linear against logarithmic time interval. The thermal conductivity is calculated from the slope of the rise in the wire's temperature against logarithmic time interval.

4.6. Nanostill

Researches in heat transfer have been carried out over the previous several decades, leading to the development of the currently used heat transfer enhancement techniques. The use of additives is a technique applied to enhance the heat transfer performance of water in the still basin. Recently, as an innovative material, nanosized particles have been used in suspension in conventional solar still. The fluids with nanosized solid particles suspended in them are called "nanofluids". The suspended metallic or nonmetallic nanoparticles change the transport properties and heat transfer characteristics of the water in the still. Thus the water temperature in the basin increases. The carbon nanotube (CNT)-based nanofluids are expected to exhibit superior heat transfer properties compared with conventional water and other type of nanofluids [22].

4.7. Measurements

The solar still made up of Aluminum, inside bottom black paint coated and addition of nanofluids is operated at ambient conditions from 6:00 a.m. to 6:00 p.m. during the months of April and May 2012. Addition of nanofluids with water in the basin increases the water temperature and thereby increasing the evaporation rate of the modified solar still. The same measurement process is repeated for various parameters to find out the enhanced performance of the solar still using nanofluids and compare the performance of them. Readings are taken for various parameters to find out the enhanced performance of the still with nanofluids and compare performance of them. In this study various operating conditions have been examined such as; different water depth, insulation thickness, ambient temperature and salt concentration with nanofluids inside the till. The mixing of nanofluids with water inside still basin helps to increase heat transfer. The suspended metallic or nonmetallic nanoparticles change the transport properties and heat transfer characteristics of the base fluid. Addition of nanofluids increases the water temperature and thereby increasing the evaporation rate. This improves the evaporation rate and in turn improves the efficiency of the still to certain extent.

5. Discussion of Results

The single basin solar still made up of Aluminium sheet is operated from 6.00 a.m. to 7.00 p.m. The measurements of the temperatures, solar radiation intensity and the production of distilled water are taken hourly to study the effect of each parameter on the single basin still productivity without using nanofluids. In this study, various operating conditions have been examined such as different water depths; insulation thickness, salt concentration, ambient temperature and productivity are measured hourly. The total productivity and solar Intensity are also measured daily. The output of the solar still varies directly with the ambient temperature. The productivity rate varies as time passes from the early morning until late afternoon. In the morning, the temperature of water is low; therefore it needs high energy to change its phase from saturated liquid to saturated vapor phase. The results show that temperature and required heat are inversely proportional. In the early afternoon the temperature of water reaches the maximum so it needs less heat to vaporize, and vice versa in the late afternoon.

The hourly output is maximum in afternoon hours when the ambient temperature is at its daily peak. The wind speed is found to be around 2 - 4 m/s. The water temperature has a direct effect on the productivity whereas the depth of water increases from 1 to 3 cm, the daily still output decreases *i.e.* inversely proportional. The solar radiation is absorbed by black painting inside the bottom of the basin and thus increases the temperature of the water. The black paint absorbs all the incident radiation falling on it. Due to this the amount of distillate collected in this still made up of aluminium sheet is higher and hence the increase in efficiency by 20%. Aluminium has higher thermal conductivity which increases the heat transfer rate. Due to this distillate collected is higher for the still made up of aluminium sheet and hence the increase in efficiency by 20% when compared with the still made up of Galvanized Iron sheet for the same basin area.

The efficiency increases when the insulation thickness increases. This is due to the decrease in the heat loss from the still to the surroundings. The insulation material should be dimensionally and chemically stable at high temperatures, and resistant to weathering and dampness from condensation. Usually, glass-wool insulation 10 cm thick is recommended. It would be between if the insulation also could contribute to the structural rigidity of the collector, but more rigid insulating materials are often less stable than glass-wool. Temperatures in flat-plate solar collectors can be high enough to melt some foam insulations, such as Styrofoam. And some foam give off corrosive frames at high temperatures, which could damage the absorber plate. The efficiency increases by 20% by providing insulation for a thickness of 10 cm. The theoretical value of absorbtivity multiplied with transmissivity is a function of the solar intensity and the ambient condition.

Figure 6 shows that the productivity increases with time until reaching the maximum value in the afternoon. At the maximum, the incident solar radiation is larger than heat losses early in the afternoon. When the radiation from the sun increases the heat gained by the solar stills also get increased thereby the heat stored in the solar stills also raised inspite of that the water gets heated quickly and starts to vaporize finally the vapour are condensed and potable water are collected.

Figure 6 shows that the output water level is very low at period of 9:00 a.m. to 10:00 a.m. the water collected at that time is 80 ml but the yield is increased at the peak time of the day between 14:00 p.m. to 15:00 p.m. to the water level of about 360 ml/hr which shows that at a certain period of time the output gain is more than the initial State. **Figure 6** also shows that the output water level is low at 13:00 p.m. due to clouds formation and they cover the sun and there is reduction in output.

Figure 7 shows the temperature variation along with the time. The wet bulb temperature indicates the amount

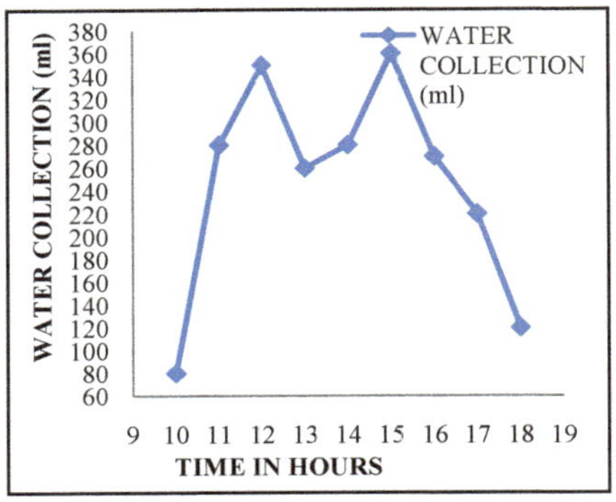

Figure 6. Time vs water collections.

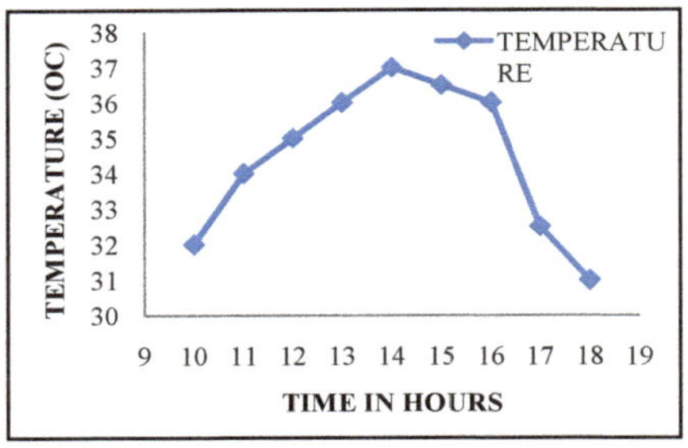

Figure 7. Time vs temperature.

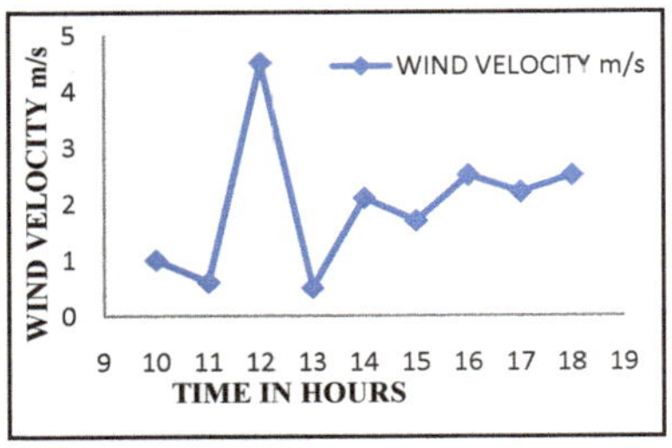

Figure 8. Time vs wind velocity.

of moisture content in the atmospheric air. Wind is caused by the uneven heating of the earth surface.

Depends upon the time period of flow of wind the output varies as shown in **Figure 8**. It can be concluded that as the solar intensity increases, the heat loss decreases and the water and ambient temperature difference increases considerably due to the increase of the water temperature through conduction process between the black base and the water. As the ambient temperature increases, the efficiency increases.

The same measurement process is repeated for various parameters to find out the enhanced performance of the solar still by mixing nanofluids with water inside still basin which helps to increase heat transfer. Addition of nanofluids increases the water temperature and thereby increasing the evaporation rate and in turn increases the efficiency of the still 60%.

The average daily output is found to be 6 litres/day for the basin area of 1 m^2. The optimized glass cover angle is 10°. The efficiency is calculated as 100% higher when compared with stills being used worldwide. In this paper the solar still made up of aluminium is fabricated and tested for both the conditions *i.e.* with and without use of nanofluids. The measurements are taken separately and the experimental data are compared here as shown in **Table 2**. The distillate collected is higher for the still using nanofluids when compared with the conventional fluid. The yield during the period between 6:00 a.m. to 7:00 p.m. is also high in the solar still made up of aluminium sheet.

Table 2. Time vs water collection comparision.

Time duration	Water collection	Water collection
	Ordinary still	Still with nanofluids
Hour	ml/m^2	ml/m^2
06 a.m. - 07 a.m.	20	100
07 a.m. - 08 a.m.	50	250
08 a.m. - 09 a.m.	80	300
09 a.m. - 10 a.m.	100	400
10 a.m. - 11 a.m.	280	500
11 a.m. - 12 a.m.	350	500
12 p.m. - 13 p.m.	360	550
13 p.m. - 14 p.m.	280	600
14 p.m. - 15 p.m.	360	650
15 p.m. - 16 p.m.	300	550
16 p.m. - 17 p.m.	220	450
17 p.m. - 18 p.m.	120	350
18 p.m. - 19 p.m.	30	200

The readings are plotted on a graph in **Figure 9** and the readings are compared. When comparing both the solar stills, the solar still using nanofluid inside the basin yields higher output for the same solar radiation. The average daily output of 6 liters/day is achieved when compared to that of 3 litres/day in an ordinary solar still painted with black paint. The solar radiation from the sun is higher at the mid noon period depend upon the time period and also the radiation the yield of pure water is depended on these factor.

For a solar still made up of aluminum sheet using nanofluids, graphs are drawn for productivity and time for various salt concentrations of 0%, 10% and 20% and for the depth of water level 1cm. It reveals the lower the salt concentrations the higher will be the productivity as shown in **Figure 10**.

For a solar still made up of Aluminum sheet without using nanofluids, graphs are drawn for productivity and time for different depths of water level 1 cm, 3 cm and 5 cm for salt concentrations of 10%. It reveals an increase in the productivity for minimum depths of water level. It also shows that the lower the salt concentrations the higher the productivity as shown in **Figure 11.**

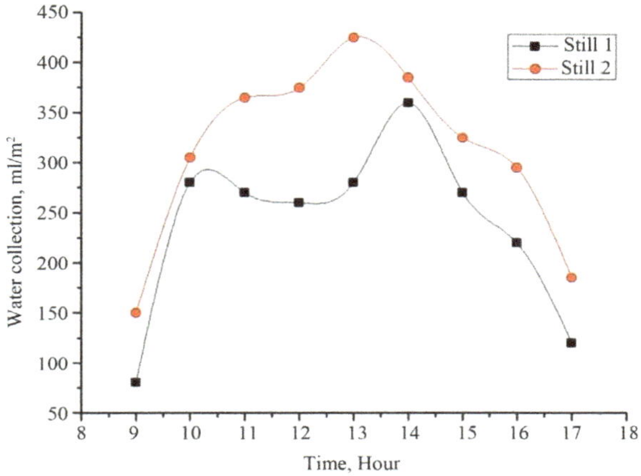

Figure 9. Comparision of yield output.

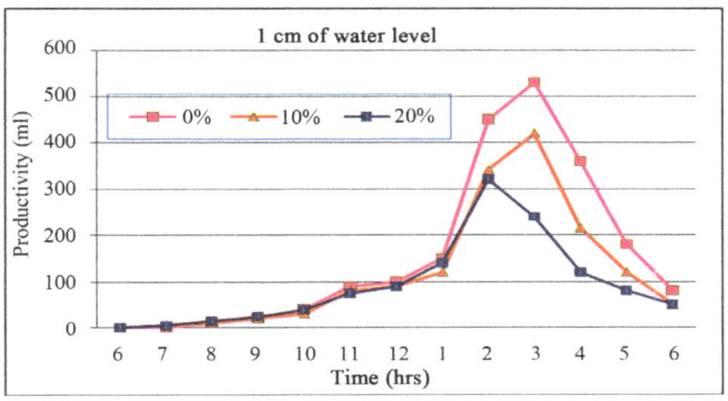

Figure 10. Productivity Vs Time for various concentrations.

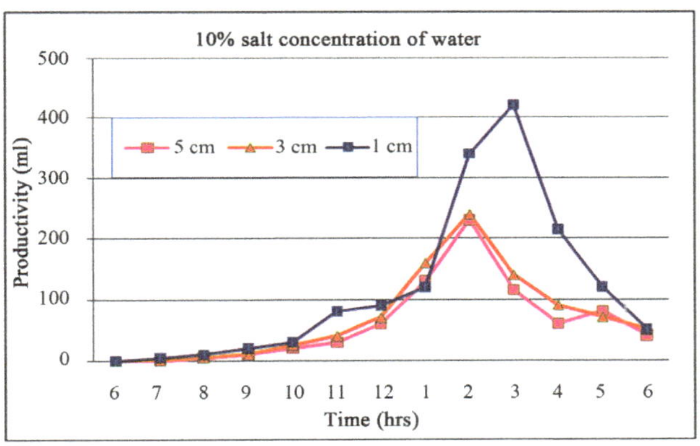

Figure 11. Productivity vs time for various water level.

The mixing of nanofluids with water inside still basin helps to increase heat transfer. The suspended metallic or nonmetallic nanoparticles change the transport properties and heat transfer characteristics of the base fluid. Addition of nanofluids increases the heat transfer rate and thereby increasing the water temperature. This increases the evaporation rate and in turn improves the efficiency of the still to certain extent. Thus the productivity is higher for the solar still using nanofluids in compared with the still using conventional fluids like water.

5.1. Thermal Analysis

The energy received by the saline water in the still (from the sun and base) is equal to the summation of energy lost by convective heat transfer between water and glass, radiative heat transfer between water and glass, evaporative heat transfer between water and glass, and energy gained by the saline water:

$$I(t) + A_b\alpha_b = m_b C_{pb}\left(\frac{dT_b}{dt}\right) + Q_{cb-w} + Q_{loss} \quad (1)$$

$$Q_{cb-w} = h_{cb-w} A_b (T_b - T_w) \quad (2)$$

$$Q_{loss} = UA_b (T_b - T_a) \quad (3)$$

Substitute Equations (2) and (3) in Equation (1) to solve the thermal equation.

The energy received by the saline water in the still $I(t)$ solar radiation and Q_{cb-w} convective heat transfer between basin and water are equal to the summation of energy lost by Q_{cw-g} convective heat transfer between water and glass, Q_{rw-g} radiative heat transfer between water and glass, Q_{ew-g} evaporative heat transfer between water and glass and energy gained by the saline water:

$$I(t) A_w\alpha_w + Q_{cb-w} = Q_{cw-g} + Q_{rw-g} + Q_{ew-g} + m_w C_{pw}\left(\frac{dT_w}{dt}\right) \quad (4)$$

α_w absorptivity of the water

Substitute to solve the thermal equation.

$$Q_{cw-g} = h_{cw-g} A_w (T_w - T_g) \quad (5)$$

The convective heat transfer coefficient between water and glass was given by,

$$h_{cw-w} = 0.884\left\{T_w + T_g + \frac{(P_w - P_g)(T_w + 273)}{268.9\times10^3 - P_w}\right\}^{1/3} \quad (6)$$

Here P_w, P_g - Partial pressure of glass and water.

$$Q_{rw-g} = h_{rw-g} A_w (T_w - T_g) \quad (7)$$

The radiative heat transfer coefficient between water and glass was given by,

$$h_{rw-g} = \varepsilon_{eff}\sigma\left[(T_w + 273)^2 + (T_g + 273)^2\right]\left((T_w + T_g + 546)\right) \quad (8)$$

T_g, T_w—Temperature of Glass and Water (°C)

σ -Stefan–Boltzmann constant (W/m²·K⁴)

$$Q_{ew-g} = h_{ew-g} A_w (T_w - T_g) \quad (9)$$

The evaporative heat transfer coefficient between water and glass was given by,

$$h_{ew-g} = \left[\left(16.273\times10^{-3}\right)\right]h_{cw-g}\left[\frac{P_w - P_g}{T_w - T_g}\right] \quad (10)$$

Energy gained by the glass cover (from sun and convective, radiative and evaporative heat transfer from water to glass) is equal to the summation of Q_{rw-g} energy lost by radiative and Q_{cg-sky} convective heat transfer between glass and sky and energy gained by glass:

$$\begin{aligned} &I(t)A_g\alpha_g + Q_{cw-g} + Q_{rw-g} + Q_{ew-g} \\ &= Q_{rg-sky} + Q_{cg-sky} + m_g C_{pg}\left(\frac{\mathrm{d}T_g}{\mathrm{d}t}\right) \end{aligned} \quad (11)$$

$$\varepsilon_{eff} = \frac{1}{\left(\frac{1}{\varepsilon_w} + \frac{1}{\varepsilon_g} - 1\right)} \quad (12)$$

ε_w, ε_g —Emissivity of water and glass Substitute Equations (5), (7), (9) and (13) in Equation (11) to solve the thermal equation.

$$Q_{rg-sky} = h_{rg-sky}A_g\left(T_g - T_{sky}\right) \quad (13)$$

The radiative heat transfer coefficient between glass and sky is given by,

$$h_{rg-sky} = \varepsilon\sigma\left[\frac{\left(T_g + 273\right)^4 - \left(T_{sky} + 273\right)^4}{T_g - T_{sky}}\right] \quad (14)$$

The effective sky temperature,

$$T_{sky} = T_a - 6 \quad (15)$$

The changes in basin temperature $\left(\mathrm{d}T_b\right)$, increase in saline water temperature $\left(\mathrm{d}T_w\right)$ and glass temperature $\left(\mathrm{d}T_g\right)$ are used in the above Equations.

The daily efficiency $\left(\eta_d\right)$ is obtained by summing up the hourly condensate production $\left(m_w\right)$, multiplied by the latent heat of vaporization (L), and divided by the daily average solar radiation $I(t)$ over the still area (A).

The overall Thermal Efficiency of still (η) is given

$$\eta = \frac{\sum m_w \cdot h_{fg}}{\sum A_s \cdot I} \quad (16)$$

where, m_w = mass of the distilled water collected as output in kg/s.

A = Area of the basin in m^2

$I(t)$ = Solar radiation with respect to time W/m^2 and

L = Latent heat of vaporization (h_{fg}) in KJ/s:

$$\begin{aligned} L = 2.4935\times10^6&\left[\left(1 - 9.4779\times10^{-4}T\right) + \left(1.3132\times10^{-7}T^2\right)\right. \\ &\left.-\left(4.7974\times10^{-9}T^{-3}\right)\right]\left(\text{Lower than } 70^\circ\text{C}\right) \end{aligned} \quad (17)$$

$$\begin{aligned} &L = 3.1615\times10^6\left[\left(1 - 7.6160\times10^{-4}T\right)\right] \\ &\left(\text{Higher than } 70^\circ\text{C}\right) \end{aligned} \quad (18)$$

The still efficiency is defined as the ratio of heat energy used for vaporizing the water in the basin to the total solar Intensity of radiation absorbed by the still.

5.2. Cost Estimation

The overall cost of the aluminium solar still experimental setup is given in **Table 3**.

5.3. Economic Analysis

The payback period of the solar still setup depends on overall cost of fabrication, maintenance cost, operating cost and cost of feed water.

The overall fabrication cost is Rs. 13,000 ($260).

The maintenance cost, operating cost and cost of feed water are negligible.

The overall cost of the project = Rs.: 13,000 ($260).

Cost of water produced per day = Daily yield × Cost of water per litre = 7.5 × Rs. 15 = Rs. 120 ($2).

The payback period is less than 1 year.

6. Conclusion

A single basin solar still made up of aluminium sheet is fabricated and tested for both the conditions with and without nanofluids. The distilled water production rate of a single basin solar still can vary with the design of the solar still, absorbing materials, depth of water and salt concentrations inside the still. The aluminium has higher thermal conductivity which increases the heat transfer so that the solar still made up of aluminium sheet yields more output of 6 litres/day. The efficiency of the solar still is increased by 55%. The efficiency of the solar still is increased further 20% by providing insulation to the still which reduces heat loss.

The efficiency of the solar still is 15% increased further by painting black colour in inside bottom of the still

Table 3. Cost estimation.

Sl. No.	Description	Amount in Rs.
1	Aluminum sheet	1600
2	Plywood	1600
3	Glass	550
4	Supply tank with accessories	400
5	Collecting tank	600
6	Thermo cool	150
8	Nanofluids	7600
9	Overhead charges	500
	Total	Rs. 13,000

to absorb more heat. The efficiency of solar still can be increased further by mixing nanofluids with water inside the still. Addition of nanofluids in the basin surface increases the thermal conductivity by 40% which in turn increases water temperature by increasing heat transfer rate and thereby increasing the evaporation rate and increases the efficiency by 60%.

The modified innovative nanostill has an enhanced performance when compared with the still using conventional fluids like water and more flexible with climatic conditions. The system will serve a family of 4 members. This gives the total water consumption to be around 7.5 litres/day/m^2. This cost-effective design is expected to provide the rural communities an efficient way to convert the brackish water in to potable water.

REFERENCES

[1] V. Velmurugan and K. Srithar, "Performance Analysis of Solar Stills on Various Factors Affecting the Productivity," *Renewable Energy and Sustainable Energy*, Vol. 15, No. 2, 2011, pp. 1294-1304.

[2] G. N. Tiwari, H. N. Singh and R. Tripathy, "Present Status of Solar Distillation," *Solar Energy*, Vol. 75, No. 5, 2003, pp. 367-373.

[3] O. O. Badran and M. M. Abu-Khader, "Evaluating Thermal Performance of a Single Slope Solar Still," *Heat and Mass Transfer*, Vol. 43, No. 10, 2007, pp. 985-995.

[4] S. Abdallah, O. Badran and M. M. Abu-Khader, "Performance Evaluation of a Modified Design of a Single-Basin Solar Still," *Desalination*, Vol. 219, No. 219, 2008, pp. 222-230.

[5] V. Velmurugan, K. K. J. Naveen, H. T. Noorul and K. Srithar, "Performance Analysis in Stepped Solar Still for Effluent Desalination," *Energy*, Vol. 34, No. 9, 2011, pp. 1179-1186.

[6] K. K. Murugavel, Kn. K. S. K. Chockalingam and K. Srithar, "Progresses in Improving the Effectiveness of the Single Basin Passive Solar Still," *Desalination*, Vol. 220, No. 1-3, 2008, pp. 677-686.

[7] M. K. Gnanadason, P. S. Kumar, G. Sivaraman and J. E. S. Daniel, "Design and Performance Analysis of a Modified Vacuum Single Basin Solar Still," *Smart Grid and Renewable Energy*, Vol. 2, No. 4, 2011, pp. 388-395.

[8] Y. Hwang, J. K. Lee, C. H. Lee, Y. M. Jung, S. I. Cheong, C. G. Lee, B. C. Ku and S. P. Jang, "Stability and Thermal Conductivity Characteristics of Nanofluids," *Thermochimica Acta*, Vol. 455, No. 1-2, 2007, pp. 70-74.

[9] Y. Xuan and Q. Li, "Heat Transfer Enhancement of Nanofluids," *International Journal of Heat and Fluid Transfer*, Vol. 21, No. 1, 2000, pp. 58-64.

[10] G. M. Koilraj, P. Senthilkumar and G. Sivaraman, "Design and Performance Analysis of a Vacuum Single Basin Solar Still," *International Journal of Advanced Engineering Technology*, Vol. 2, No. 4, 2011, pp. 174-181.

[11] M. K. Kalidasa, Kn. K. S. K. Chockelingam and K. Srither, "An Experimental Study on Single Basin Double Slop Simulation Solar Still within Layer of Water in the Basin," *Desalination*, Vol. 220, No. 1-3, 2008, pp. 687-693.

[12] R. Tripathi and G. N. Tiwari, "Thermal Modeling of Passive and Active Solar Stills for Different Depths of Water by Using Concept of Solar Fraction," *Solar Energy*, Vol. 80, No. 8, 2006, pp. 956-967.

[13] S. Lee, S. U. S. Choi, S. Li and J. A. Eastman, "Measuring Thermal Conductivity of Fluids Containing Oxide Nanoparticles," *Journal of Heat Transfer*, Vol. 121, No. 2, 1999, pp. 280-289.

[14] Y. J. Hwang, Y. C. Ahn, H. S. Shin, C. G. Lee, G. T. Kim, H. S. Park and J. K. Lee, "Investigation on Characteristics of Thermal Conductivity Enhancement of Nanofluids," *Current Applied Physics*, Vol. 25, No. 4, 2005, pp. 609-735.

[15] M. J. Assael, C. F. Chen, N. Metaxa and W. A. Wakeham, "Thermal Conductivity of Suspensions of Carbon Nanotubes in Water," *International Journal Thermophysics*, Vol. 2, No. 25, 2004, pp. 971-985.

[16] H. Xie, J. Wang, T. Xi, Y. Liu, F. Ai and Q. Wu, "Thermal Conductivity Enhancement of Suspensions Containing Nanosized Alumina Particles," *Journal of Applied Physics*, Vol. 4, No. 91, 2002, pp. 4568-4572.

[17] C. H. Li and G. P. Peterson, "Experimental Investigation of Temperature and Volume Fraction Variations on the Effective Thermal Conductivity of Nanoparticle Suspensions (Nanofluids)," *Journal of Applied Physics*, Vol. 8, No. 99, 2006, pp. 84-94.

[18] L. Jiang, L. Gao and J. Sun, "Production of Aqueous Colloidal Dispersion of Carbon Nanotubes," *Journal of Colloid Interface Science*, Vol. 260, No. 1, 2003, pp. 89-94.

[19] T. K. Hong, H. S. Yang and C. J. Choi, "Study of the Enhanced Thermal Conductivity of Fe Nanofluids," *Journal of Applied Physics*, Vol. 6, No. 97, 2005, pp. 1-4.

[20] S. A. Putnam, D. G. Cahill, P. V. Braun, Z. Ge and R. G. Shimmin, "Thermal Conductivity of Nanoparticle Suspensions," *Journal of Applied Physics*, Vol. 8, No. 99, 2006, pp. 84-88.

[21] E. Natarajan and R. Sathish, "Role of Nanofluids in Solar Water Heaters," *International Journal Advanced Manufacturing Technology*, Vol. 8, No. 170, 2009, pp. 1876-1882.

[22] M. K. Gnanadason, P. S. Kumar, G. Jemilda and S. R. Kumar, "Effect of Nanofluids in Vacuum Single Basin Solar Still," *International Journal of Scientific and Engineering Research*, Vol. 3, No. 1, 2012, pp. 2229-5518.

Modeling, Control & Fault Management of Microgrids

Mehdi Moradian[1], Faramarz Mahdavi Tabatabaei[2], Sajad Moradian[3]

[1]Department of Electrical Engineering, Sahand University of Technology, Tabriz, Iran; [2]Saman Gostar Company (Distributor of SANTERNO, Italy), Tehran, Iran; [3]Department of Electrical Engineering, Iran University of Science and Technology, Tehran, Iran.

ABSTRACT

In this paper, modeling and decentralize control principles of a MicroGrid (MG) whom equipped with three Distributed Generation (DG) systems (consist of: Solar Cell System (SCS), MicroTurbine System (MTS) and Wind Energy Conversion System (WECS)) is simulated. Three arrangement of load changing have investigated for the system. In first one the system doesn't have transfer of power between MG and grid. In other two arrangements system have transfer of power between MG and utility grid. Of course in third case transfer of power between DG resources is considerable. Case study system is equipped by energy storage devices (battery bank) for each DG's separately by means of increasing the MG reliability. For WECS and SCS, MPPT control and for MTS, voltage and frequency (V&F) controller has designed. The purpose of this paper is load respond in MG and storage process of surplus energy by consider of load changing. MATLAB/Simulink and its libraries (mainly the Sim Power Systems toolbox) were employed in order to develop a simulation platform suitable for identifying MG control requirements. This paper reported a control and operation of MG in network tension by applying a three phase fault.

Keywords: Microgrid; Decentralize Control; Wind Energy Conversion System; Microturbine; Solar Cell

1. Introduction

Microgrid concept widely developed in countries such as USA, Canada, Japan and UK. It has been investigated and implemented [1,2]. The increase in researches due to benefits of this type of networks including provide the reliability and security of network and loads, high efficiency, environmentally friendly and self-healing [3]. In today's power systems, very large problems including electricity production cost and also reduce of fossil fuel, on the other hand, the increasing pollution created from burning oil and gas and dramatically growth of demands has been increase the greenhouse gases in the air which is considered as a big threat for ozone layer. Because electricity costs much less and in some cases is zero (for DG Resources), so in today's power systems used distributed energy resources (e.g. wind and sun are free resources to generate electricity). Another advantage of smart grid which it should be mentioned is loss reduction caused by power transmission line. Because one of the goals in smart grids is producing power by distributed energy resources and removed the power plant as much as possible. So with this, the line power flowing and its corresponding losses can be reduced to acceptable level. In recent years, many researches about structure, control and implementation of smart grids has been worked in laboratories. Control of MG and performance of energy storages have close relationships, which in this paper have been reviewed. By energy storage MG control can be more easily and the system reliability can be increased. A MG is a connection of distributed energy resources like: wind energy conversion, microturbine, fuel cell, PV arrays, Combined Heat and Power (CHP) and energy saving factors such as flywheel, batteries or Uninterruptable Power Supply (UPS) and power capacitors in low voltage power systems [4].

Basic structure of a typical MG is shown and discussed in [5].

Here is assumed that distributed generation sources have the ability to respond the loads. After disappearing network fault, synchronization operation performed and isolated network connected to Utility Power Source again [6].

In reference [7] a very simple scheme of a MG with three DG resources has been studied. In the article that was published in 2010, there is no mention and have not been analyzed the: DG resources structure, controllers of each micro source, fault occurrence in the grid and reaction of MG against the sudden event and also transferred power between DG's.

Note that the decentralized control means that on each DG resources has been an independent controller and each of this resources performed control operations independently. May be the type of applied controller is different and even similar. Fault events that may lead to islanding of a distribution system are discussed in [8]. The work described in this paper regards the simulation and control of the MG equipped by three distributed energy storage device and local loads. Used distributed generations in this paper are consists of: WECS, SCS (with MPPT control strategies), and MTS (with voltage and frequency control strategy). The robustness of the tested control strategies were studied for disturbances taking place in the utility network, followed by a forced islanding of the MG. Experimental tests for islanding and synchronization were presented in [9]. Islanding of the MG can take place by unplanned events like faults in the utility network or by planned actions like maintenance requirements.

2. Structure of Distributed Generation Systems

In **Figure 1** the general structure of a micro-distribution network is shown. The input power producing by distributed generation resources converted to electrical energy for network and load requirements. Control tasks are divided into two parts:

1) Input-Side Controller: which should be possible to take the maximum power from the input source. Naturally, the protection of input side converter must be in considered.

2) Grid-Side Controller: that can follow these tasks: a) Input active power control derived for network; b) Control of the reactive power transferred between network and micro-grid; c) DC link voltage control; d) Synchronization of network; e) Assurance of power quality injected to the network. Generally the network controller position is VSI, which both amplitude and phase of the output voltage are controlled. All items listed above are the basic features for the grid-side controller that these converters should have. Studied network in this paper consists of two distributed generation sources, which is briefly explains their structure.

3) Microgrids Internal Structure: According to given system studied in this article is including distribution generating resources of PV and MT. The structure of this two system and their relevant controlling parts are shown below. It is noteworthy that PV and Fuel Cell (FC) systems have similar hardware structures [10,11].

3.1) Photovoltaic and FC systems: As previously noted the PV and FC hardware structure is similar. Although voltage or current by FC and PV is low, but by binding a set of them together can increase the production levels and also can increase or decrease the voltage level by using DC-DC converters such as boost converter for increasing the voltage level.

Non-linear relationship between V-I obtain from the below equation [12].

$$V_{pv} = \frac{N}{\lambda}\ln\left(\frac{I_{SC} - i_{pv} + MI_o}{MI_o}\right) - \frac{N}{\lambda}R_S i_{pv} \quad (1)$$

Which in it I_{SC} is short circuit current, I_o is reverse saturation current, R_S is series resistance and λ is constant factor which is depends on the type of materials used in cell.

In this paper a silicon solar panel, ($M = 1, N = 36$) has been used. Sample model is constructed by Iranian Optical Fiber Fabrication Company (OFFC) that related table of its coefficients and parameters comes to Table A1 [13].

According to related values Equation (1) is written as follows:

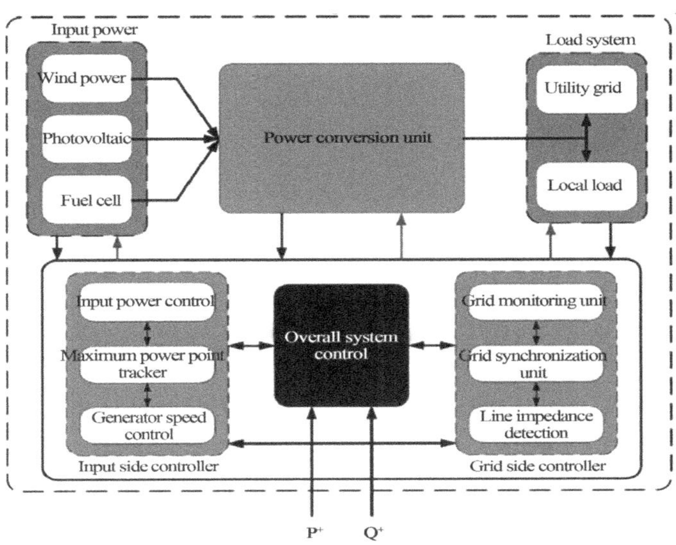

Figure 1. Topology of smart systems control.

$$V_{pv} = 1.767 \ln\left(\frac{I_{SC} - i_{pv} + 0.00005}{0.00005}\right) - i_{pv} \quad (2)$$

Non-linear characteristics of V-I and P-I are shown in **Figure 2.** P_{MP}, V_{MP} are known as power and voltage of maximum power in PV cell. These Curves are:

By changing temperature, the coefficients will changes [12]. Two samples of these changes are estimated in Equations 3-a (70˚C) and 3-b (−20˚C).

$$V_{pv} = 1.69 \ln\left(\frac{3.005 - i_{pv} + 0.00024}{0.00024}\right) - i_{pv} \quad (3\text{-a})$$

$$V_{pv} = 1.82 \ln\left(\frac{2.83 - i_{pv} + 0.00001}{0.00001}\right) - i_{pv} \quad (3\text{-b})$$

For displaying MPPT technique in PV we act the way [14]: cell voltage with corresponding maximum power production by considering the open circuit voltage for different temperature show a dependency.

$$V_{MP} = MV_{OC} \quad (4)$$

This equation shows MPPT technique which in it M_V called voltage factor that for OFFC its value considered 0.71. This method for maximum power estimating is simple and fast.

Equivalent circuit to PV cell block shown in **Figure 3(a)** which the related equation to non-linear V-I relationship is placement. Also a delay function to limit the current of rapid response to voltage controlled source and to improving the convergence responses is used.

For VMPPT its related equivalent circuit shown in **Figure 3(b)**. This block will calculate the open circuit voltage (By using I_{SC} and Equation (2)), then comprised it with PV output voltage and produces the fire command for the PWM block. The delay shown here is the same reason as in **Figure 3(a)**.

Now we want to see the performance of this system in connected to the grid-connected mode and with applying a controller in AC part according to what is seen in **Figure 4**. After increasing the voltage level by boost, we paralleled it by energy storage devices (That somehow we can call them UPS). We do this in order to increase reliability of system. Now it is necessary that DC voltage produced by inverter becomes AC. The purpose of the grid-side controller is to maintain DC link voltage at a constant value regardless production power range.

Vector control in a rotating reference frame with the line voltage vector is used. The purpose in this controller is regulation of DC voltage and reactive power control. Using the Park conversions, voltage equations can be controlled to reference frame d-q. The idea of control is taken from [10].

Figure 5 shows simulated model of grid-side controller. PI standard controllers are used in order to regulate the line current in rotational synchronous frame in internal control loop and DC voltage in external loop.

i_d is active part of current and i_q is reactive section on current. In order to obtain a transformation from active power, the value of current reference i_q (reactive part) considered as zero. PLL used in figure is to synchronize converter frequency with main grid. It is assumed that the harmonics produced by switching is zero.

3. Structure And Mts Control Model

Recently microturbines have been much attention because of their small size, relative low cost, repair and cheap maintenance and relatively simple control. Different dynamic models have been discussed for micro-turbines by Rowen, Hannet, Saha and Nern for combustion gas turbine [15-17]. In 1993 mathematical method of gas turbine by Rowen was developed [15]. While in 1993, Prime Mover Working Group by considering the control of speed, acceleration, fuel and temperature made this model wider [16]. MT used in this article is a small combustion turbine with an installed capacity 25 to 500 KW and a high rotation speed (between 50,000 to 120,000 rpm). This model includes the speed governor,

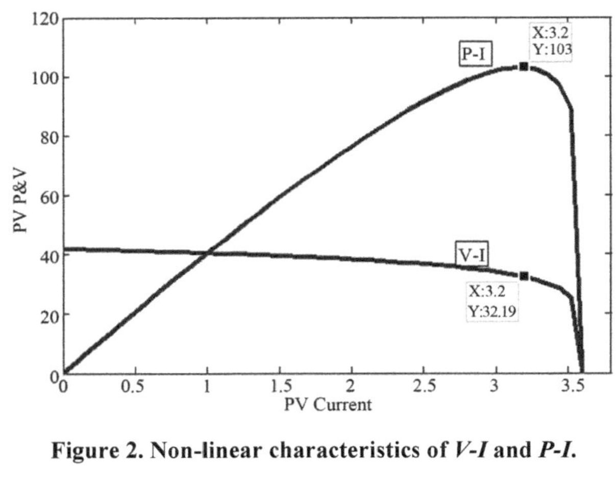

Figure 2. Non-linear characteristics of V-I and P-I.

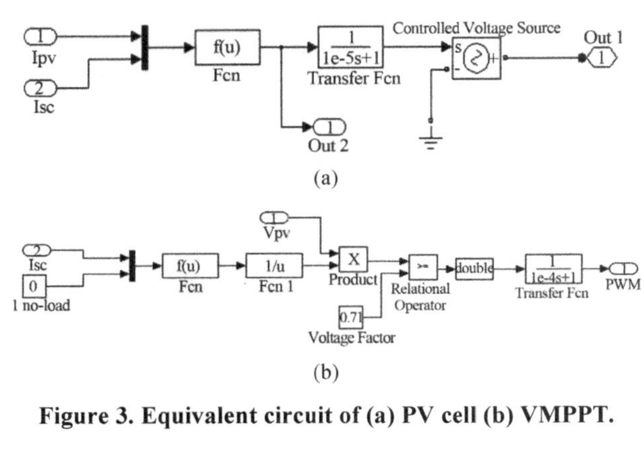

Figure 3. Equivalent circuit of (a) PV cell (b) VMPPT.

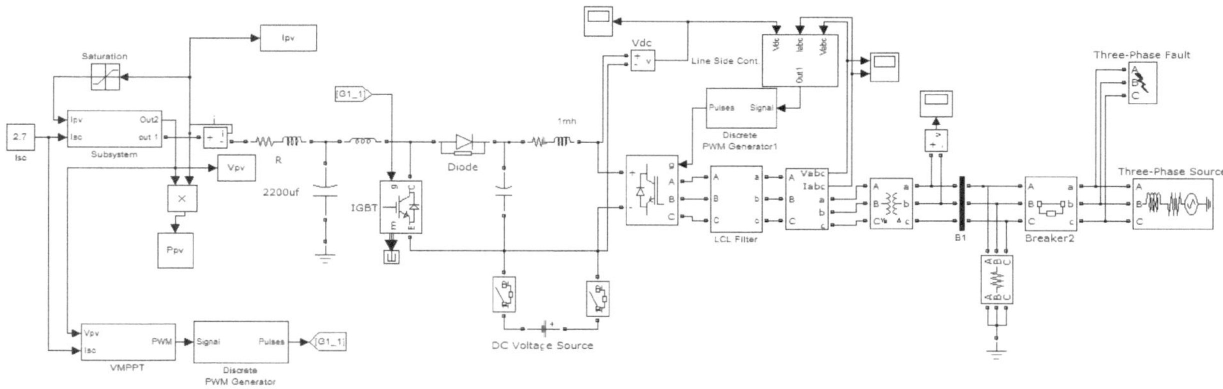

Figure 4. SCS in grid-connected mode and applying grid-side controller.

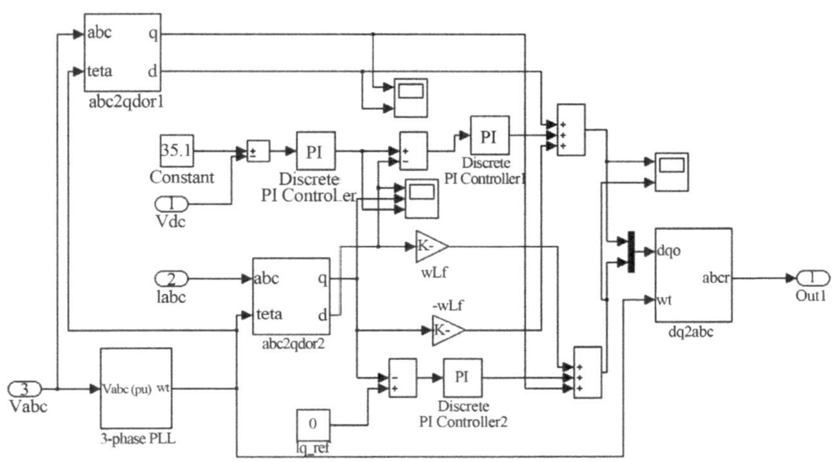

Figure 5. Equivalent circuit of grid-side controller.

acceleration control block, fuel system control and temperature control. Single-shaft turbine model is considered. Power producer with a Permanent Magnet Synchronous Generator (PMSG) has two poles and smooth poles rotor. Because of high speed shaft, generators of an AC voltage source will be a high frequency (frequency angular higher than 100,000 rad/sec) [17]. Since turbines moves at high speed, so AC generator is a high-frequency generator which cannot be directly coupled the AC network [18]. One way to model a system of distributed generation MT, based on all classification system are three following separate parts [19,20]:

1) Module 1: mechanical system of turbine and fuel.

2) Module 2: PMSG and AC/DC rectifier and energy storage devices.

3) Module 3: AC/DC voltage source inverter, PWM controller.

Mechanical Model and MT Control Functions:

Based on Rowen and Hannet model, we examine the MT model. Dynamic equations of MTS in [15] are investigated.

According to the principle of energy conversion and ignore the inverter losses, total of instantaneous powers in output of AC terminal must be equal to the instant powers in dc terminal like.

$$V_{dc}I_{dc} = v_a i_a + v_b i_b + v_c i_c \tag{5}$$

Which I_{DC} and V_{DC} are dc link voltage and current.

VSI simplified model shows in **Figure 6(a)**. The inverter which used in this essay is hysteresis model.

Diagram block of V&F controlling model presented in **Figure 6(b)**. $V_{d_{ref}}$ and $V_{q_{ref}}$ are reference amounts. In order to have unit power factor, the amount of $V_{q_{ref}}$ is zero and $V_{d_{ref}}$ is 1 p.u. Voltage and frequency (V&F) control has to regulate the voltage value at the Point of Common Coupling (PCC) and also the frequency of the whole grid.

Now MT model of distributed generation in grid-connected mode is shown in **Figure 7**. The produced frequency by inverter will have the equal amount of 50 Hz corresponding to the network.

LCL Filters in this paper is designed by the idea in [21,22].

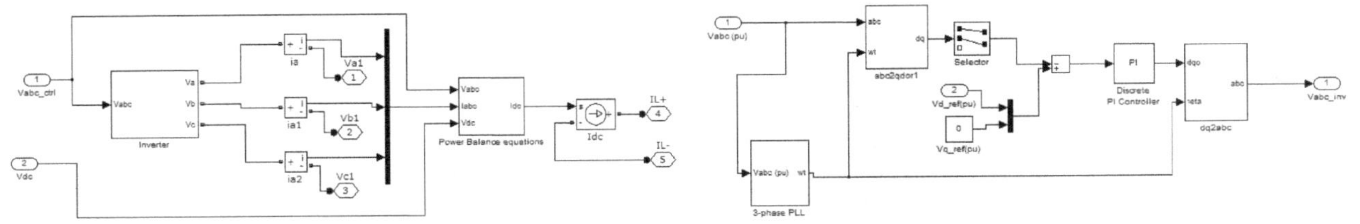

(a) (b)

Figure 6. Bolck diagram of MTS components; (a) VSI simplified model; (b) Diagram block of V&F controlling model.

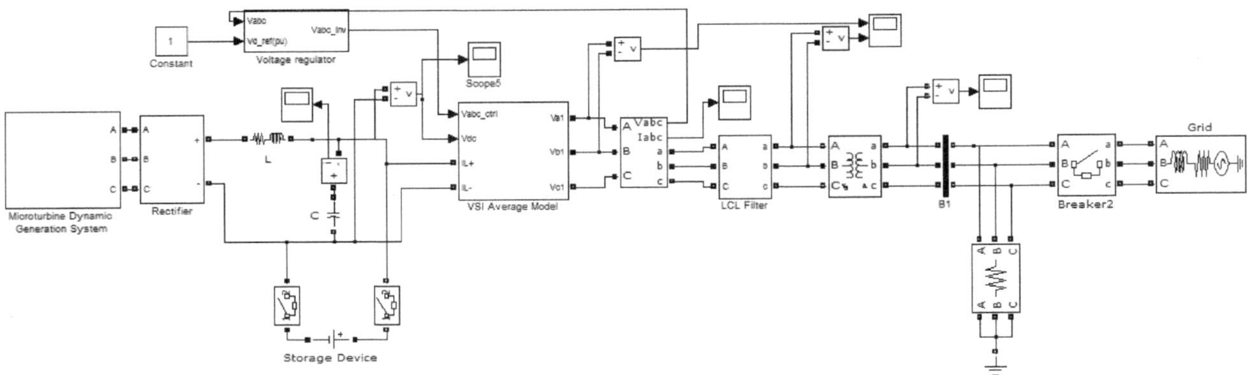

Figure 7. The general model of MTS in grid-connected mode.

Conceptual and technical solution of MG is presented in [23,24].

4. Structure and WECS Control Model

Electrical wind generators are the equipment who converts wind to electrical energy. Different types of generators are used in wind turbines. For example small sized wind turbines are equipped by DC generators with capacity up to 90 kw (from 10 to 90).

In wind turbines modern systems three phase AC generators are customs [25].

General kind of AC generators who are used in modern wind turbines are:

1) Squirrel Cage Induction Generator (SCIG)
2) Wired Rotor Induction Generator (WRIG)
3) DoubleFed Induction Generator (DFIG)
4) Synchronous Generators with output excitation (SG)
5) Permanent Magnet Synchronous Generator (PMSG)

Synchronous Generator is a kind of generators who are used in some researches [26,27]. These generators could connect to wind turbine without any gearbox. These benefits are attractive by consider of maintenance and limited shelf life. Synchronous generators could to excite by electric or with permanent magnet rotor. By considering the above reasons, used generator in this paper is kind of PMSG.

5. Simulation of WECS

This system is modeled by equations of wind turbine as could be seen in Equations below. In this paper a variable speed wind turbine is used. Wind speed 12 m/sec is considered. The parameters value of PMSG is shown in Table A2. If the speed of wind was variable, WECS should be used the Buck/Boost converter. In this case the trigger signal should produce for two switches. This performance cause to system be complicated.

Equations for wind turbine are shown in below [28].

$$P_m = 0.5 \times C_p(\lambda, \beta) \times \rho \times A \times v_w^3 \tag{6}$$

$$C_p(\lambda, \beta) = C_1\left(\frac{C_2}{\lambda_i} - C_3\beta - C_4\right)e^{\frac{-C_5}{\lambda_i}} + C_6\lambda \tag{7}$$

$$\lambda = \frac{Rw}{v_w} \tag{8}$$

$$\frac{1}{\lambda_i} = \frac{1}{\lambda + 0.08\beta} + \frac{0.035}{\beta^3 + 1} \tag{9}$$

Output mechanical power in watt is shown in Equation (6). In this equation, ρ air density in (kg/m^3), C_p, performance coefficient, v_w wind speed in m/sec, λ, tip speed ratio, β, pitch angle, A, turbine swept area.

In Equation (7), the coefficients C_1 to C_6 are: $C_1 = 0.5176, C_2 = 116, C_3 = 0.4, C_4 = 5, C_5 = 21$ and $C_6 = 0.0068$ [28].

In Equation (8), R is rotor radius in meter; w, angular speed in rad/sec the output torque of wind turbine is input of used PMSG.

In order to acquiring the output maximum power in WECS, we use the MPPT algorithm **Figure 8**.

Inverter's in each DG's are modeled base on SANTERNO products [29].

In this algorithm, the initial value adjusted for DC references voltage. Correspondingly, voltage and current will be measured. After the measurement, DC output power (Po) would be calculated. In next step, the reference voltage ought to be altered as much as dc variations (ΔV_{dc}). By this way:

$$V_{ref}(k) = V_{ref}(k-1) + \Delta V_{dc} \tag{10}$$

Then dc power will calculated with $P(k) = V_{dc}(k) \times I_{dc}(k)$. If $P(k) \succ P_o$, the system output isn't in maximum point, so accordingly the reference voltage have to rise a quantum of ΔV_{dc} and power should compare with earlier amount $(P(k) \succ P(k-1))$. This process continued till receiving to maximum point. Now if $(P(k) \prec P(k-1))$, the reference voltage should decrease.

Value and parameters of boost model is shown in Table A3.

In **Figure 9** MPPT block of WECS is shown. This block should produce the trigger signal of switches in DC link for tracking the maximum power.

In **Figure 10(a)** Torque-Speed characteristics of WT is shown. Note that all of curves are in 12 m/sec wind speed.

As you see in **Figure 10(b)**, maximum amount of power coefficient in used turbine is 0.41 and $\lambda = 7.71$.

As we said, output power of WT is 7100 watt that is shown in **Figure 10(d)**.

Produced torque of WT in system run time is 184 N.m.

DC link Voltage in battery bank terminal is shown in **Figure 10(e)**. Voltage ripple is small. But it regulate in 243 volt. Battery bank capacity should be proportion to

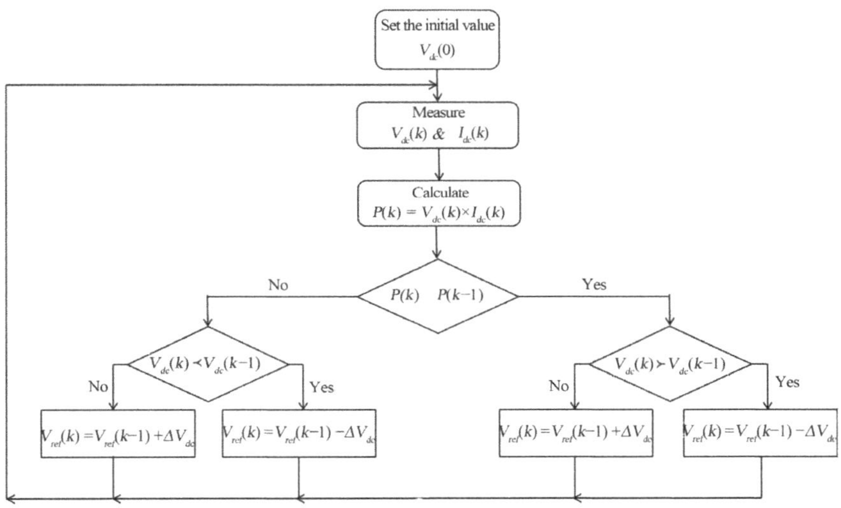

Figure 8. WECS MPPT flowchart.

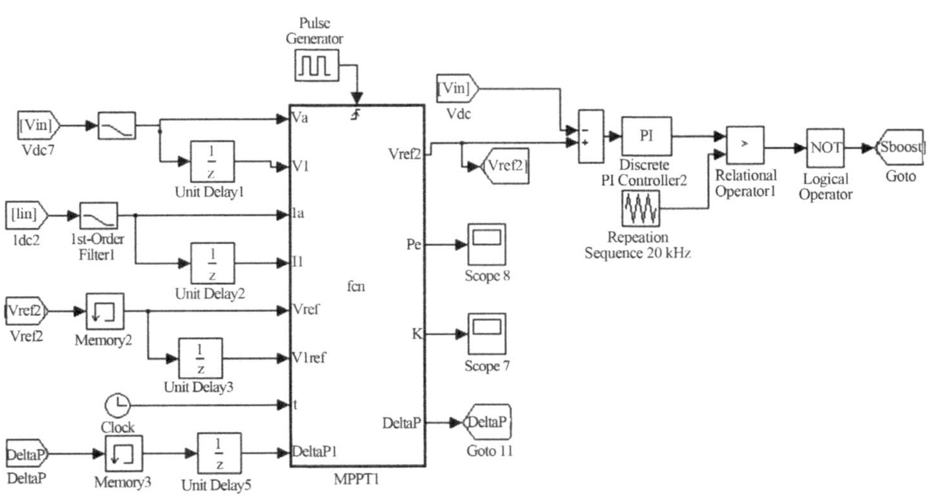

Figure 9. MPPT simulated block for WECS.

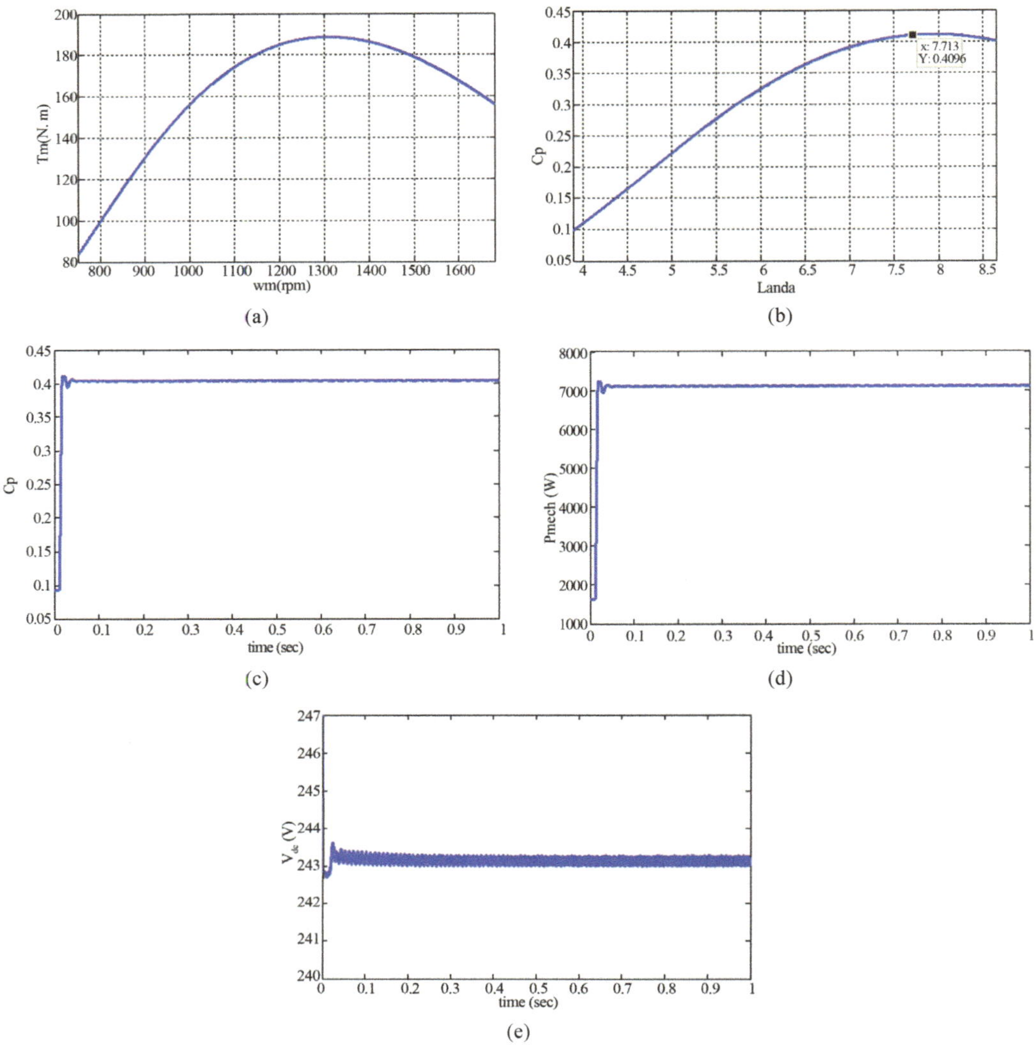

Figure 10. Output of WECS; (a) Torque-speed characteristics of WT; (b) $C_p - \lambda$ curve of WT; (c) Changing of power coefficient in time; (d) Produced mechanical power; (e) DC link Voltage in battery bank terminal.

produced power and connected load to the system. This signal is input of three phase inverter. In **Figure 11** grid-connected WECS is simulated. In t = 0.25 sec three phase fault (phase-phase) applied and the system goes to islanding mode.

Figure 12 has shown the output voltages before/after LCL filter.

6. Simulation

As you see in **Figure 13**, each DG sources protect local (three-phase and balance) load, the considered load for MTS is 375 KVA, WECS is 7100 W and SCS load, 110 W. Each DG sources is equipped with an energy storage device. Because if in specifics circumstances the producers cut, storage resources can continue to support the loads. To review the islanding mode and fault management in this networks, 3 phase fault (phase-phase) applied to system and breaker (Point of Common Coupling) in moments will stop applying the fault and load and DG sources goes to islanding mode. Output shown in this case shows that in this moment DG resources are responsive to load. When the network fault disappears, part of autonomous again can be connected to the utility grid. Some of the values required in the above system are given in Appendix.

Simulation process is for three cases: 1) the case that load capacity and produced power are equal. Here the system doesn't have any transfer of power between MG and utility grid. 2) The case that loads capacity is less than produced power. Here system has power transfer between grid and MG. 3) The case that MTS load capacity is less than its local load, SCS and WECS local loads are bigger than its produced capacity. Here the system

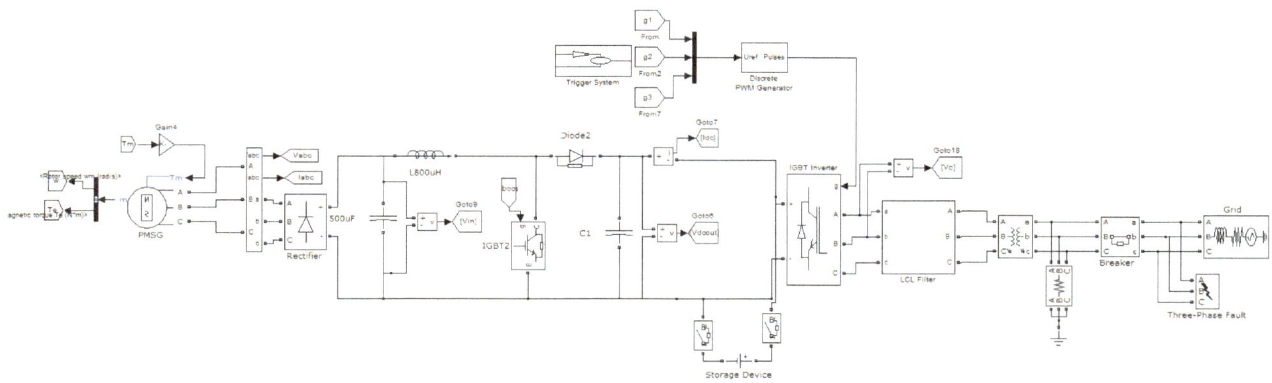

Figure 11. WECS in grid-connected mode.

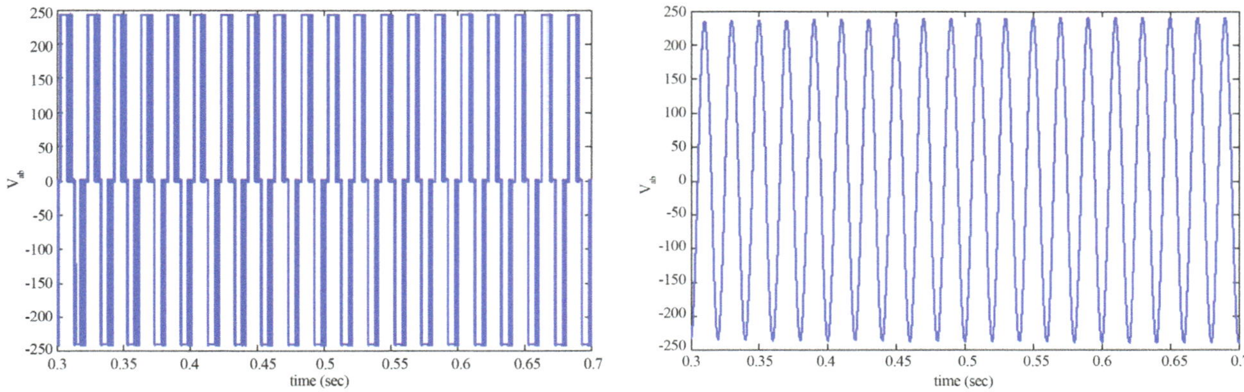

(a) (b)

Figure 12. Output voltage of WECS (a) Before; (b) After LCL filter.

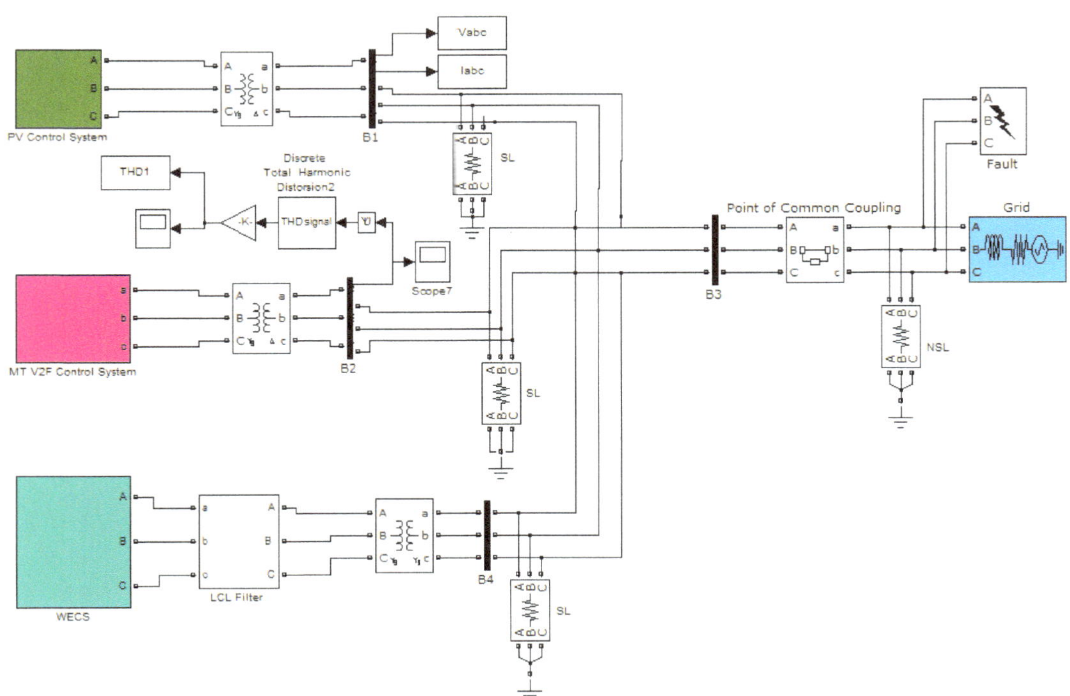

Figure 13. Case study system.

has power transfer between MG and grid. Also transfer power between DG resources is considerable.

Output curve of used PV system is shown in **Figure 3**. The points Corresponds to maximum power is shown in the figure.

7. Simulation Result

In this part simulated migrogrids result are displayed and investigates. First case: amount of produced power by each resources are equal with the loads toward each DG: in this case SL's respond's by each micro sources and we don't have any power transfer between MG and grid. At t = 0.25 sec three phase fault applied to the system and breaker in PCC guide the system to islanding mode operation. In below curves the time before 0.25 sec is for grid-connected operation and after 0.25 sec is for islanding mode. In first case NSL isn't connected to grid or it can support by utility grid. Load capacities are equal to produce capacity. Waveforms of the system in this case are shown in figures below.

In simulated system, SL's are sensitive loads and NSL is non-sensitive load of system.

By using simulation, produced power waveform of PV is as follows (**Figure 14(a)**).

Three phase voltage of SCS is 32 v (**Figure 14(b)**). it means that MPPT process is done by the applied controller.

In **Figure 15** the voltage and current curves shows. In this case, the system can continue to stable operation and in islanding mode the loads could responds by DG sources.

In **Figure 16(b)** total harmonic distortion in system output is shown. In the moment of t = 0.25 sec three phase fault applied. THD rise up to 2.035 percent (in fault time).

It can be seen in **Figure 17**, the network can respond to loads and we don't have any transferred power between grid and MG.

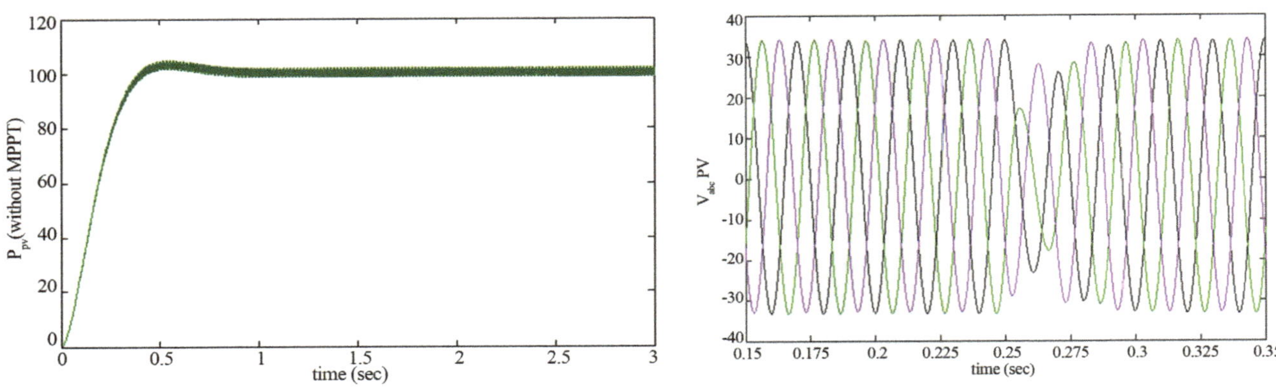

(a) (b)

Figure 14. Output of SCS; (a) Pulled output power from the PV in DC link; (b) Three phase voltages of SCS.

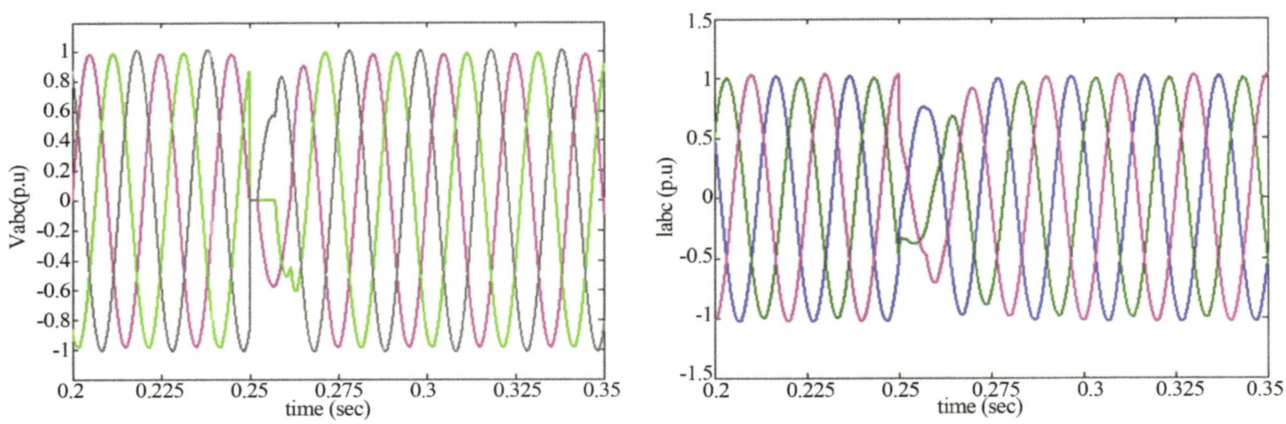

(a) (b)

Figure 15. Voltage and current of line and loads (in p.u.); (a) Three phase voltage of line (in p.u.); (b) Line and loads three phase current (in p.u.)

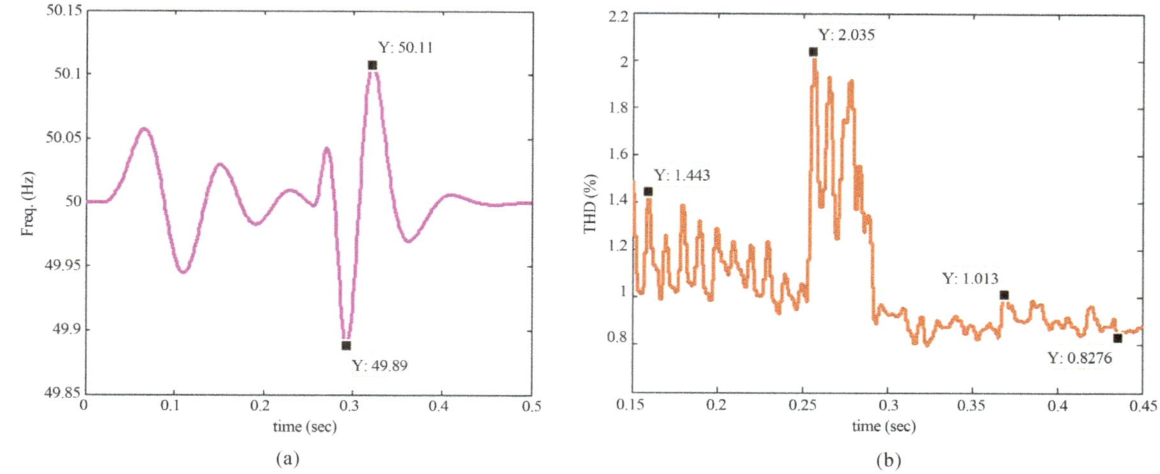

Figure 16. Frequency and THD of system; (a) System frequency changing curve; (b) Total harmonic distortion in line voltage.

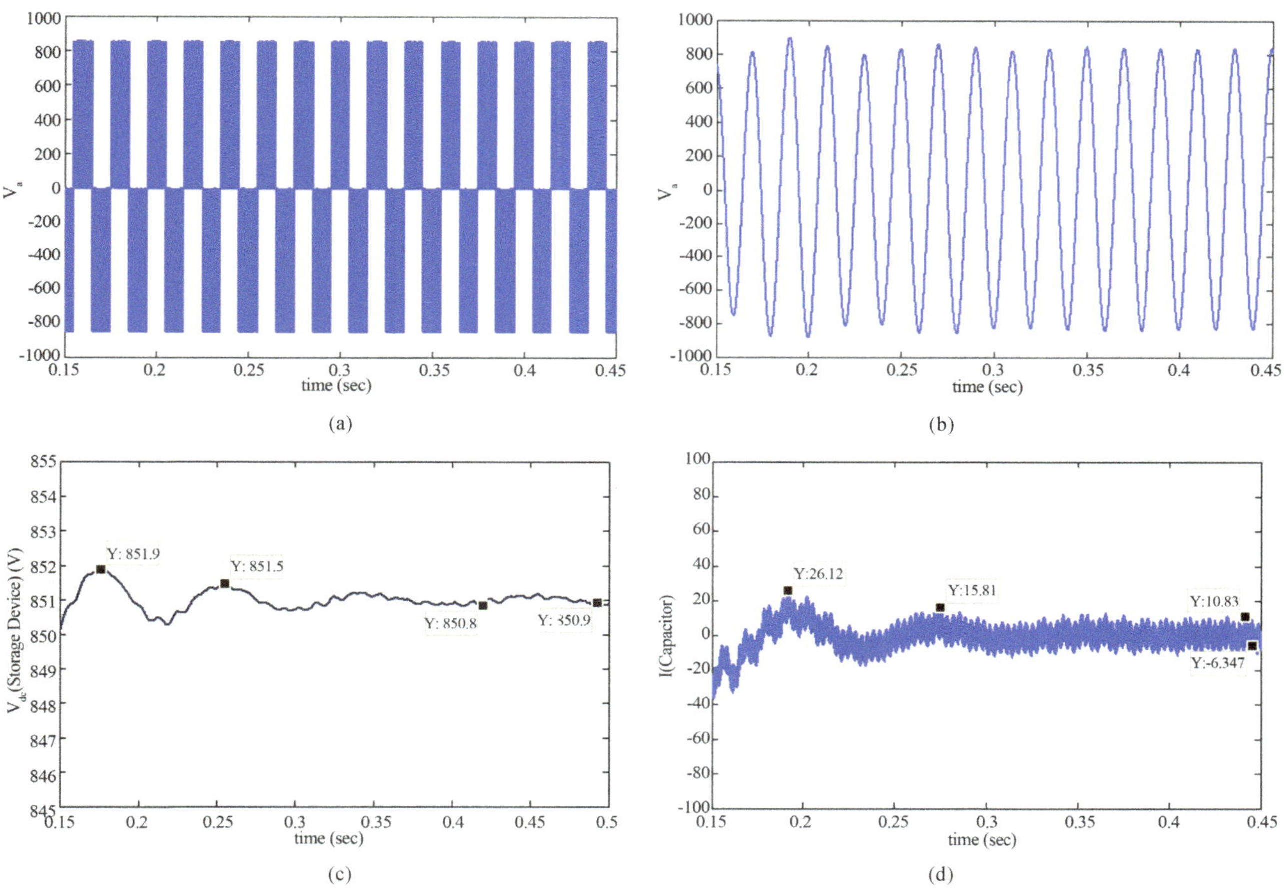

Figure 17. Terminal voltage and current toward the MT; (a) Output voltage of MT system before LCL filter; (b) Output MT system after LCL filter; (c) Terminal voltage in the battery bank; (d) Current of shunt capacitor in DC link.

Second case: non-sensitive load with 7130 watt capacity is connected to utility system. And sensitive loads are decrease by size 1130 w (SCS load 30 w, MTS load 1 kw, WECS load 100 w are decreases). We want to see and investigate the effect of these changes in the system. The outputs are like these:

In **Figure 18**, level of energy storage will increase after operation of breaker in PCC and fault occurrence time.

Injected current to network by each DG's are showed in **Figure 19**. In third case, the PV load rise up to 130 w

(30 w more than its nominal load) and WECS load rise up to 7200 w (100 w more than its nominal load and MTSs load decrease in amount of 1130 w. in this case NSL rise up to 7 kw (1 kw more than its first case). it showed that additional part of loads in the network supplies by MTS. Output of this case is showed in **Figure 20**.

8. Conclusion

In this paper a microgrid with three DG resources equipped by energy storage devices and grid side controllers has simulated. Control principles and modeling of the system has investigated. Output of the system displayed for three load arrangement. By means of showing load management and support of loads in developed systems, Fault management and control vision has showed. And energy storage operation in the moment of load respond and when loads have the changes has displayed. We use different controllers such as: MPPT controller (for SCS

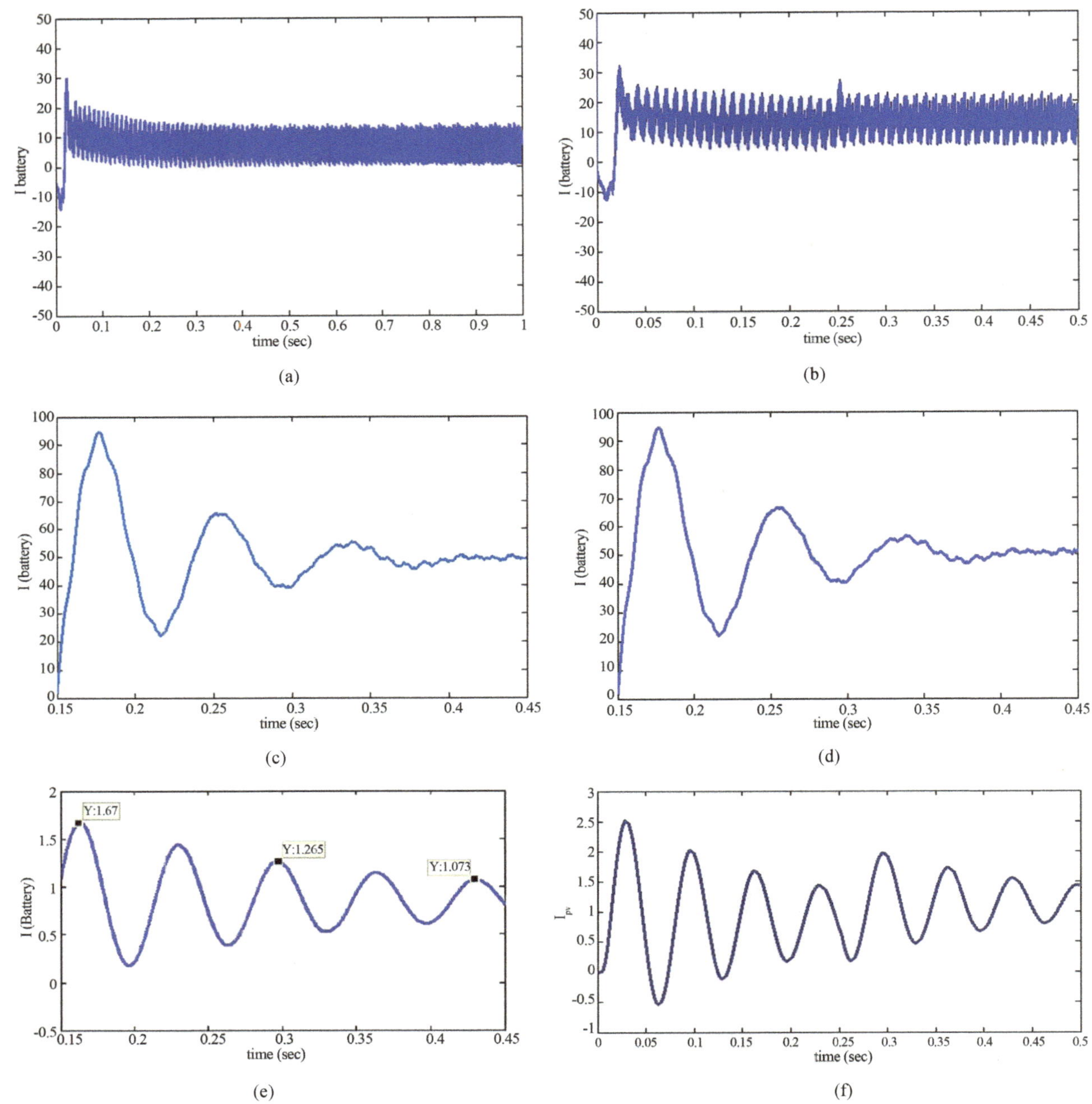

Figure 18. Storage operation in load changing condition; (a) Battery bank current of WECS in first case; (b) Battery bank current of WECS in second case; (c) Battery bank current of MTS in first case; (d) Battery bank current of MTS in second case; (e) Battery bank current of SCS in first case; (f) Battery bank current of SCS in second case.

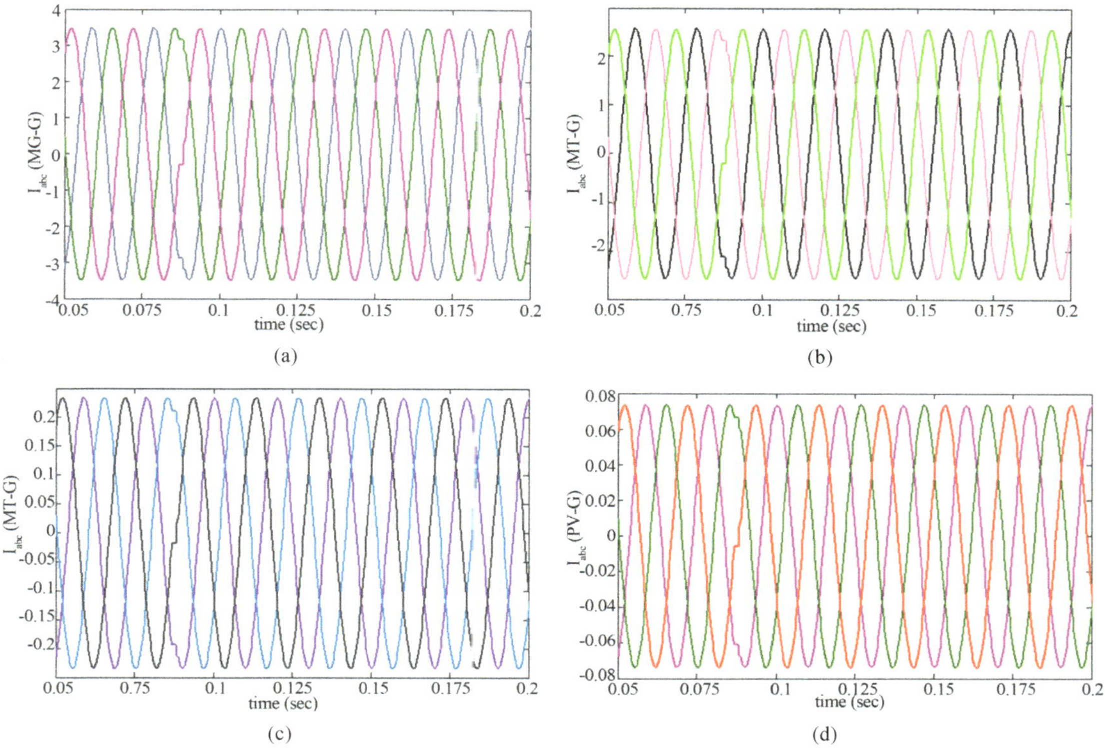

Figure 19. Second case injected current of DG's for the grid-connected operation; (a) Transferred current between MG and grid in grid connected mode; (b) Injected current from MTS to grid; (c) Injected current from WECS to grid; (d) Injected current from SCS to grid.

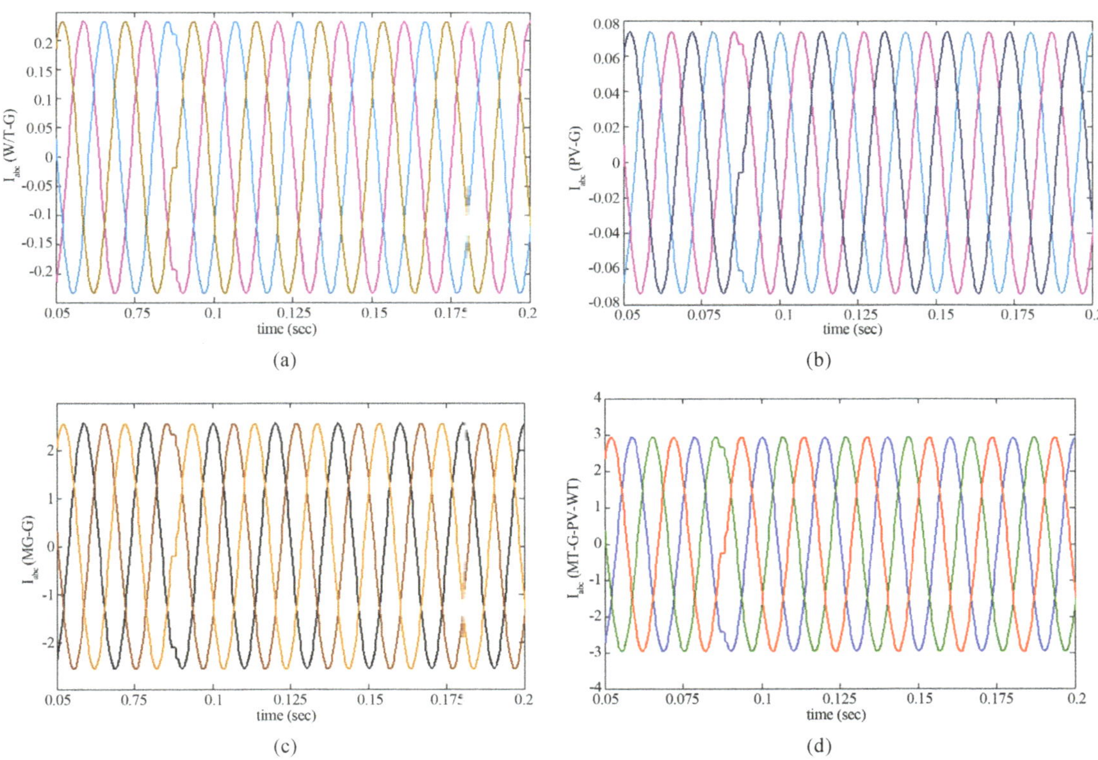

Figure 20. Third case injected current of DG's for the grid-connected operation; (a) Injected current from MTS to WECS; (b) Injected current from MTS to SCS; (c) Injected current from MTS to grid; (d) Injected current from MTS to WECS, SCS and grid.

and WECS) and V&F controller (for MTS) in order to research about decentralize control operation and showing the effect of this kind of control. In Grid-connected and islanding mode, additional product of DG resources have stored in battery banks. Of course it could be seen that in all conditions the system can continue to stable operation and loads are in good respond condition.

REFERENCES

[1] B. Lasseter, "Microgrids (Distributed Power Generation)," *Proceedings of the IEEE PES Winter Meeting*, Vol. 1, 2001, pp. 146-149.

[2] N. Hatziargyriou, H. Asano, R. Iravani and C. Marnay, "Microgrids: An Overview of Ongoing Research, Development, and Demonstration Projects," *IEEE Power Energy Magazine*, Vol. 5, No. 4, 2007, pp. 78-94.

[3] M. Pipattanasomporn, H. Feroze and S. Rahman, "Multi-Agent Systems in a Distributed Smart Grid: Design and Implementation," *Power Systems Conference and Exposition*, Seattle, 15-18 March 2009, pp. 1-8.

[4] Public Power Corporation, "Microgrids—Large Scale Integration of Micro-Generation to Low Voltage Grids," Technical Annex, 2002.

[5] P. Piagi and R. H. Lasseter, "Autonomous Control of Microgrids," *IEEE PES Meeting*, Montreal, June 2006.

[6] C. L. Moreira, F. O. Resende and J. A. P. Lopes, "Using Low Voltage MicroGrids for Service Restoration," *IEEE Transactions on Power Systems*, Vol. 22, No. 1, 2007, pp. 395-403.

[7] R. Zamora and A. K. Srivastava, "Controls for Microgrids with Storage: Review, Challenges, and Research Needs," *Elsevier*, Vol. 14, No. 7, 2010, pp. 2009-2018.

[8] F. Katiraei, M. R. Iravani and P. W. Lehn, "Microgrid Autonomous Operation during and Subsequent to Islanding Process," *IEEE Transactions on Power Delivery*, Vol. 20, No. 1, 2005, pp. 248-257.

[9] D. Georgakis, S. Papathanassiou, N. Hatziargyriou, A. Engler and C. Hardt, "Operation of a Prototype Microgrid System Based on Micro-Sources Equipped with Fast-Acting Power Electronics Interfaces," *Proceedings of IEEE 35th PESC*, Aachen, Vol. 4, 2004, pp. 2521-2526.

[10] F. Blaabjerg, R. Teodorescu, M. Liserre and A. V. Timbus, "Overview of Control and Grid Synchronization for Distributed Power Generation Systems," *IEEE Transactions on Industrial Electronics*, Vol. 53, No. 5, 2006, pp. 1398-1409.

[11] M. Uzunoglu, O. C. Onar and M. S. Alam, "Modeling, Control and Simulation of a PV/FC/UC Based Hybrid Power Generation System for Stand-Alone Applications," *Renewable Energy*, Vol. 34, No. 3, 2009, pp. 509-520.

[12] Z. M. Salameh, B. S. Borowy and A. R. A. Amin, "Photovoltaic Module-Site Matching Based on the Capacity Factors," *IEEE Transactions on Energy Conversion*, Vol. 10, No. 2, 1995, pp. 326-332.

[13] http://www.solarserver.com/yellow-pages/companies/company-search/optical-fiber-solar-cell-fabrication-company.html

[14] M. A. Masoum, H. Dehbonei and E. F. Fuchs, "Theoretical and Experimental Analyses of Photovoltaic Systems with Voltage- and Current-Based Maximum Power-Point Tracking," *IEEE Transactions on Energy Conversion*, Vol. 22, No. 8, 2002, p. 62.

[15] W. I. Rowen, "Simplified Mathematical Representations of Heavy Duty Gas Turbines," *Journal of Engineering for Power*, Vol. 105, No. 4, 1983, pp. 865-869.

[16] L. N. Hannet and A. Khan, "Combustion Turbine Dynamic Model Validation from Tests," *IEEE Transactions on Power Systems*, Vol. 8, No. 1, 1993, pp. 152-158.

[17] A. K. Saha, S. Chowdhury, S. P. Chowdhury and P. A. Crossley, "Modeling and Performance Analysis of a Microturbine as a Distributed Energy Resource," *IEEE Transactions on Energy Conversion*, Vol. 24, No. 2, 2009, pp. 529-538.

[18] Working Group on Prime Mover and Energy Supply Models for System Dynamic Performance Studies, "Dynamic Models for Combined Cycle Plants in Power System Studies," *IEEE Transactions on Power Systems*, Vol. 9, No. 3, 1994, pp. 1698-1708.

[19] I. Zamora, J. S. Martin, A. Mazon, J. S. Martin and V. Aperribay, "Emergent Technologies in Electrical Micro-Generation," *International Journal of Emerging Electric Power Systems*, Vol. 3, No. 2, 2005, pp. 1553-1779.

[20] C.-M. Ong, "Dynamic Simulation of Electric Machinery Using Matlab/Simulink," Prentice Hall, Upper Saddle River, 1998.

[21] M. Malinowski, S. Stynski, W. Kolomyjski and M. P. Kazmierkowski, "Control of Tree-Level PWM Converter Applied to Variable Speed-Type Turbine," *IEEE Transactions on Industrial Electronics*, Vol. 56, No. 1, 2009, pp. 69-77.

[22] M. Liserre, F. Blaabjerg and S. Hansen, "Design and Control of an LCL Filter-Based Three-Phase Active Rectifier," *IEEE Transactions on Industry Applications*, Vol. 4, No. 5, 2005, pp. 1281-1291.

[23] J. A. P. Lopes, C. L. Moreira and A. G. Madureira, "Defining Control Strategies for MicroGrids Islanded Operation," *IEEE Transactions on Power Systems*, Vol. 21, No. 2, 2006, pp. 916-924.

[24] R. H. Lasseter and P. Piagi, "Microgrid: A Conceptual Solution," *PESC*'04, Aachen, 20-25 June 2004.

[25] T. Ackermann, "Wind Power in Power Systems," John Wiley & Sons, Chichester, 2005.

[26] A. J. G. Westlake, J. R. Bumby and E. Spooner, "Damping the Power-Angle Oscillations of a Permanent-Magnet Synchronous Generator with Particular Reference to Wind Turbine Applications," *IEE Proceedings of Electric Power Applications*, Vol. 143, No. 3, 1996, pp. 269-280.

[27] L. Dambrosio and B. Fortunato, "One Step Ahead Adaptive Control Technique for a Wind Turbine-Synchronous Generator System," *Proceedings of the 32nd Intersociety*

Energy Conversion Engineering Conference, Honolulu, 27 July-1 August 1997, pp. 1970-1975.

[28] A. H. M. A. Rahim, M. A. Alam and M. F. Kandlawala, "Dynamic Performance Improvement of an Isolated Wind Turbine Induction Generator," *Computers and Electrical Engineering*, Vol. 35, No. 4, 2009, pp. 594-607.

[29] Carraro Group. www.santerno.com/company/company-profile/

Appendix

Carrier frequency in VMPPT PWM generator, 3000 Hz and in grid-side controller, 5000 Hz, boost converter parameters: $L = 0.0034H$, $C = 0.00561F$. PI coefficients in grid-side controller: $K_{pV_{dc}} = 0.05$, $K_{iV_{dc}} = 3$, $K_{pI_d} = 2.5$, $K_{iI_d} = 700$, $K_{pIq} = 2.5$, $K_{iIq} = 700$.

Table A1. Values and coefficients used in pv cell.

Current temp. coefficient	$\alpha = 0.002086$	[A/°C]
Voltage temp. coefficient	$\beta = 0.0779$	[V/°C]
Reverse saturation current	$I_0 = 0.5\times10^{-4}$	[A]
Short circuit cell current	$I_{ph} = I_{SC} = 0.5\times10^{-4}$	[A]
Cell resistance	$R_S = 0.0277$	[Ω]
Cell material coefficient	$\lambda = 0.049$	[1/V]

Table A2. Synchronous generator parameters amounts.

Parameters	Amount	Unit
Stator phase resistance R_s	0.0485	Ω
Stator inductances (L_d, L_q)	0.395	mH
Inductive flow by permanent magnet	0.1194	Wb
Moment of inertia (J)	0.0027	$kg\cdot m^2$
Nominal power	14	kw
Pairs of poles (P)	4	-

Table A3. Boost converter coefficient values.

Parameters	Amount	Unit
Low voltage capacitor C1	500	µF
High voltage capacitor Co	4700	µF
Inductance	800	µH
Switching frequency	20	KHz

8

Parasitic Effects on the Performance of DC-DC SEPIC in Photovoltaic Maximum Power Point Tracking Applications

Nur Mohammad[1], Muhammad Quamruzzaman[1], Mohammad Rubaiyat Tanvir Hossain[2], Mohammad Rafiqul Alam[1]

[1]Department of Electrical & Electronic Engineering, Chittagong University of Engineering & Technology, Chittagong, Bangladesh;
[2]Department of Electrical & Electronic Engineering, Bangladesh University of Engineering & Technology, Dhaka, Bangladesh.

ABSTRACT

This paper presents an analysis of the effect of parasitic resistances on the performance of DC-DC Single Ended Primary Inductor Converter (SEPIC) in photovoltaic maximum power point tracking (MPPT) applications. The energy storage elements incorporated in the SEPIC converter possess parasitic resistances. Although ideal components significantly simplifies model development, but neglecting the parasitic effects in models may sometimes lead to failure in predicting first scale stability and actual performance. Therefore, the effects of parasitics have been taken into consideration for improving the model accuracy, stability, robustness and dynamic performance analysis of the converter. Detail mathematical model of SEPIC converter including inductive parasitic has been developed. The performance of the converter in tracking MPP at different irradiance levels has been analyzed for variation in parasitic resistance. The converter efficiency has been found above 83% for insolation level of 600 W/m^2 when the parasitic resistance in the energy storage element has been ignored. However, as the parasitic resistance of both of the inductor has increased to 1 ohm, a fraction of the power managed by the converter has dissipated; as a result the efficiency of the converter has reduced to 78% for the same insolation profile. Although the increasing value of the parasitic has assisted the converter to converge quickly to reach the maximum power point. Furthermore it has also been observed that the peak to peak load current ripple is reduced. The obtained simulation results have validated the competent of the MPPT converter model.

Keywords: Photovoltaic (PV) Renewable Energy Systems; DC-DC Converter; Parasitic Resistance; Maximum Power Point Tracking (MPPT); Single Ended Primary Inductance Converter (SEPIC)

1. Introduction

The Photovoltaic (PV) energy is one of the promising alternative renewable energy resources that can be used to minimize the existing electricity crisis in the world considerably. PV energy is getting increasing importance as a renewable source due to the advantages such as the absence of fuel cost, little maintenance, no carbon emission at all and no noise due to absence of moving parts. The output current vs. voltage curve of PV array shows a non-linear I-V characteristic that depends on environmental conditions such as solar irradiance and temperature. In this curve, there is a certain operating point at which the PV array produces maximum power which is known as maximum power point (MPP). Therefore, to maximize the PV array output power at any irradiance and temperature, maximum power point tracking (MPPT) is used in the PV system. The maximum power theory is based on impedance matching. By adjusting the duty cycle of the converter, the equivalent load impedance as seen by the PV source is matched with its own impedance. Using an appropriate MPPT algorithm, the duty cycle of the converter is adjusted continuously to track the MPP. The MPPT scheme using the basic switch mode converter topologies have their own advantages and disadvantages. The main requirement of any converter used in the MPPT system is that it should have a low input current ripple. Buck or its derived topologies give pulsating currents on the PV array side. On the other hand, as compared with buck topologies, the boost or its derived topologies present low current ripple on the PV side, but high ripple in the load current. The requirement of load voltage, either lower or higher than the array voltage, can be realized by means of buck-boost converters. But, still the PV array current and load current are pulsating in nature. Furthermore, load voltage is inverted with buck-boost converter. These aforementioned restrictions of the conventional converters have motivated researchers and system designers to investigate the feasibility of employing single switch fourth order converters, namely single-ended primary inductance converter (SEPIC) [1-3], Cuk converter [4-9] and Zeta converter [10,11] to provide alternative solution. The term fourth order means these converters have four energy storage elements—two inductors and two capacitors to transfer energy from input side to output side. In general, these fourth order converters have wide range of input-to-output conversion ratio, better adaptability of integrating transformers for galvanic isolation, and non-pulsating input and output current. For example SEPIC converter provides the buck-boost conversion function without polarity reversal unlike buck-boost, in addition to the low ripple input current. The authors in [12,13] have suggested that SEPIC topology is highly suitable for multiple-input DC-DC converter because of its non-pulsating input current, grounded switch and non-inverting output voltages. SEPIC may also be preferred for battery charging systems because the diode placed on the output stage works as a blocking diode preventing a reverse current going to PV source from the battery [14]. Additionally, the Electromagnetic Interference (EMI) behavior of a SEPIC topology is much better than a step-down or flyback topology and avoids the problems with leakage inductance and snubbers [15]. The inductor used in the SEPIC has a certain amount of non-zero dc parasitic resistance, as it is usually a winding of several turns of long metallic wire. Similarly, the capacitor has also a small equivalent series parasitic resistance. But the parasitic resistance of capacitor may be neglected comparing to that of inductor may be neglected without loss of generality [16]. Apart from adding ohmic losses, these parasitic resistances add current damping and affect the ripple attenuation [17]. Although considering ideal components significantly simplifies model development, but neglecting the parasitic effects in models may sometimes lead to failure in predicting the fast-scale instabilities [18]. Therefore, it is important to take the effects of parasitics into consideration for improving the efficiency, dynamic performance, stability and robustness of the converter. In case of SEPIC converter used for MPPT applications, a detail investigation is necessary to observe and analyze the effects of parasitics on the overall performance of the converter which is still not reported in the literature. In this paper, the effects of these parasitic resistances on the overall performance of these converters especially in MPPT applications are analyzed.

2. PV Array Characteristics

An ideal solar cell may be modeled by a current source connected in parallel with a diode; the current source represents the generated photocurrent when the sunlight hits the solar panel, and the diode represents the p-n transition area of the solar cell. In practice no solar cell is ideal and a shunt resistance R_{sh} and a series resistance R_s component are incorporated in the model according to its behavior. The equivalent circuit of solar cell comprising desecrates components is shown in **Figure 1**.

From the above electrical equivalent circuit of solar cell, it is evident that the V is the voltage across the load resistance R and the current I which is flowing through this load can be written as Equation (1).

$$I = I_L - I_D - I_{sh} \tag{1}$$

where I_L light generated current, I_D is the diode current I_{sh} is the current which is shunted through R_{sh}.

By the Shockley diode Equation [19], the current diverted through the diode is given by Equation (2).

$$I_D = I_0\left[\exp\left(\frac{q(V+IR_s)}{nkT}\right)-1\right] \tag{2}$$

Here T is the absolute temperature in Kelvin. q is the charge of a electron, k is the Boltzmann's constant, n is the diode ideality factor which depends on the certain PV technology and I is the reverse saturation current in amperes. Substituting these into the Equation (1), produces the characteristic Equation (3), of a typical solar cell, this relates solar cell parameters to the output current and voltage.

$$I = I_L - I_0\left[\exp\left(\frac{q(V+IR_s)}{nkT}\right)-1\right]-\frac{V+IR_s}{R_{sh}} \tag{3}$$

Sometimes, to simplify the model, the effect of the shunt resistance is not considered, that is R_{sh} is infinite, so the expression of (3), simplify to as Equation (4).

$$I = I_L - I_0\left[\exp\left(\frac{q(V+IR_s)}{nkT}\right)-1\right] \tag{4}$$

A PV panel is composed of many solar cells, which are connected in series and/or parallel so the output current and voltage of the PV panel are high enough for a certain application. Taking into account the simplification of Equation (4), the output current-voltage characteristic of a PV panel is expressed by Equation (5), where, N_p *and* N_s are the number of solar cells in parallel and series respectively.

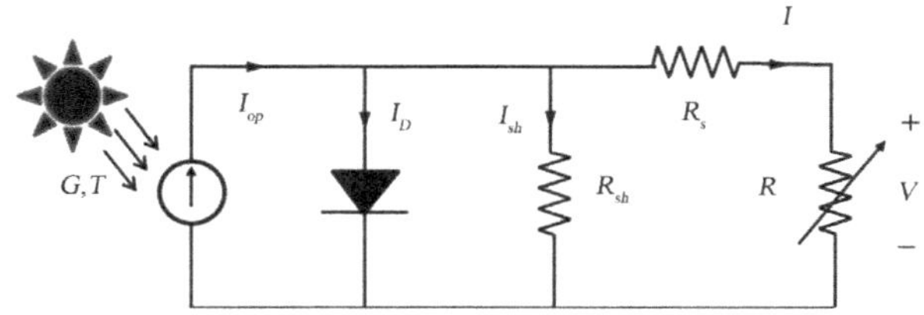

Figure 1. Equivalent circuit of a solar cell.

$$I = N_p I_L - N_p I_0 \left[\exp\left(\frac{q(V + IR_s)}{nkTN_s} \right) - 1 \right] \quad (5)$$

The intensity of solar irradiance (insolation) is the most dominant environmental factor which is strongly affecting the electrical characteristics of solar panel according to the Equation (5). The effect of the irradiance on the voltage-current (V-I) and voltage-power (V-P) characteristics of STF100P6 solar panel under various irradiances level is best depicted in **Figure 2**. From this Figure it is clear that under higher irradiance, the PV cell produces higher output currents because the light generated current is proportionally generated by the flux of photons.

The maximum power point (MPP) decreases with decreasing irradiance and this is indicated on each (V-P) curve in **Figure 3**.

3. Maximum Power Point Tracking (MPPT)

Usually there are two major approaches adopted for maximizing power extraction from PV sources. First one is the mechanical tracking of the solar panel. In this case the panel is attempted to position in any terrain at an angle of ninety degree with the direction incoming ray of sun. This issue is beyond our topic of discussion. The second one is the electrical MPPT where electrical operating point is forced at the peak power point continuously by adjusting the duty cycle of the DC-DC converter inserted between PV array and load. However several MPPT methods have been summarized in the literatures [20,21]. The methods vary in complexity, sensors required, tracking efficiency, convergence speed, cost, and in other respects. Some of the well-known techniques are Perturb & Observe (P & O), Incremental Conductance, Fractional Open-Circuit, Fractional Short-Circuit, Fuzzy Logic and Neural Network based algorithms. Among them the Perturb and Observe (P & O) algorithm is most commonly used in practice due to its fast tracking speed, low cost and eases of implementation by the majority of authors [22-27]. It is an iterative method of obtaining MPP. It measures the PV array voltage and current, and then perturbs the operating point of PV generator to encounter the change direction. The maximum point is reached when $dP/dV = 0$. There are many varieties, from simple to complex. But the most basic form explained in **Figure 4**; that has been adopted in this paper. To seek MPP, the operating voltage of the PV array perturbed, by a small amount, dV and the resulting change in power, dP is measured. If the power increases due to the perturbation then the next perturbation of the operating voltage is continued in the same direction. However, if the power decreases, the subsequent perturbation should be reversed.

The process is repeated until the MPP is reached. The system may oscillates about the MPP if the meteorological conditions change abruptly. But in reality insolation and temperature has slower dynamics and never gives a big jump in a very short time. A summary of the algorithm is presented in **Table 1**.

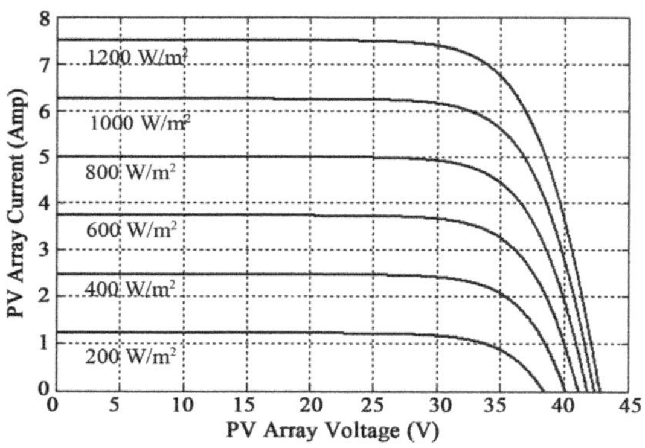

Figure 2. I-V characteristics of the solar PV array due to change in insolation at 25˚C.

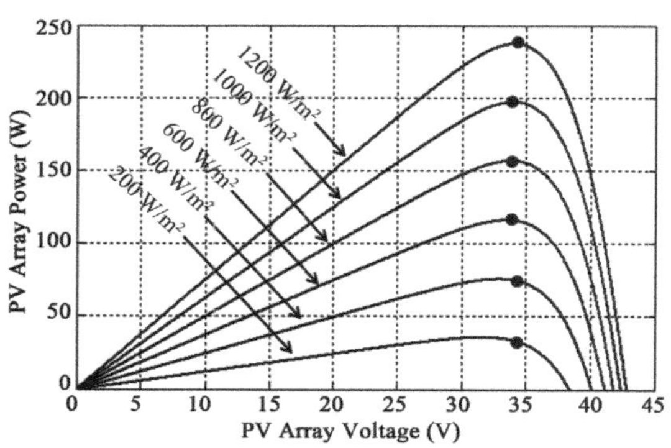

Figure 3. P-V characteristics of the solar PV array due to change in insolation at 25˚C.

4. Operation and Modeling of SEPIC

The single-ended primary-inductor converter (SEPIC) as indicated in the MPPT implementation circuit of **Figure 5** has the ability to operate from an input voltage that is

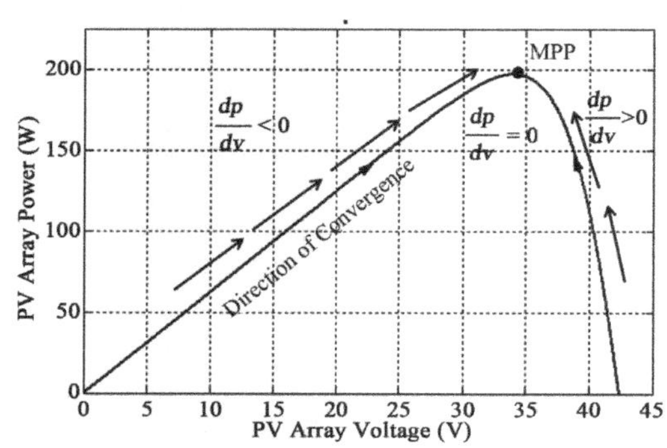

Figure 4. Maximum power point tracking technique.

Table 1. Summary of perturb and observe algorithm.

Perturbation	Change in power	Next perturbation
Positive	Positive	Positive
Positive	Negative	Negative
Negative	Positive	Negative
Negative	Negative	Positive

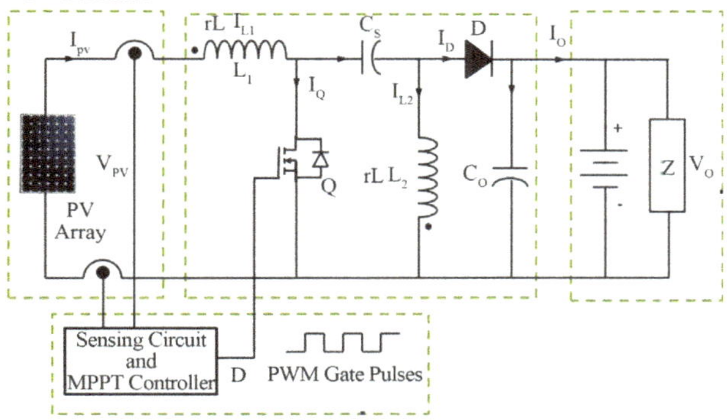

Figure 5. MPPT implementation circuit using SEPIC.

greater or less than the regulated output voltage without polarity reversal. Aside from being able to function as both a buck and boost, its minimal current ripple improves average value of current and voltage as well. The converter exchanges energy between the capacitors and inductors in order to convert from one level of voltage to another. The primary means of transfer energy from the input to the output of the converter is through a series capacitor which is known as SEPIC capacitor or coupling capacitor, C_s.

The voltage rating of this capacitor must be greater than the maximum input voltage. The output voltage is controlled by adjusting duty cycle of control switch, Q. The control switch is typically a MOSFET, which offer much higher input impedance and lower voltage drop and do not require biasing resistors. The first input inductor L_1 together with the MOSFET control switch resembles a simple boost topology, whereas the shunt inductor L_2, location is similar to a buck-boost topology. The Diode, D should have fast recovery time and low forward-voltage drop. Its peak current rating is greater than or equal to the peak inductor current and reverse breakdown voltage must be greater than the output voltage. When the control switch is turned on, diode is turn off and the output current is solely supplied by the output capacitor, C_o. Thus the selected output capacitor must be capable of handling the maximum rms current. To understand the voltages at the various nodes and current through different branches, the circuit is assumed in an average state with continuous conduction mode. The initial state of a SEPIC before the switch closes is shown in **Figure 6(a)**. The coupling capacitor has charged to V_{in}. The output voltage is zero, and no current is flowing in any of the components. When the switch closes as shown in **Figure 6(b)**, V_{in} is applied across the first inductor and current, i_{L1} ramps up through it.

The voltage, V_{in} across the coupling capacitor is also applied across the second inductor L_2 and current i_{L2} ramps up through it in similar way that of L_1. The diode is reverse biased and the output capacitor is left to provide the load current in this time. The fact that when the switch is on, both L_1 and L_2 are charged up and disconnected from the load and both the capacitors discharge.

The associated voltage across the inductor L_1 and the diode D as well as the instantaneous current through each components of the converter are best sketched in **Figure 7**. When the switch is off, the current through L_1 has no place to go but through the C_s to the load as the diode is forward biased. The current through shunt inductor L_2 must also go to the output. Both inductors provide current to the load capacitor and load. This switching state of the converter circuit is shown in **Figure 6(c)**. In order for the current to continue to flow through L_1, the voltage on the switch boosts up to $V_{in}+V_o+V_D$. The current flowing through the capacitor C_s charges it up in this time to enable it to transfer this energy to shunt inductor L_2 until the switch closes again for the next cycle. There is an energy balance between the coupling capacitor C_s and shunt inductor L_2, which helps determine the value of the capacitor C_s. As the capacitor C_s blocks direct current and hence the average current through it is zero, making inductor L_2 the only source of load current. Therefore, the average current through inductor L_2 is the same as the average

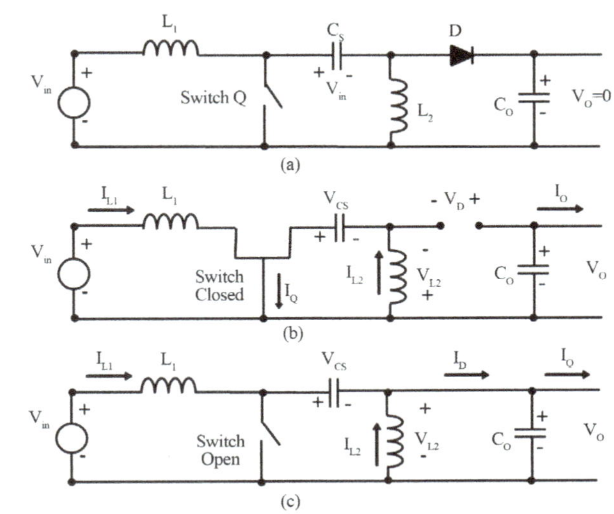

Figure 6. SEPIC circuit state (a) Before switch is on; (b) When switch on; and (c) When switch off.

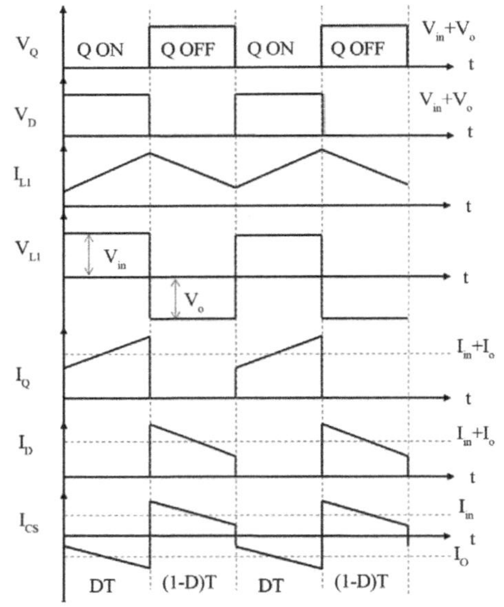

Figure 7. Voltages and current of each component of the SEPIC in CCM.

load current and hence independent of the input voltage. Looking at average voltages across L_1 which is actually zero for a full cycle, the energy balance volts-hertz [28] equation can be written as:

$$V_{L1} = D \cdot V_{in} + (1 - D) \cdot (-V_o) = 0 \tag{6}$$

$$\Rightarrow V_o = \frac{DV_{in}}{1 - D} \tag{7}$$

For ideal converter

$$\Rightarrow I_o = \frac{1 - D}{D} I_{in} \tag{8}$$

The above voltage and current relation prove that the SEPIC has the ability to operate from an input voltage that is greater or less than the regulated output voltage without polarity reversal unlike the conventional buck bust converter.

4.1. Determining Peak Inductor Current

The graph of i_{L1} is shown in **Figure 8**. For PV applications, it is desired to have low ripple in i_{L1} to keep the solar panel operating at its MPP. At discharging phase of inductor, L_1 can be written as:

$$\frac{\mathrm{d}i_{L1}}{\mathrm{d}t} = \frac{-V_o}{L_1} \tag{9}$$

$$\Delta I_1 = \frac{V_o}{L_1}(1 - D)T \tag{10}$$

$$\frac{V_o(1 - D)}{L_1 f} \tag{11}$$

4.2. Determining Peak Inductor Current

One of the first steps in designing any PWM switching regulator is to decide how much inductor ripple current, ΔI_L, to allow. Too much increases EMI, while too little may result in unstable PWM operation. A good rule for determining the inductance is to allow the peak-to-peak ripple current to be approximately 40% of the maximum input current at the minimum input voltage [29]. The ripple current flowing in equal value inductors L_1 and L_2 is given by:

$$\Delta I_L = I_{in} \times 40\% = I_o \times \frac{V_o}{V_{in}} \times 40\%$$

The value of the first inductor L_1 is derived from the following fundamental relation of

$$\frac{\mathrm{d}i_{L1}}{\mathrm{d}t} = \frac{V_{in}}{L_1}$$

$$\Rightarrow L_1 = \frac{V_{in}}{\Delta i L_1 \cdot f} D \tag{12}$$

Similarly the value of the inductor L_2 is derived from the following fundamental relation of

$$\frac{\mathrm{d}i_{L1}}{\mathrm{d}t} = \frac{-V_o}{L_2} \tag{13}$$

$$\Rightarrow L_2 = \frac{-DV_{in}}{(1 - D)} \frac{1}{\Delta i L_2} (1 - D)T$$

$$\Rightarrow L_2 = \frac{-V_{in}}{\Delta i L_2 \cdot f} D \tag{14}$$

Ignoring the sign from Equation (12) and Equation (14) and considering the magnitude of current ripple $\Delta I_{L1} = \Delta I_{L1}$, it can be said that, $L_1 = L_2$. That is, both the inductors have to have same level of inductance. It proves that they have induced same level of voltage with opposite polarity. Physically the windings are constructed with the same number of turns on the similar ferrite iron core. To ensure the inductor does not saturate, the peak current in the inductor is given [30] by:

$$I_{L1} = I_o \times \frac{V_o}{V_{in}} \times \left(1 + \frac{40\%}{2}\right) \tag{15}$$

$$I_{L2} = I_o \times \left(1 + \frac{40\%}{2}\right) \tag{16}$$

5. Dynamic Performance Analysis

For the sake of simplicity, we consider here that the inductor is the only non-ideal component, and that it is equivalent to an inductor and a resistor in series. This assumption is acceptable because an inductor is made of one long wound piece of wire, so it is likely to exhibit a non-negligible parasitic resistance (r_L). Furthermore, current flows through the inductor both in the on and the off states in continuous conduction mode of operation. Using the state-space averaging method, we can write:

$$V_{in} = V_L + V_Q \quad (17)$$

where V_L and V_Q are respectively the average voltage across both of the inductor. If we consider that the converter operates in steady-state, the average current through the inductor is constant [31]. The average voltage across the inductor is:

$$V_L = L\frac{dI_L}{dt} + r_L I_L = r_L I_L \quad (18)$$

When the switch is in the on-state, $V_Q = 0$ and when it is off, the diode is forward biased. Therefore, $V_Q = V_{in} + V_o$. So, the average voltage across the switch is:

$$V_Q = D\cdot 0 + (1-D)\cdot(V_{in}+V_o) = (1-D)\cdot(V_{in}+V_o) \quad (19)$$

The average inductor current in terms of average output current is:

$$I_L = \frac{D}{1-D} I_o \quad (20)$$

Assuming the output current and voltage have negligible ripple. For the purely resistive load of R Equation (28) becomes:

$$I_L = \frac{V_o D}{R\cdot(1-D)} \quad (21)$$

Using the previous equations, the input voltage becomes:

$$V_{in} = r_L \frac{V_o D}{R\cdot(1-D)} + (1-D)\cdot(V_{in}+V_o) \quad (22)$$

$$\Rightarrow \frac{V_0}{V_{in}} = \frac{1}{\frac{r_L}{R}\frac{D}{(1-D)} + \frac{(1-D)}{D}} \quad (23)$$

If the r_L is zero, Equation (23) represents the ideal case. However, as r_L increases, the voltage gain of the converter decreases. The effect of r_L increases with the increase of duty cycle which is displayed graphically in **Figure 9**. As the inductor becomes less ideal, the possible gain drops off sharply from the theoretical value, especially as the duty cycle increases above 90%. The normal operating cell temperature (NOCT) rating of 48°C was assumed throughout the simulation. Only the change of intensity of solar irradiation is taken into consideration. The complete model of the proposed system in MATLAB/SIMULINK [32] environment is given in **Figure 10**. Specification of PV module [33] and converter parameters are optimized and calculated values are summarized in **Table 2**.

The voltage and power in the output of the converter for the case of ideal inductor SEPIC and when the inductor parasitic resistance is not taken into consideration is shown in **Figures 11-14**, respectively. In **Figure 11**, it has been

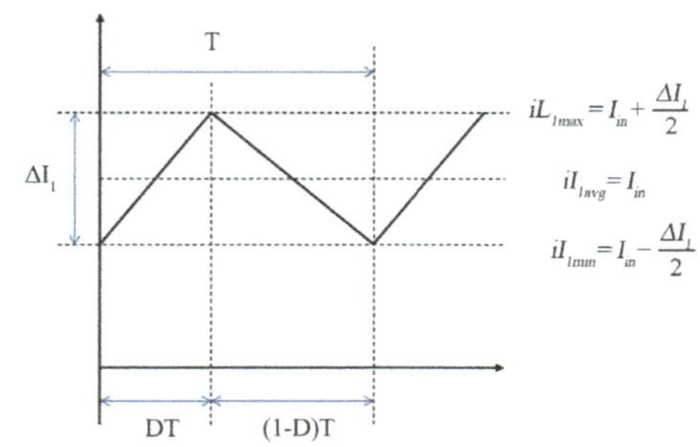

Figure 8. Inductor current waveform for CCM.

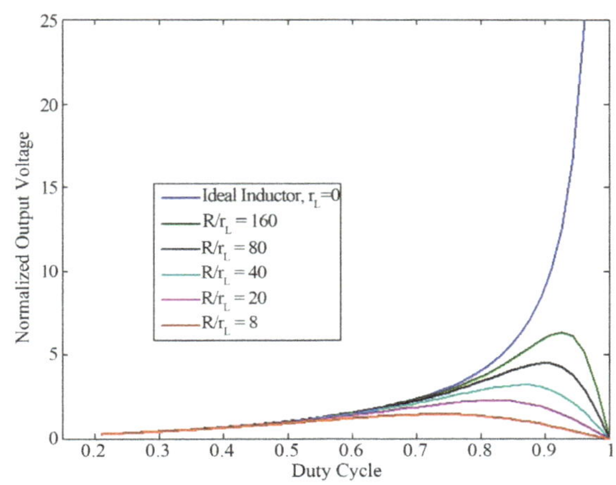

Figure 9. Normalized voltage of SEPIC with the duty cycle when the parasitic resistance of inductor varies.

Table 2. Specification of PV module and converter ratings.

Parameters	Specification
Maximum power (P_m)	100 W
Open circuit voltage (V_{oc})	21.5 V
Short circuit current (I_{sc})	6.22 A
Voltage at maximum power (V_m)	17.30 V
Current at maximum power (I_m)	5.8 A
Short-circuit current temp coefficient	6.928 mA/°C
Open-circuit voltage temp coefficient	−0.068 V/°C
Module size	36 cells (4 × 9)
Inductor L_1	250 μH
Inductor L_2	250 μH
Inductor parasitic resistance r_L	1.00 Ω
Capacitor C_s	47 μF
Capacitor C_o	47 μF

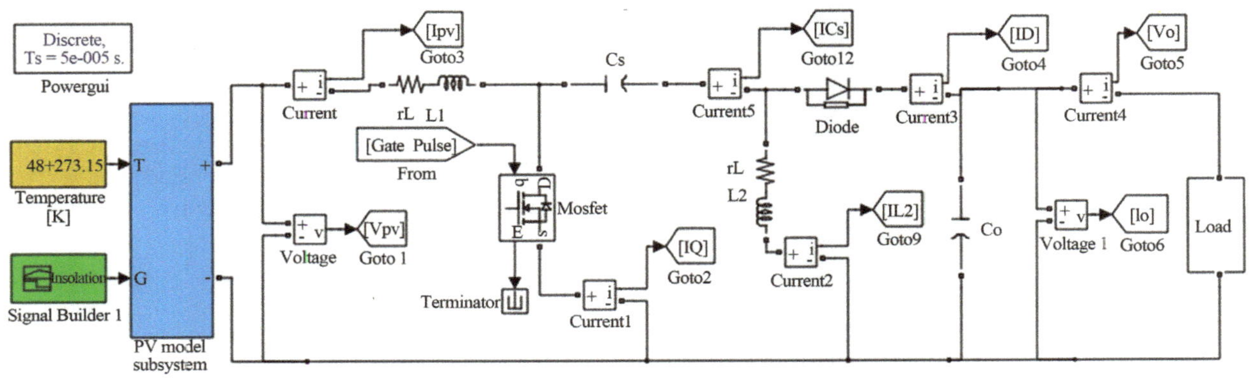

Figure 10. The detailed simulink model of SEPIC based MPPT system.

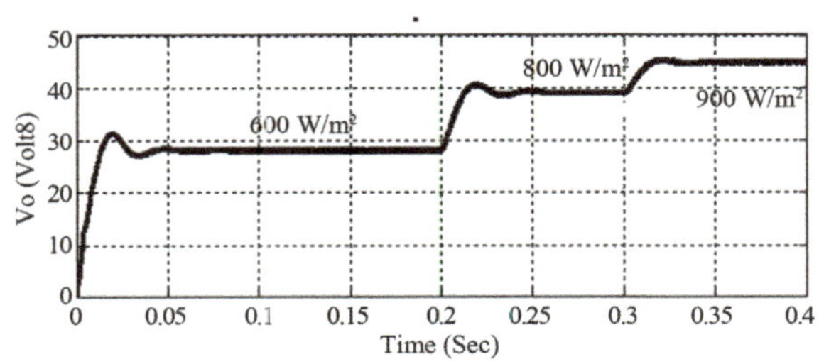

Figure 11. Higher output voltage swing for step up change of insolation for ideal inductor.

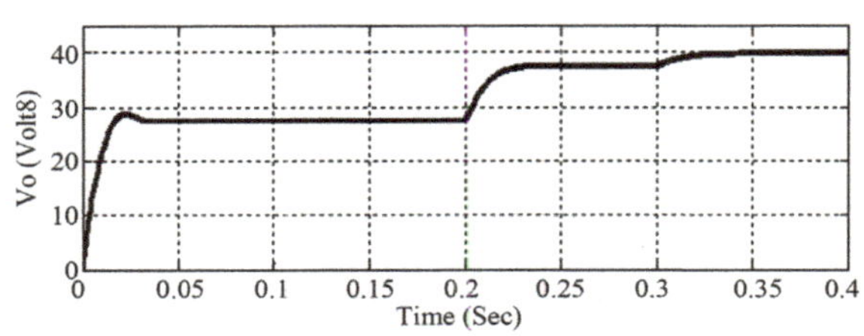

Figure 12. Reduced output voltage swing for the presence of inductor parasitic resistance.

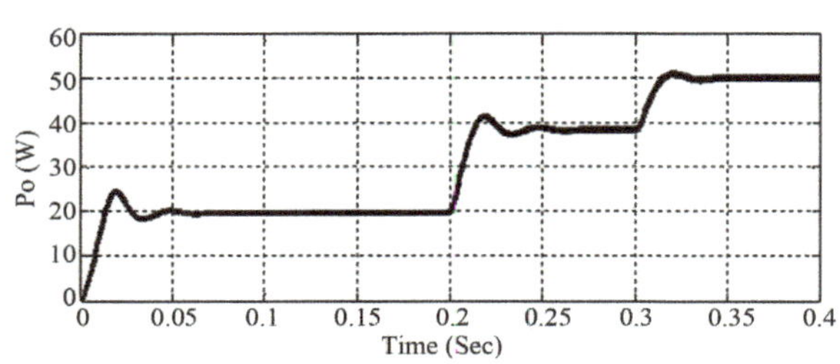

Figure 13. Output power for step up change of insolation for ideal inductor.

observed that voltage get back its stable level at 28 V, 39 V and 45 V for insolation step of 600 W/m^2, 800 W/m^2 and 900 W/m^2 respectively after some considerable delay.

The respective maximum power levels are shown in **Figure 13** has stable in 20 W, 38 W and 50 W accordingly. This happens because the algorithm provides the optimum duty ratio perturbation as the PV arrays are very

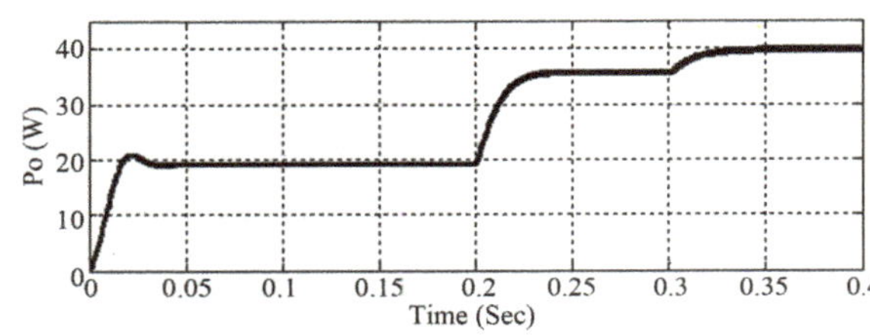

Figure 14. Parasitic resistance assists to converge MPPT.

sensitive to rapid environmental changes. Similar phenomena are observed in **Figure 12**, but here change of voltage swing is relatively less before get back its stable position. The variations in the voltage and power of the converter are almost terminated when the solar radiation is almost constant. For example this can be seen between 0.05 sec. and 0.2 sec. in the time scale. The converter efficiency has been found more than 83% for insolation level of 600 W/m^2. It's stable at 28 V, 38 V and 40 V for insolation level of 600 W/m^2 800 W/m^2 and 900 W/m^2 respectively. For the same level of last two insolation profile as taken into consider for first the case it has been found that 5% and 15% less than the previous on due to effect of parasitic of inductor. The respective power levels are shown in **Figure 14**, which are stabled at 20 W, 38 W and 50 W. The variations in the voltage and power of the converter are almost terminated when the solar radiation is almost constant but in this case both of these get back its stability very quickly for inductor damping. For example this can be seen between 0.03 sec. and 0.2 sec. in the time scale. However the converter efficiency has been found less than 78% for insolation level of 600 W/m^2. The output current for step down change of insolation profile of 1000 W/m^2, 850 W/m^2 and 700 W/m^2 for ideal inductor as well as when the effect of inductor parasitic resistance is considered is shown in **Figures 15** and **16** respectively.

It is clear from these figures that the average value of output current is reduced to 17.54% for the presence of 1 ohm parasitic resistance in the energy storage element.

The effect of parasitic in current ripple is presented in **Figures 17** and **18**. The calculated value of the peak to peak output current ripple is 5 mA when the parasitic resistance has been ignored. The average value of output current is 1.24 A. On the other hand, the peak to peak output current ripple is 4 mA when the parasitic resistance of 1 ohm has been considered and the average value of output current is reduced to 1.0225 A. This is due to account for adding ohmic losses of parasitic elements. Apart from adding ohmic losses these resistances add current damping and affect the ripple attenuation as well.

6. Conclusion

The effects of parasitics on the overall performance of the converter have been analyzed in this paper and have been found that the inductor parasitic resistances have large effects on the converter efficiency and ripple. Firstly both of the two inductors are assumed ideal and hence power is transmitted with fewer losses from the input side to the load. The loss which has been encounter was due to the switching loss of the converter. The converter efficiency has been found above 83% for insoletion level of 600 W/m^2 in this case. However, as r_L increases, the possible gain drops off sharply from the theoretical value, especially as the optimum duty cycle of around 90%. A fraction of the power managed by the converter is dissipated by these parasitic resistances. The voltage as well as current gain of the converter decreases compared to the ideal case. The converter efficiency has been found less than 78% for the insolation level of 600 W/m^2. So Inductors with lower series resistance allow less energy to be dissipated as heat, resulting in greater efficiency and a larger portion of the input power being transferred to the load.

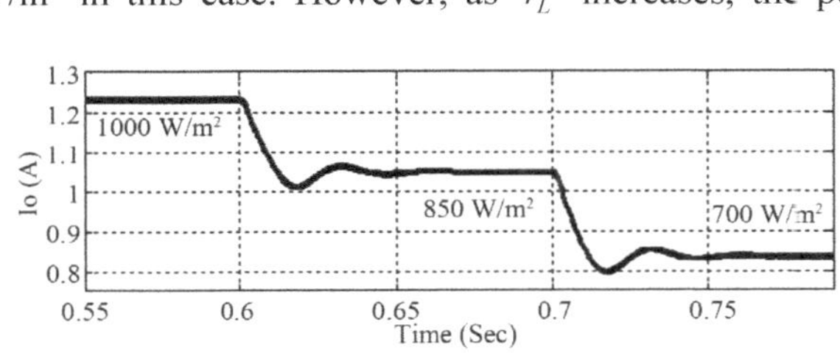

Figure 15. Output current for step down change of insoletion for ideal inductor.

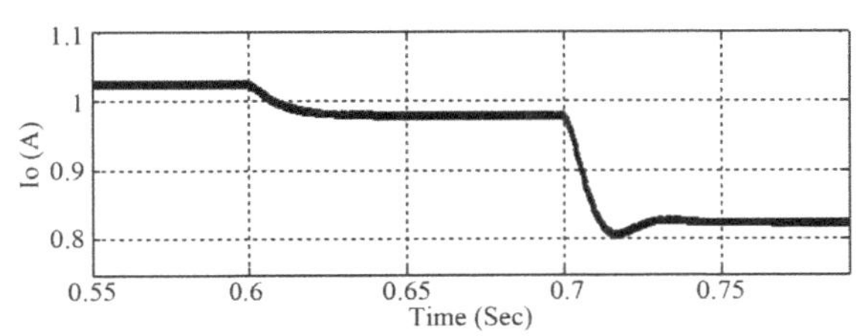

Figure 16. Consequence of inductor parasitic resistance on output current.

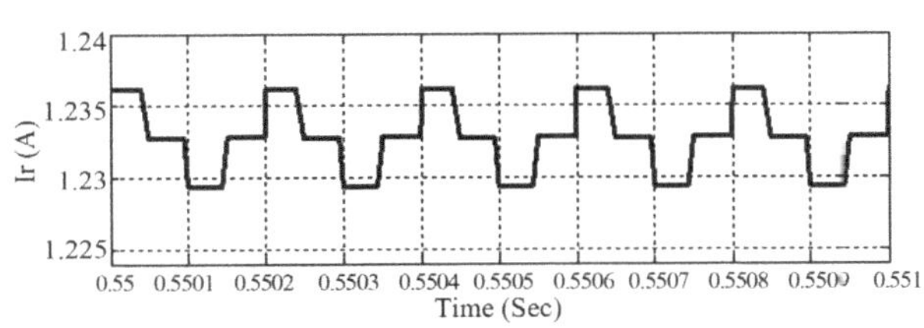

Figure 17. Output current ripple for ideal inductor.

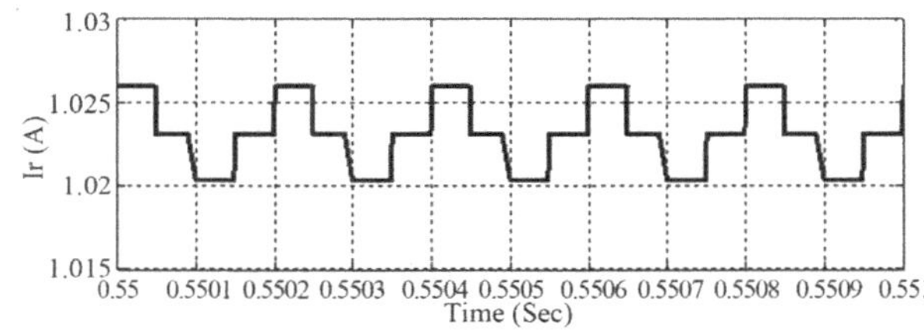

Figure 18. Consequence of parasitic resistance on output current ripple.

7. Acknowledgements

This work has been supported partially by the post graduate research grant of the Chittagong University of Engineering & Technology (CUET), Bangladesh. I also express my gratitude to my honorable teacher, Muhammad Quamruzzaman, Assistant Professor, Department of Electrical & Electronic Engineering, who gave me inspiration, cordially guidance, and valuable suggestions which helped me to complete my research work.

REFERENCES

[1] M. Veerachary, "Power Tracking for Nonlinear PV Sources with Coupled Inductor SEPIC Converter," *IEEE Transactions on Aerospace and Electronic Systems*, Vol. 41, No. 3, 2005, pp. 1019-1029.

[2] J. Hu, A. D. Sagneri, J. M. Rivas, Y. Han and D. J. Perreault, "High-Frequency Resonant SEPIC Converter with Wide Input and Output Voltage Ranges," *IEEE Transactions on Power Electronics*, Vol. 27, No. 1, 2012, pp. 189-200.

[3] S. B. Kjaer, J. K. Pdersen and F. Blaabjerg, "Modeling and Control of PV Charger System with SEPIC Converter," *IEEE Transactions on Industrial Electronics*, Vol. 56, No. 11, 2009, pp. 4344-4353.

[4] R. M. Middlebrook and S. Cuk, "A General Unified Approach to Modeling Switching Converter Power Stages," *Proceedings of IEEE Power Electronical Special Conference*, Vol. 4, 1976, pp. 18-34.

[5] A. Safari and S. Mekhilef, "Simulation and Hardware

Implementation of Incremental Conductance MPPT with Direct Control Method Using Cuk Converter," *IEEE Transactions on Industrial Electronics*, Vol. 58, No. 4, 2011, pp. 1154-1161.

[6] S. C. Tan, "Constant-Frequency Reduced-State Sliding Mode Current Controller for Cuk Converters," *IET Power Electronics*, Vol. 1, No. 4, 2008, pp. 466-477.

[7] J. Knight, S. Shirsavar and W. Holderbaum, "An Improved Reliability Cuk Based Solar Inverter with Sliding Mode Control," *IEEE Transactions on Power Electronics*, Vol. 21, No. 4, 2006, pp. 1107-1115.

[8] D. B. Costa and C. M. C. Duarte, "The ZVS-PWM Active-Clamping Cuk Converter," *IEEE Transactions on Industrial Electronics*, Vol. 51, No. 1, 2004, pp. 54-60.

[9] B. R. McGee and R. M. Nelms, "Using a Cuk Converter to Interface a Pulsed Load to a Fuel Cell," *Energy Conversion Engineering Conference*, 29-31 July 2004, pp. 313-318.

[10] D. Martins and G. de Abreu, "Application of the Zeta Converter in Switch-Mode Power Supplies," *Applied Power Electronics Conference and Exposition*, San Diego, 7-11 March 1993, pp. 214-220.

[11] D. C. Martins, F. S. Campos and I. Barbi, "Zeta Converter with High Power Factor Operating in Continuous Conduction Mode," 18*th International Telecommunications Energy Conference*, Boston, 6-10 October 1996, pp. 107-113.

[12] S. Yu, R. Zhao and A. Kwasinski, "Design Considerations of a Multiple-Input Isolated Single Ended Primary Inductor Converter (SEPIC) for Distributed Generation Sources," *Energy Conversion Congress and Exposition* (*ECCE*), Phoenix, 17-22 September 2001, pp. 3960-3967.

[13] A. El Shahat, "Stand-Alone PV System Simulation for DG Applications," *Journal of Automation & Systems Engineering*, Vol. 6, No. 1, 2012, pp. 55-72.

[14] R. Zhao and A. Kwasinski, "Multiple-Input Single Ended Primary Inductor Converter (SEPIC) Converter Distributed Generation Applications," *Energy Conversion Congress and Exposition*, 20-24 September 2009, pp. 1847-1854.

[15] N. Kalja and S. P. Rao, "SEPIC Converters Solve Automotive Power Needs," *Maxim Power Electronics Technology*, Vol. 35, No. 4, 2009, p. 16.

[16] S. Maniktala, "Switching Power Supplies A to Z," Elsevier, London, 2003.

[17] S. K. Mazumder, A. H. Nayfeh and D. Boroyevich, "Theoretical and Experimental Investigation of the Fast- and Slow-Scale Instabilities of a DC-DC Converter," *IEEE Transactions on Power Electronics*, Vol. 16, No. 2, 2001, pp. 201-216.

[18] A. Davoudi and J. Jatskevich, "Parasitics Realization in State-Space Average-Value Modeling of PWM DC-DC Converters," *IEEE Transactions on Circuits and Systems*, Vol. 54, No. 9, 2007, pp. 1003-10012.

[19] S. M. Szee, "Semiconductor Physic and Devices," 3rd Edition, John Willy and Sons, Chichester, 2003.

[20] V. Salas, E. Olias, A. Barrado and A. Lazaro, "Review of the Maximum Power Point Tracking Algorithms for Stand-Alone Photovoltaic Systems," *Journal of Solar Energy Materials & Solar Cells*, Vol. 90, No. 11, 2006, pp. 1555-1578.

[21] T. Esram and P. L. Chapman, "Comparison of Photovoltaic Array Maximum Power Point Tracking Techniques," *IEEE Transactions on Energy Conversion*, Vol. 22, No. 2, 2007, pp. 439-449.

[22] Z. Salameh and D. Taylor, "Step-Up Maximum Power Point Tracker for Photovoltaic Arrays," *Solar Energy*, Vol. 44, No. 1, 1990, pp. 57-61.

[23] K. Hussein, I. Muta, T. Hoshino and M. Osakada, "Maximum Photovoltaic Power Tracking: An Algorithm for Rapidly Changing Atmospheric Conditions," *Proceedings of IEEE Generation, Transmission and Distribution Conference*, Vol. 142, No. 1, 1995, pp. 59-64.

[24] N. Femia, G. Petrone, G. Spagnuolo and M. Vitelli, "Optimizing Sampling Rate of P&O MPPT Technique," *Proceedings of IEEE Power Electronical Special Conference*, Vol. 3, 2004, pp. 1945-1949.

[25] C.-X. Liu and L.-Q. Liu, "An Improved Perturbation and Observation MPPT Method of Photovoltaic Generate System," 4*th IEEE Conference on Industrial Electronics and Applications*, Xi'an, 25-27 May 2009, pp. 2966-2970.

[26] (IP&O) of MPPT Control for Photovoltaic Power Systems," in Proc. IEEE Photovoltaic Spec. Conf., 2005, pp. 1788-1791.

[27] D. Sera, T. Kerekes, R. Teodorescu and F. Blaabjerg, "Improved MPPT Algorithms for Rapidly Changing Environmental Conditions," *Power Electronics and Motion Control Conference*, Portoroz, 30 August-1 September 2006, pp. 1614-1619.

[28] C. Hua and C. Shen, "Study of Maximum Power Tracking Techniques and Control of DC-DC Converters for Photovoltaic Power System," 29*th Annual IEEE Power Electronics Specialists Conference*, New York, 17-22 May 1998, pp. 86-93.

[29] N. Mohan, M. T. Undeland and P. W. Robbins, "Power Electronics, Converters, Application and Design," 2nd Edition, John Willey & Sons, Chichester, 2003.

[30] M. H. Rashid, "Power Electronic Circuits, Devices and Applications," 2nd Edition, Tata McGraw-Hill Publishing Company Limited, New Delhi, 2009.

[31] V. Michal, "Modulated-Ramp PWM Generator for Linear Control of the Boost Converter's Power Stage," IEEE Transactions on Power Electronics, Vol. 27, No. 6, 2012, pp. 2958-2965.

[32] M. Simulink, "User's Manual," The Math Works, Inc., New York. http://www.mathworks.com

[33] "Poly-Crystalline Silicon Photovoltaic Modules Datasheet: Malaysia," 2012.

Towards Implementation of Smart Grid: An Updated Review on Electrical Energy Storage Systems

Md Multan Biswas[1*], Md Shafiul Azim[2], Tonmoy Kumar Saha[2], Umama Zobayer[3], Monalisa Chowdhury Urmi[3]

[1]Department of Electrical and Electronic Engineering, Bangladesh University of Engineering and Technology, Dhaka, Bangladesh;
[2]Department of Electrical and Electronic Engineering, Khulna University of Engineering and Technology, Khulna, Bangladesh;
[3]Department of Electrical and Electronic Engineering, Stamford University Bangladesh, Dhaka, Bangladesh.

ABSTRACT

A smart grid will require, to greater or lesser degrees, advanced tools for planning and operation, broadly accepted communications platforms, smart sensors and controls, and real-time pricing. The smart grid has been described as something of an ecosystem with constantly communication, proactive, and virtually self-aware. The use of smart grid has a lot of economical and environmental advantages; however it has a downside of instability and unpredictability introduced by distributed generation (DG) from renewable energy into the public electric systems. Variable energies such as solar and wind power have a lack of stability and to avoid short-term fluctuations in power supplied to the grid, a local storage subsystem could be used to provide higher quality and stability in the fed energy. Energy storage systems (ESSs) would be a facilitator of smart grid deployment and a "small amount" of storage would have a "great impact" on the future power grid. The smart grid, with its various superior communications and control features, would make it possible to integrate the potential application of widely dispersed battery storage systems as well other ESSs. This work deals with a detailed updated review on available ESSs applications in future smart power grids. It also highlights latest projects carried out on different ESSs throughout all around the world.

Keywords: Battery; Distributed Generation; Hybrid Energy Storage Systems; Power Quality; Smart Grid

1. Introduction

The world's electricity systems face a number of challenges, including aging infrastructure, continuous increase in demand, the integration of growing numbers of variable renewable energy DGs and electric and hybrid electric vehicles, the need to improve the security of power supply, and the need to lower carbon dioxide (CO_2) emissions [1-3]. These challenges must be addressed also with regard to each region's unique technical, economic, and commercial regulatory environment [4]. Smart grid technologies offer ways not just to meet these challenges but also to develop a sustainable energy supply that is more energy efficient and more affordable.

Compared to other industries, our electrical grid has been largely bypassed by technological innovation until relatively recently, owing to the fact that historically it has been heavily regulated and modeled to keep the lights on and costs low. Partly for this reason, its modernization by means of information-technology tools and techniques has been somewhat of a back-burner priority. Like the telecom and internet revolutions that preceded it, technology holds the key to the smart grid and its realization. The smart grid and the technologies embodied within it are an essential set of investments that will help bring our electric grid into the 21st century using megabytes of data to move megawatts of electricity more efficiently, reliably, and affordably [5-9].

The smart grid will:

- Provide power quality for the digital economy;
- Accommodate all generation and storage options;
- Enable new products, services, and markets;
- Optimize asset utilization and operate efficiently;
- Enable active participation by consumers;
- Anticipate and respond to system disturbances;
- Operate resiliently against attack and natural disaster.

Storage is perhaps the most important smart grid advanced component because of its key role in complementing renewable generation. With the proper amount and type of storage broadly deployed and optimally controlled, renewable generation can be transformed from an energy source into a dispatchable generation source

*Corresponding author.

[10-12]. And with the addition of energy storage, more wind and solar generation can be added to a typical power system that employs a large percentage of slow-response fossil and nuclear generation. It is feasible that, the penetration of renewables can be significantly above 20 percent with the addition of sufficient energy storage technologies [13].

This paper is organized as follows. Section 2 presents a brief description of smart grid technology with its essential features. The necessity and prospect of energy storage systems in future smart power grid are broadly discussed in Section 3. Finally, Section 4 elaborately reviews the available energy storage technologies that are considered for use in smart grid applications with updated technologies, which is followed by some concluding remarks in Section 5.

2. Smart Grid: Technology Description

Though there has been much debate over the exact definition of smart grid, it actually comprises a broad range of technology solutions that optimize the energy value chain. Depending on where and how a specific utility operates across that chain, it can benefit from deploying certain parts of a smart grid solution set. Smart grid is a large electricity network that uses digital and other advanced technologies to improve efficiency, reliability, and security of the electric system: from large generation, through the delivery systems to electricity consumers and a growing number of distributed-generation and energy storage resources [14-16]. Smart grids co-ordinate the needs and capabilities of all power generators, grid operators, end-users, and electricity market stakeholders to operate all parts of the system as efficiently as possible, minimising costs and environmental impacts while maximising system stability, reliability, and resilience. Smart grids are an evolving set of technologies that will be deployed at different rates in a variety of settings around the world, depending on local commercial attractiveness, compatibility with existing technologies, regulatory developments, and investment frameworks. **Figure 1** demonstrates the evolutionary character of smart power network [4].

Advanced metering infrastructure (AMI) is an approach to integrating consumers based upon the development of open standards. It provides consumers with the ability to use electricity more efficiently and provides utilities with the ability to detect problems on their systems and operate them more effectively [17,18]. AMI enables consumer-friendly efficiency concepts like "Prices to Devices" to work like this: Assuming that energy is priced on what it costs in near real-time—price signals are relayed to "smart" home controllers or end-consumer devices like thermostats, washer/dryers, and refrigerators—the home's major energy-consumers [19-21]. The devices, in turn, process the information based on consumers' learned wishes and power accordingly. The house or office responds to the occupants, rather than vice-versa. Because this interaction occurs largely in the background with minimal human intervention, there's a dramatic savings on energy that would otherwise be consumed [14]. Far more than "smart meters," a fully-functioning smart grid will feature sensors throughout the transmission and distribution grid to collect data, real-time two-way communications to move that data between utilities and consumers, and the computing power necessary to make that intelligence actionable and transactive as shown in **Figure 2**. Indeed, only by bringing the tools, techniques, and technologies that enabled the internet to the utility and the electric grid is such a transformation possible [8].

Smart Grid's Principal Characteristics

A smart power grid brings the power of networked, interactive technologies into an electricity system, giving utilities and consumer's unprecedented control over energy use, improving power grid operations, and ultimately reducing costs to consumers. In brief, the main features of smart grids are [9,22-26]:

1) A smart grid accommodates not only large, centralised power plants, but also the growing array of customer-sited distributed energy resources.

2) Not all commercial enterprises, and certainly not all residential customers, need the same quality of power. A smart grid supplies varying grades (and prices) of power.

3) Correctly designed and operated markets efficiently create an opportunity for consumers to choose among competing services.

4) A smart grid applies the latest technologies to optimize the use of its assets. For example, optimized capacity can be attainable with dynamic ratings, which allow assets to be used at greater loads by continuously sensing and rating their capacities.

5) Provides resiliency to disturbances, attacks, and natural disasters. Self-healing actions result in reduced interruption of service to consumers and help service providers better manage the delivery infrastructure.

6) Consumers help balance supply and demand, and ensure reliability by modifying the way they use and purchase electricity.

3. Energy Storage—A Key Enabler of Smart Grid

Liberalization of the power market and widespread use of distributed energy resources (DERs), in particular DG and ESSs, could enable smart grids to have a significant influence on electricity market prices and ancillary services [27]. As focused in [27], the charge/discharge of

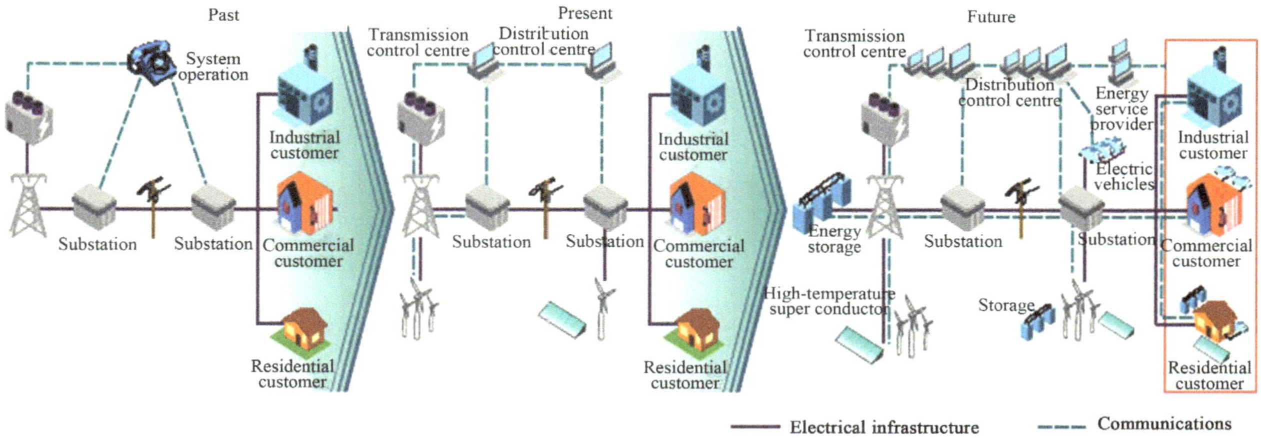

Figure 1. Smarter electricity systems.

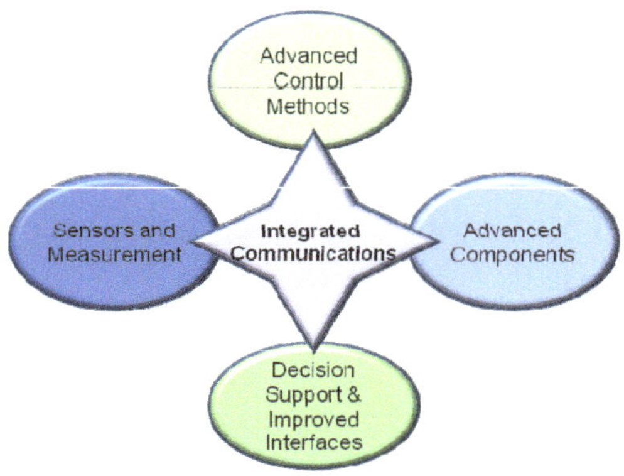

Figure 2. Advanced communication and control in smart grids.

Figure 3. Distributed energy storage in a Windfarm.

various ESSs can be controlled in order to guarantying proper applications. Further, in modern power distribution systems, where a significant amount of the total electricity demand is met by renewable generation, ESSs can mitigate the uncertainties of energy sources (such as solar and wind) and can store the energy during high renewable production and/or low price periods, and deliver when either necessary or convenient [28-31]. Based on the ESSs technologies, in [27] the applications of ESSs are classified in instantaneous, short-, mid- and long-term. Instantaneous and short-term applications are involved in real time regulations, for example aiming at ancillary services provision or integration of electric drive vehicles batteries in the networks [32,33].

Energy storage systems improve the efficiency and reliability of the electric supply system by reducing the requirements for spinning reserves to meet peak power demands [34]. Storage devices can also provide frequency regulation to maintain the balance between the network's load and power generated, and they can achieve a more reliable power supply for high tech industrial facilities as shown in **Figure 3**.

High voltage power electronics, such as switches, rectifiers, inverters, and controllers, allow electric power to be precisely and rapidly controlled to support long distance transmission [35,36]. Thus, energy storage and power electronics hold substantial promise for transforming the electric power industry [28]. This capability will allow the system to respond effectively to disturbances and operate more efficiently, thereby reducing the need for additional infrastructure.

4. Available Energy Storage Systems

The adoption of smart power grid devices throughout utility networks will effect tremendous change in grid operations and usage of electricity over the next two decades. Increased deployment of energy storage devices in the distribution grid will help make this process happen more effectively and improve system performance. The energy storage systems (ESSs) applicable in power networks can be divided into two major categories [29] as shown in **Figure 4**. The first category being large-scale storage systems that can be used in utility transmission applications. The second group includes small-scale storage systems sited at the consumer's premises. Some of the available energy storage systems are potentially discussed in the following sub-sections.

4.1. Battery

Battery is one or more electrochemical cells that convert chemical energy into electrical energy. Batteries have been using for energy storage purpose for over one-hundred years and possess some very important, unique, and desirable features. Battery energy storage systems (BESSs) are modular, quiet, and non-polluting [37]. Batteries are manufactured in a wide variety of capacities ranging from less than 100 watts to modular configurations of several megawatts. As a result, batteries can be used for various utility applications in the areas of generation, transmission and distribution, and customer service and they can be installed relatively quickly. Battery has convenient size and voltage characteristics. The operating principle of a typical battery is shown in **Figure 5**.

A grid-scale BESS consists of a battery bank, control system, power electronics interfacing circuit for ac-dc power conversion, protective circuitry, and a transformer to convert the BESS output to the desired transmission or distribution system voltage level [38]. The battery bank consists of numerous batteries connected in a combination series-parallel configuration to provide the desired energy and power capabilities for the application.

The use of Lead Acid batteries for energy storage dates back to mid-1800s. There have been developed several new battery technologies to store more energy, last longer and less cost than the Lead Acid battery. Battery technologies differ widely in terms of their energy and power densities, energy efficiencies, cycle-life, availability, and operating conditions. Some of these new battery technologies are Lithium Ion, Lithium Polymer, Nickel Metal Hydride (Ni-MH), Vanadium Redox (VRB), Nickel Cadmium (Ni-Cd), Sodium Sulfur (NaS), and Zinc Bromide [39-41]. **Table 1** summarizes the characteristic parameters of different batteries [27,28, 42-44].

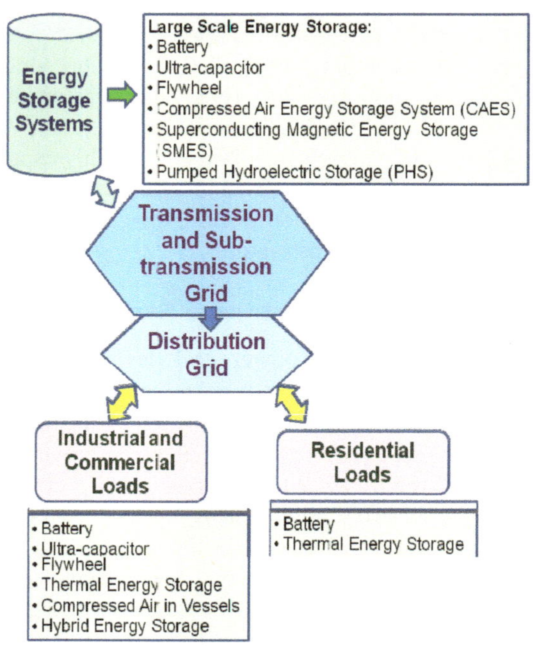

Figure 4. Energy storage systems in smart power networks.

4.2. Ultra-Capacitors

Ultra-capacitors (UCs), also known as super-capacitors, or electric double-layer capacitors (EDLCs), store energy in the electrical double layer at an electrode/electrolyte interface. There are no chemical reactions involved in the UC's energy storage mechanism. The main parts of an electrochemical capacitor are electrodes and electrolyte. The electrical energy E accumulated in ultracapacitors is related to the capacitance C or the stored charges Q and voltage V by following formula:

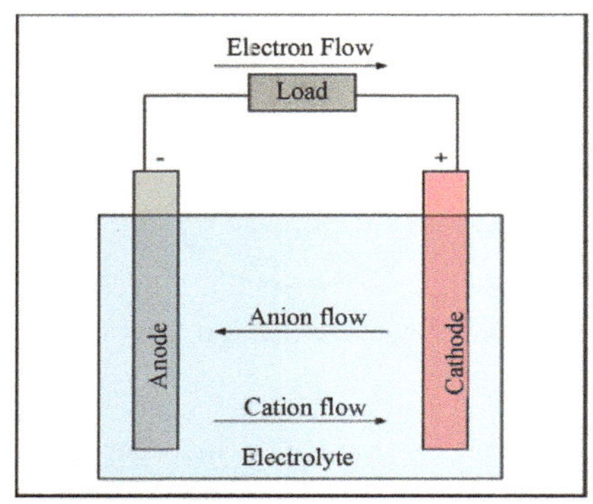

Figure 5. Working principle of a battery.

Table 1. Comparison of different types of batteries.

Type	Energy Density (Wh/kg)	Energy Efficiency (%)	Power Density (W/Kg)	Cycle Life (Cycles)	Self Discharge (%/Month)
Lead-Acid	30 - 40	70 - 90	180	200 - 2000	3 - 4
Li-Ion	100 - 250	75 - 90	1800	500 - 2000	5 - 10
Li Polymer	130 - 200	70	3000	>1200	4 - 8
Ni-MH	30 - 80	70	250 - 1000	500 - 100	30
Ni-Cd	40 - 60	60 - 90	140 - 180	500 - 2000	10 - 15
NaS	150	80 - 90	120 - 150	2500	-
VRB	25 - 40	80	100 - 150	>16,000	<1
Zinc Bromide	70	70	-	1000	-

$$E = \frac{1}{2}CV^2 = \frac{1}{2}QV \quad (1)$$

And,

$$Q = CV = \frac{A\varepsilon}{d}V \quad (2)$$

As for the conventional capacitor, the capacitance C is proportional to the area A of the plates and the permitivity of the dielectric ε and is inversely proportional to the distance d between the plates. UCs are designed to have a very high electrode surface area and use high permitivity dielectric. Due to the high permeability and close proximity of the electrodes, UCs have a low voltage withstand capability (usually 2 - 3 V) [30]. The electrode surface area is maximized by using porous carbon as the current collector, allowing a relatively large amount of energy to be stored at the collector surface. Therefore, UCs attain very high capacitance ratings. Larger UCs have capacities up to 5000 farads [45].

Ultracapacitors store energy by physically separating unlike charges. UCs have a long cycle life due to the fact that there are no chemical changes on the electrodes ideally in normal operation and UCs have superior efficiency. UCs also provide exceptional power density, since the charges are physically stored on the electrodes. Conversely, energy density is low since the electrons are not bound by chemical reactions [30]. The UC is temperature resistant with an operating range between −40°C to +65°C and is also shock and vibration resistant [46, 47].

In order to be a viable alternative in a large scale energy storage system they will need to be able to handle multiple kV. The ability of modular, non-polluting, quiet, quick charge and discharge capability, long life (10 to 12 years), and very high cycle life makes the UC a very desirable energy storage device. They can be use for short term ride through capabilities as well as voltage regulation, frequency control and other power quality issues [47,48]. UCs are currently available in many sizes. Two 3000 F capacitors are shown in **Figure 6**. There is even a startup company that claims to be able to create ultra capacitors with higher energy densities than lead-acid, Nickel metal hydride and even lithium ion batteries [49]. There have also been advances in the design of the ultracapacitors using nano-tube technology to improve the surface area of the capacitor. This "nano-tube ultracapacitor" would improve the ultracapacitor's energy density to be compatible with again that of a chemical battery [46,50,51].

4.3. Flywheel

During the past decade, flywheel energy storage systems (FESSs) have been rediscovered by the power industry due to their advantages in comparison with other energy storage systems. FESSs have found an important technical role on the application of enhancing the electric power quality, grid voltage and frequency support, and unbalanced load compensation. By virtue of their high dynamics, long lifetime, and good efficiency, FESSs are well suited for short-term storage systems [52-54].

FESS consists of a flywheel coupled to permanent magnets synchronous machine (PMSM) as shown in **Figure 7**. Flywheels store energy in the form of momentum in a rotating wheel or cylinder. A FESS stores energy through accelerating a rotor up to a high rate of speed and maintaining the energy in the system as inertial energy. Advanced composite materials are sometimes used for the rotor to lower its weight while allowing for the extremely high speeds. The flywheel releases the energy by reversing the process and using the motor as a generator. As the flywheel releases its stored energy, the

Figure 6. 3000 farad ultracapacitors.

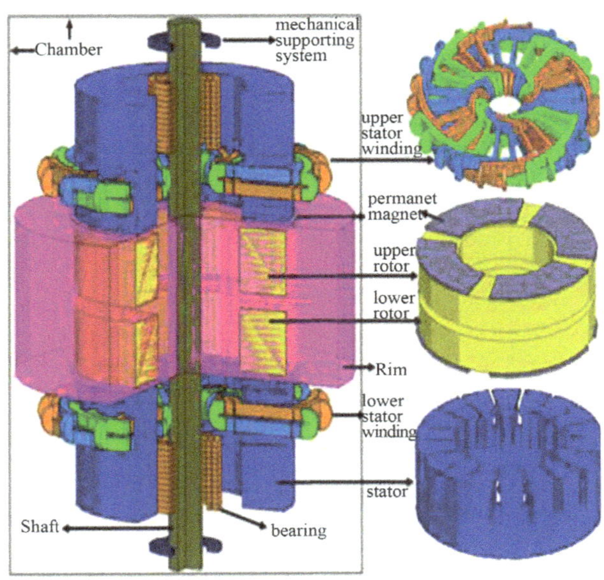

Figure 7. Flywheel energy storage system.

flywheel's rotor slows until it is fully discharged. The energy that can be stored depends on its rotational velocity ω and moment of inertia I, as in (3). They have high power density and more energy can be stored if the flywheel rotates at higher rotational speed [55]. Power electronics are used to ensure that output voltage has appropriate amplitude and frequency characteristics.

$$E = \frac{1}{2} I \omega^2 \quad (3)$$

Flywheel energy storage systems using superconducting magnetic bearing (SMB) together with a permanent magnet bearing (PMB) is one of the most promising electro-mechanical energy storage systems consisting long life, high energy density, high efficiency, with no pollution or toxic material disposal problems, and low rotational loss by non-contact superconductor bearing [55-57].

Flywheels have seen most commercial success targeted for power delivery capabilities typically in the 150 kW-1MW range [50,58]. Recently, a 20 MW FESS plant has been successfully established in Stephentown, New York by Beacon Power Corp. under a pilot project of the Department of Energy (DOE), New York State Energy Research and Development Authority (NYSERDA) and currently 40 MW FESS plant project is under development [59].

4.4. Superconducting Magnetic Energy Storage (SMES)

Superconducting magnetic energy storage (SMES) systems store energy in the magnetic field produced by current flowing through a superconducting coil. The SMES principle is based on inductive energy storage in the magnetic field produced by current flowing through a superconducting coil.

A SMES system consists of four major subsystems [60], plus miscellaneous equipment for system control, data collection, and so on. The major subsystems are: 1) superconducting coil with magnet (SCM); 2) power conditioning system (PCS) that controls the flow of current into and out of the coil to charge and discharge the SMES; 3) cryogenic system (CS) that maintains the coil at a low enough temperature to maintain superconductivity; and 4) the control unit (CU), as shown in **Figure 8**.

For a SMES system, the inductively stored energy (E) and the rated power (P) are commonly the given specifications for SMES devices, and can be expressed as follows:

$$E = \frac{1}{2} L I^2 \; ; P = \frac{dE}{dt} = LI \frac{dI}{dt} = VI \quad (4)$$

where L is the inductance of the coil, I is the dc current flowing through the coil, and V is the voltage across the coil. During SMES operation, the magnet coils have to remain in the superconducting status. A refrigerator in the cryogenic system maintains the required temperature for proper superconducting operation [61].

Among the different variants of flexible alternating current transmission system (FACTS) devices and ESSs currently available, static synchronous compensators (STATCOM) integrated with SMES has been proposed as the most adequate for participating of the primary frequency control because of SMES has high efficiency (95% to 98% [61]) and rapid response to power demand [62-64]. Depending on the control loop of its power conversion unit and switching characteristics, the SMES system can respond very rapidly (MWs/milliseconds). The ability of injecting/absorbing real or reactive power can increase the effectiveness of the control, and enhance system reliability and availability. Comparing with other storage technologies, the SMES technology has a unique advantage in two types of applications: Power system transmission control and stabilization, and power quality improvement [65].

Several SMESs in the range of kWh to MWh scale have been already implemented for compensation of load/generation fluctuation as well as energy storage [66]. Recently, 10 MVA/20 MJ SMES prototype has been tested at an actual power system including hydro power generators in order to compensate the fluctuating power load for a metal rolling factory [67].

4.5. Compressed Air Energy Storage System (CAES)

Compressed air energy storage system (CAES) is a hybrid technology of long term power storage and generation. A CAES plant mainly consists of 1) compressor train; 2) motor-generator unit; 3) gas turbine; and 4) underground compressed air storage tank (see **Figure 9**). During low-cost off-peak load periods, air is compressed and stored in large underground salt caverns [68]. Upon demand, the process is reversed; the compressed air is returned to the surface; this air is used to burn natural gas in the combustion chambers. The resulting combustion gas is then expanded in the two-stage gas turbine to spin

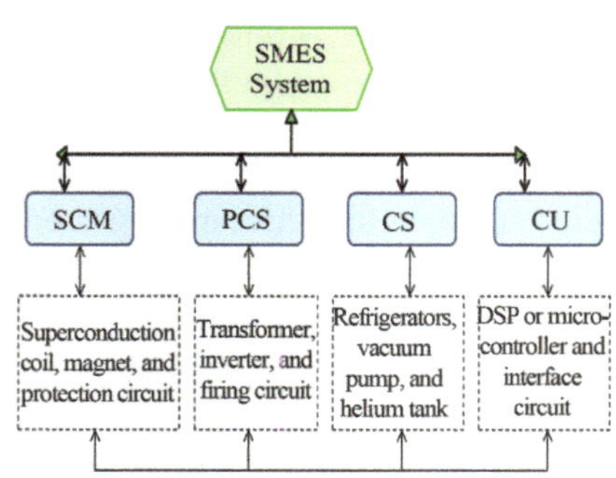

Figure 8. Block diagram of an SMES system.

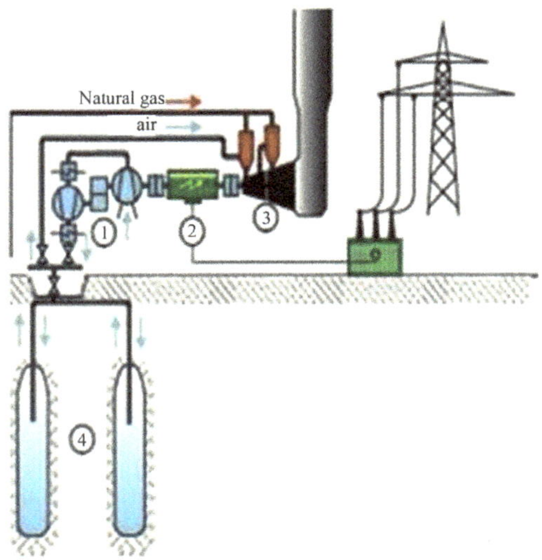

Figure 9. Major components of a CAES plant.

the generator and produce electricity.

CAES can be placed near consumers and in large, small, and micro scale smart power networks. As in many countries, the potentials of installing new hydro-storage plants with pumping facilities are limited and many other energy storage devices are far from being economic, CAES may be an attractive investment opportunity for such purposes. The first such plant was built in Huntorf, Germany, in 1978 with a capacity of 290 MW [68]. In 1991, the first U.S. CAES facility was built in McIntosh, Alabama, by the Alabama Electric Cooperative and EPRI, and has a capacity rating of 110 MW [69]. Liu *et al.* [70] has proposed a novel hybrid-fuel CAES system for China. The design is based on using standard, industry proven equipment components to deliver a reliable and economic compressed air energy cycle. The resulting CAES plant consists of 410 MW generation, 205 MW of compression, and 2050 MW of storage.

In the past few years, research has been conducted to improve the efficiency of the turbines and heat transfer mechanisms used to pump and retrieve compressed air [71-73]. In an adiabatic CAES, the air's heat energy is stored separately and recovered before the compressed air is expanded in an air turbine. Such plants are currently under development and promise higher efficiencies and zero direct CO_2-emissions [71].

4.6. Pumped Hydroelectric Storage

Pumped hydroelectric storage (PHS) is the oldest kind of large-scale energy storage technology [74]. Since 1904, they are in active operation and new ones are still being built because of their operational flexibility and ability to provide rapid response to changes in system loading or spot price of electricity.

Conventional pumped hydroelectric storage consists of two large reservoirs, one is located at base level and the other is situated at a different elevation. Water is pumped to the upper reservoir where it can be stored as potential energy. Upon demand, water is released back into the lower reservoir, passing through hydraulic turbines which generate electrical power as high as 1000 MW. Pumped hydroelectric storage has huge energy and power capacity. Recently [75-77] examined the impact of pumped storage units together with large renewable penetration. They focused on reducing system operating costs and maximizing usage of renewable energy based on unit commitments and dispatches. Currently, efforts aimed at increasing the use of pumped hydro storage are focused on the development of underground facilities. Modern pumped storage plants are often designed to have fast start, load ramping and unloading capabilities. They can respond to load changes within seconds.

4.7. Hybrid Energy Storage System

The selection of the energy storage system for a particular application in power grid sometimes depends on a suitable combination of the power and energy ratings, energy, power density, cost, weight, volume, and operating temperature etc of ESSs. These requirements may not be achievable from a single energy storage source. To implement such optimal applications, hybrid energy storage devices (HESDs) have been proposed. In a HESD, two or more different energy storage systems with complementary characteristics can be combined together electronically. For future grid application proposed HESDs are listed next, with the energy-supplying device listed first, followed by the power-supplying device:

1) battery and ultracapacitor [78-81];
2) battery and flywheels [82];
3) CAES and battery or ultracapacitor [83];
4) battery and SMES [84].

Figure 10 demonstrates the relationship between power operational range and discharge time at rated power for various energy storage systems such as battery, ultracapacitors, flywheel, SMES, CAES, and pumped hydro. **Table 2** summarizes the characteristic parameters of different energy storage technologies [30,42-44,59,83].

5. Concluding Remarks

In this paper the potential application of different ESSs and their prospect have been discussed and analyzed in detail for future smart power grids. Energy can be stored both for short and long durations. Various storage systems are available for these purposes. The importance of storage systems in electricity grids is finally receiving the attention of system planners as more storage options participation of storage is increasing. The design of smart

Table 2. Comparison of various energy storage systems.

Type	Energy Density (Wh/kg)	Energy Efficiency (%)	Power Density (W/Kg)	Life Time (Cycles)	Self Discharge (%/Month)	Environmental Effect
Battery	25 - 250	60 - 90	100 - 3000	150 - 3000	3 - 30	Toxic
Ultra Capacitor	<50	95	4000	10 k - 100 k	High	Benign
Flywheel	100 - 130	95	1000	>125,000	High	Benign
SMES	30 - 80	95	Very High	-	Negligible	Benign
CAES	10 - 30	50	Fair	40 Years	Negligible	Benign
PHS	0.3	65 - 80	-	75 Years	Negligible	Benign

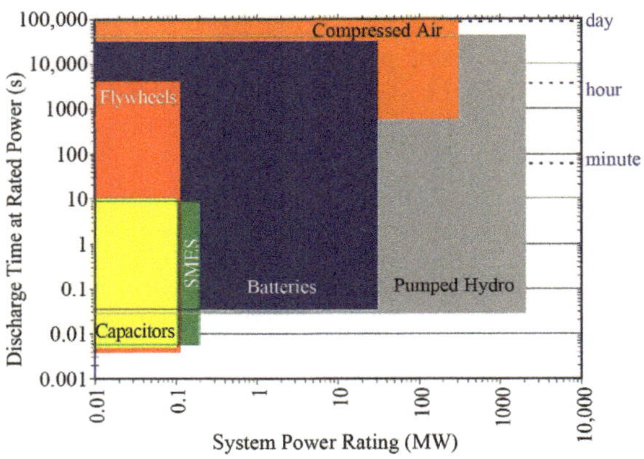

Figure 10. The relationship between power rating and discharge time for different storage technologies.

grids in the future will take advantage of storage to deal with more dynamic loads and sources. As market rules are adjusted to take advantage of the benefits of bulk and distributed storage devices, the overall capabilities and reliability of more complex electricity networks should continue to improve as fully integrated smart grids.

REFERENCES

[1] K. Moslehi and R. Kumar, "A Reliability Perspective of the Smart Grid," *IEEE Transactions on Smart Grid*, Vol. 1, No. 1, 2010, pp. 57-64.

[2] P. Zhang, F. Li and N. Bhatt, "Next-Generation Monitoring, Analysis, and Control for the Future Smart Control Center," *IEEE Transactions on Smart Grid*, Vol. 1, No. 2, 2010, pp. 186-192.

[3] S. M. Amin and B. F. Wollenberg, "Toward a Smart Grid: Power Delivery for the 21st Century," *IEEE Power Energy Magazine*, Vol. 3, No. 5, 2005, pp. 34-41.

[4] International Energy Agency, "Technology Roadmap: Smart Grids," 2011. http://www.iea.org/publications/freepublications/publication/name,3972,en.html

[5] US Department of Energy, "How the Smart Grid Promotes a Greener Future," 2010. http://energy.gov/oe/downloads/how-smart-grid-promotes-greener-future

[6] H. Gharavi and R. Ghafurian, "Smart Grid: The Electric Energy System of the Future," *IEEE Proceedings*, Vol. 99, No. 6, 2011, pp. 917-921.

[7] J. Xia and Y. Wang, "Secure Key Distribution for the Smart Grid," *IEEE Transactions on Smart Grid*, Vol. 3, No. 3, 2012, pp. 1437-1443.

[8] A. Molderink, V. Bakker, M. G. C. Bosman, J. L. Hurink and G. J. M. Smit, "Management and Control of Domestic Smart Grid Technology," *IEEE Transactions on Smart Grid*, Vol. 1, No. 2, 2010, pp. 109-119.

[9] A. Bose, "Smart Transmission Grid Applications and Their Supporting Infrastructure," *IEEE Transactions on Smart Grid*, Vol. 1, No. 1, 2010, pp. 11-19.

[10] B. P. Roberts and C. Sandberg, "The Role of Energy Storage in Development of Smart Grids," *IEEE Proceedings*, Vol. 99, No. 6, 2011, pp. 1139-1144.

[11] M. S. Whittingham, "History, Evolution, and Future Status of Energy Storage," *IEEE Proceedings*, Vol. 100, No. , 2012, pp. 1518-1534.

[12] G. D. Rodriguez, "A Utility Perspective of the Role of Energy Storage in the Smart Grid," 2010 *IEEE Power and Energy Society General Meeting*, Minneapolis, 25-29 July 2010, pp. 1-2.

[13] US Department of Energy, "Energy Storage—A Key Enabler of the Smart Grid," 2009. http://www.netl.doe.gov/smartgrid/refshelf.html#White%20Papers

[14] US Department of Energy, "The Smart Grid: An Introduction, 2008. http://energy.gov/oe/downloads/smart-grid-introduction-0

[15] F. Li, W. Qiao, H. Sun, H. Wan, J. Wang, *et al.*, "Smart Transmission Grid: Vision and Framework," *IEEE Transactions on Smart Grid*, Vol. 1, No. 2, 2010, pp. 168-177.

[16] M. D. Hopkins, A. Pahwa and T. Easton, "Intelligent Dispatch for Distributed Renewable Resources," *IEEE Transactions on Smart Grid*, Vol. 3, No. 2, 2012, pp. 1047-1054.

[17] S. S. S. R. Depuru, L. Wang, V. Devabhaktuni and N. Gudi, "Smart Meters for Power Grid—Challenges, Issues, Advantages and Status," 2011 *IEEE/PES Power Systems Conference and Exposition* (*PSCE*), Phoenix, 20-23 March 2011, pp. 1-7.

[18] D. Rua, D. Issicaba, F. J. Soares, P. M. R. Almeida, R. J. Rei and J. A. P. Lopes, "Advanced Metering Infrastructure Functionalities for Electric Mobility," 2010 *IEEE PES Innovative Smart Grid Technologies Conference Europe* (*ISGT Europe*), Porto, 11-13 October 2010, pp. 1-7.

[19] H. Suleiman, K. A. Ahmed, N. Zafar, *et al*., "Inter-Domain Analysis of Smart Grid Domain Dependencies Using Domain-Link Matrices," *IEEE Transactions on Smart Grid*, Vol. 3, No. 2, 2012, pp. 692-709.

[20] Microsoft Power and Utilities, "Smart Energy Reference Architecture (SERA)," 2009.

[21] A. P. S. Meliopoulos, G. Cokkinides, R. Huang, E. Farantatos, S. Choi, *et al.*, "Smart Grid Technologies for Autonomous Operation and Control," *IEEE Transactions on Smart Grid*, Vol. 2, No. 1, 2011, pp. 1-10.

[22] A. Ipakchi and F. Albuyeh, "Grid of the Future," *IEEE Power Energy Magazine*, Vol. 7, No. 2, 2009, pp. 52-62.

[23] R. A. F. Currie, G. W. Ault, C. E. T. Foote, N. M. McNeill, *et al.*, "Smarter Ways to Provide Grid Connections for Renewable Generators," *IEEE PES General Meeting*, Minneapolis, 25-29 July 2010, pp. 1-6.

[24] Q. Pang, H. Gao and M. Xiang, "Design of Intelligent Terminal Unit for Smart Distribution Grid," *Proceedings of the IEEE China International Conference on Electricity Distribution* (*CICED*), Nanjing, 13-16 September 2010, pp. 1-6.

[25] F. Rahimi and A. Ipakchi, "Demand Response as a Market Resource under the Smart Grid Paradigm," *IEEE Transactions on Smart Grid,* Vol. 1, No. 1, 2010, pp. 82-88.

[26] E. Santacana, G. Rackliffe, L. Tang and X. Feng, "Getting Smart," *IEEE Power Energy Magazine*, Vol. 8, No. 2, 2010, pp. 41-48.

[27] K. C. Divya and J. Østergaard, "Battery Energy Storage Technology for Power Systems: An Overview," *Electric Power System Research*, Vol. 79, No. 4, 2009, pp. 511-520.

[28] M. M. Chowdhury, M. E. Haque, M. Aktarujjaman, M. Negnevitsky and A. Gargoom, "Grid Integration Impacts and Energy Storage Systems for Wind Energy Applications—A Review," *Proceedings of the IEEE PES General Meeting*, San Diego, 24-29 July 2011, pp. 1-8.

[29] S. C. Smith, P. K. Sen and B. Kroposki, "Advancement of Energy Storage Devices and Applications in Electrical Power System," *IEEE PES General Meeting*, Pittsburgh, 20-24 July 2008, pp. 1-8.

[30] S. Vazquez, S. M. Lukic, E. Galvan, L. G. Franquelo, *et al.*, "Energy Storage Systems for Transport and Grid Applications," *IEEE Transactions on Industrial Electronics*, Vol. 57, No. 12, 2010, pp. 3881-3895.

[31] A. A. Thatte and L. Xie, "Towards a Unified Operational Value Index of Energy Storage in Smart Grid Environment," *IEEE Transactions on Smart Grid*, Vol. 3, No. 3, 2012, pp. 1418-1426.

[32] W. Kempton and J. Tomic, "Vehicle-to-Grid Power Fundamentals: Calculating Capacity and Net Revenue," *Journal of Power Sources*, Vol. 144, No. 1, 2005, pp. 268-279.

[33] A. Bracale, P. Caramia and D. Proto, "Optimal Operation of Smart Grids Including Distributed Generation Units and Plug in Vehicles," *International Conference on Renewable Energies Power Quality* (*ICREPQ'11*), Las Palmas de Gran Canaria, 13-15 April 2011, pp. 553-558.

[34] M. M. Biswas, K. K. Das, I. A. Baqee, M. A. H. Sadi and H. M. S. Forhad, "Prospects of Renewable Energy and Energy Storage Systems in Bangladesh and Developing Economics," *Global Journal of Researches in Engineering* (*GJRE*), Vol. 11, No. 5, 2011, pp. 23-31.

[35] Y. M. Cheng, Y. C. Liu, S. C. Hung and C. S. Cheng, "Multi-Input Inverter for Grid-Connected Hybrid PV/Wind Power System," *IEEE Transactions on Power Electronics*, Vol. 22, No. 3, 2007, pp. 1070-1076.

[36] J. M. Carrasco, L. G. Franquelo, J. T. Bialasiewicz, E. Galvan, R. C. P. Guisado, *et al.*, "Power Electronic Systems for the Grid Integration of Renewable Energy Sources: A Survey," *IEEE Transactions on Power Electronics*, Vol. 53, No 4, 2006, pp. 1002-1016.

[37] R. B. Schainker, "Executive Overview: Energy Storage Options for a Sustainable Energy Future," *IEEE PES General Meeting*, Palo Alto, June 2004, Vol. 2, pp. 2309-2314.

[38] C. A. Hill, M. C. Such, D. Chen, J. Gonzalez, *et al.*, "Battery Energy Storage for Enabling Integration of Distributed Solar Power Generation," *IEEE Transactions on Smart Grid*, Vol. 3, No. 2, 2012, pp. 850-857.

[39] J. McDowall, "Nickel-Cadmium Batteries for Energy Storage Applications," *The* 14*th Annual Battery Conference on Applications and Advances*, Long Beach, 12-15 January 1999, pp. 303-308.

[40] J. P. Barton and D. G. Infield, "Energy Storage and Its Use with Intermittent Renewable Energy," *IEEE Transactions on Energy Conversion*, Vol. 19, No. 2, 2004, pp. 441-448.

[41] P. C. Symons, "Opportunities for Energy Storage in Stressed Electrical Supply Systems," *IEEE PES Summer Meeting*, Vancouver, July 2001, Vol. 1, pp. 448-449.

[42] D. Y. Goswami and F. Kreith, "Energy Conversion," CRC Press, Boca Raton, 2007.

[43] A. A. H. Hussein, S. Harb, N. Kutkut, J. Shen and I. Batarseh, "Design Considerations for Distributed Micro- Storage Systems in Residential Applications," 32*nd International Telecommunications Energy Conference (INTELEC)*, Orlando, 6-10 June 2010, pp. 1-6.

[44] B. R. Williams and T. Hennessy, "Energy Oasis [Vanadium Redox Battery System in Power Distribution Application]," *IEEE Power Engineer*, Vol. 19, No. 1, 2005, pp. 28-31.

[45] Nesscap Ultracapacitors Products, "(EDLC) Electric Double-Layer Capacitor," 2012. http://www.nesscap.com/product/edlc_large1.jsp

[46] D. Halber, "Researchers Fired Up over New Battery," *MIT Tech Talk*, Vol. 50, No. 16, 2006, pp. 1-5. http://web.mit.edu/newsoffice/2006/batteries-0208.html

[47] Maxwell Technologies, "An Overview of Ultracapacitor Technology," 2012. http://www.maxwell.com/ultracapacitors/

[48] R. D. Weir and C. W. Nelson, "Utilization of Poly (Ethylene Terephthalate) Plastic and Composition-Modified Barium Titanate Powders in a Matrix that Allows Polarization and the Use of Integrated-Circuit Technologies for the Production of Lightweight Ultrahigh Electrical Energy Storage Units (EESU)," US Patent No. 7466536, 2008.

[49] H. Pan, J. Li and Y. P. Feng, "Carbon Nanotubes for Supercapacitor," *Nanoscale Research Letters*, Vol. 5, No. 3, 2010, pp. 654-668.

[50] H. Zhang, G. P. Cao, Y. S. Yang and Z. N. Gu, "Comparison between Electrochemical Properties of Aligned Carbon Nanotube Array and Entangled Carbon Nanotube Electrodes," *Journal of Electrochemical Society*, Vol. 155, No. 2, 2008, pp. 19-22.

[51] R. Hebner, J. Beno and A. Walls, "Flywheel Batteries Come Around Again," *IEEE Spectrum*, Vol. 39, No. 4, 2002, pp. 46-51.

[52] G. Cimuca, C. Saudemont, B. Robyns and M. M. Radulescu, "Control and Performance Evaluation of a Flywheel Energy Storage System Associated to a Variable-Speed Wind Generator," *IEEE Transactions on Industrial Electronics*, Vol. 53, No. 4, 2006, pp. 1074-1085.

[53] G. Cimuca, S. Breban, M. Radulescu, C. Saudemont, *et al.*, "Design and Control Strategies of an Induction-Machine-Based Flywheel Energy Storage System Associated to a Variable-Speed Wind Generator," *IEEE Transactions on Energy Conversion*, Vol. 25, No. 2, 2010, pp. 526-534.

[54] M. Subkhan and M. Komori, "New Concept for Flywheel Energy Storage System Using SMB and PMB," *IEEE Transactions on Applied Superconductivity*, Vol. 21, No. 3, 2011, pp. 1485-1488.

[55] J. Lee, S. Jeong, Y. H. Han and B. J. Park, "Concept of Cold Energy Storage for Superconducting Flywheel Energy Storage System," *IEEE Transactions on Applied Superconductivity*, Vol. 21, No. 3, 2011, pp. 2221-2224.

[56] T. Ichihara, K. Matsunaga, M. Kita, I. Hirabayashi, *et al.*, "Application of Superconducting Magnetic Bearings to a 10 kWh-Class Flywheel Energy Storage System," *IEEE Transactions on Applied Superconductivity*, Vol. 15, No. 2, 2005, pp. 2245-2248.

[57] N. Hamsic, A. Schmelter, A. Mohd, *et al.*, "Stabilising the Grid Voltage and Frequency in Isolated Power Systems Using a Flywheel Energy Storage System," *The Great Wall World Renewable Energy Forum*, Beijing, 23-27 October 2006, pp. 1-6.

[58] E. Ortjohann, N. Hamsic, A. Schmelter, *et al.*, "Increasing Renewable Energy Penetration in Isolated Grids Using a Flywheel Energy Storage System," *International Conference on Power Engineering, Energy and Electrical Drives*, Setubal, 12-14 April 2007, pp. 195-200.

[59] M. Lazarewicz and J. Judson, "Performance of First 20 MW Commercial Flywheel Frequency Regulation Plant," *ESA* 2011 *Annual Meeting*, Beacon Power Corporation, San Jose, June 2011.

[60] D. Sutanto and K. W. E. Cheng, "Superconducting Magnetic Energy Storage Systems for Power System Applications," *International Conference on Applied Superconductivity and Electromagnetic Devices*, Chengdu, 25-27 September 2009, pp. 377-380.

[61] IEEE Task Force and T&D Committee, "Detailed Modeling of Superconducting Magnetic Energy Storage (SMES) System," *IEEE Transactions on Power Delivery*, Vol. 21, No. 2, 2006, pp. 699-710.

[62] V. Karasik, K. Dixon, C. Weber, B. Batchelder, G. Campbell and P. Ribeiro, "SMES for Power Utility Applications: A Review of Technical and Cost Considerations," *IEEE Transactions on Applied Superconductivity*, Vol. 9, No. 2, 1999, pp. 541-546.

[63] M. G. Molina and P. E. Mercado, "Comparative Evaluation of Performance of a STATCOM and SSSC both Integrated with SMES for Controlling the Power System Frequency," 2004 *IEEE/PES Transmission and Distribution Conference and Exposition*: *Latin America*, São Paulo, 8-11 November 2004, pp. 535-541.

[64] M. G. Molina and P. E. Mercado, "New Energy Storage Devices for Applications on Frequency Control of the Power System Using FACTS Controllers," *Proceedings of X ERLAC*, Iguazú, 18-22 May 2003, pp. 1-6.

[65] Y. Tatusuta, S. Koso, H. Abe, M. Urata, H. Ohsaki, *et al.*, "Development of SMES for Power System Control," *IEEE Transactions on Applied Superconductivity*, Vol. 14, No. 2, 2004, pp. 693-698.

[66] S. Nomura, T. Shintomi, S. Akita, T. Nitta, R. Shimada and S. Meguro, "Technical and Cost Evaluation on SMES for Electric Power Compensation," *IEEE Transactions on Applied Superconductivity*, Vol. 20, No. 3, 2010, pp. 1373-1378.

[67] Q. Li, "Status and Future Prospect of SMES for Grid Applications," Advanced Microgrid Concepts and Technologies Workshops, Brookhaven National Laboratory, Washington DC, 2012.

[68] F. Crotogino, K. U. Mohmeyer and R. Scharf, "Huntorf CAES: More than 20 Years of Successful Operation," *The Solution Mining Research Institute Spring Meeting*, Orlando, 23-25 April 2001.

[69] Electric Power Research Institute, "Compressed Air En-

ergy Storage: 1994," EPRI Brochure BR-102936.

[70] W. Liu, Y. Yang, W. Zhang, G. Xu and Y. Wu, "A Novel Hybrid-Fuel Compressed Air Energy Storage System for China's Situation," *Proceedings of the International Conference of Efficiency, Cost, Optimization, Simulation Environmental Impact Energy Systems*, Perugia, 26-29 June 2012, pp. 1-16.

[71] C. Bullough, C. Gatzen, C. Jakiel, M. Koller, *et al.*, "Advanced Adiabatic Compressed Air Energy Storage for the Integration of Wind Energy," *Proceedings of European Wind Energy Conference, EWEC*, London, 22-25 November 2004, pp. 1-8.

[72] C. Linnemann and M. W. Coney, "The Isoengine: Realization of a High-Efficiency Power Cycle Based on Isothermal Compression," *International Journal of Energy Technology and Policy*, Vol. 3, No. 1-2, 2005, pp. 66-84.

[73] Ridge Energy Storage & Grid Services L. P., "The Economic Impact of CAES on Wind in TX, OK and NM," Texas State Energy Conservation Office, Austin, 2005.

[74] P. Breeze, "Power Generation Technologies," Elsevier, Boston, 2005.

[75] P. Brown, J. Lopes and M. Matos, "Optimization of Pumped Storage Capacity in an Isolated Power System with Large Renewable Penetration," *IEEE Transactions on Power Systems*, Vol. 23, No. 2, 2008, pp. 523-531.

[76] J. G. Gonzalez, R. D. L. Muela, L. Santos and A. Gonzalez, "Stochastic Joint Optimization of Wind Generation and Pumped-Storage Units in an Electricity Market," *IEEE Transactions on Power Systems*, Vol. 23, No. 2, 2008, pp. 460-468.

[77] A. Tuohy and M. O'Malley, "Impact of Pumped Storage on Power Systems with Increasing Wind Penetration," *IEEE Power & Energy Society General Meeting*, Calgary, 26-30 July 2009, pp. 1-8.

[78] W. Henson, "Optimal Battery/Ultracapacitor Storage Combination," *Journal of Power Sources*, Vol. 179, No. 1, 2008, pp. 417-423.

[79] H. Yoo, S. K. Sul, Y. Park and J. Jeong, "System Integration and Power Flow Management for a Series Hybrid Electric Vehicle Using Super-Capacitors and Batteries," *IEEE Transactions on Applied Superconductivity*, Vol. 44, No. 1, 2008, pp. 108-114.

[80] L. Shuai, K. A. Corzine and M. Ferdowsi, "A New Battery/Ultracapacitor Energy Storage System Design and Its Motor Drive Integration for Hybrid Electric Vehicles," *IEEE Transactions on Vehicular Technology*, Vol. 56, No. 4, 2007, pp. 1516-1523.

[81] M. B. Camara, H. Gualous, F. Gustin and A. Berthon, "Design and New Control of DC/DC Converters to Share Energy between Supercapacitors and Batteries in Hybrid Vehicles," *IEEE Transactions on Vehicular Technology*, Vol. 57, No. 5, 2008, pp. 2721-2735.

[82] O. Briat, J. M. Vinassa, W. Lajnef, S. Azzopardi and E. Woirgard, "Principle, Design and Experimental Validation of a Flywheel-Battery Hybrid Source for Heavy-Duty Electric Vehicles," *IET Electric Power Applications*, Vol. 1, No. 5, 2007, pp. 665-674.

[83] S. Lemofouet and A. Rufer, "A Hybrid Energy Storage System Based on Compressed Air and Supercapacitors with Maximum Efficiency Point Tracking (MEPT)," *IEEE Transactions on Industrial Electronics*, Vol. 53, No. 4, 2006, pp. 1105-1115.

[84] T. Ise, M. Kita and A. Taguchi, "A Hybrid Energy Storage with a SMES and Secondary Battery," *IEEE Transactions on Applied Superconductivity*, Vol. 15, No. 2, 2005, pp. 1915-1918.

Demand and Response in Smart Grids for Modern Power System

Muhammad Qamar Raza[1], Muhammad Usman Haider[1], S. Muhammad Ali[2], Muhammad Zeeshan Rashid[3], Farooq Sharif[1]

[1]Department of Electrical Engineering, The University of Faisalabad, Faisalabad, Pakistan; [2]Department of Electrical Engineering, COMSATS Institute of Information Technology, Abbottabad, Pakistan; [3]Department of Electrical Engineering, Islamia University of Bahawalpur, Bahawalpur, Pakistan.

ABSTRACT

Micro-grid plays a vital role in fulfilling the increasing demand by using distributed renewable energy resources. Demand and response technique can be broadly classified under the setup DR deployed (e.g. ISO's/RTO's). Demand response program can be implemented to improve power system quality, reliability and increasing demand. In modern power industry, strategic player can take more benefit from more emphasized DR study in terms of social benefit (uninterrupted power supply to consumers) and economy. This paper proposes the distributed micro-grid control and implemented control setup implemented demand response algorithm, which provides better power system reliability. This paper presents contingencies control demand and response for micro-grid. The main advantage of implementation of demand and response algorithms in Micro-grids provides reliable power supplies to consumers. The proposed micro-grid TCP/IP setup provides a chance to respond the contingencies to recover the shed to active condition. Micro-grid controller implements demand and response algorithm reasonable for managing the demand of the load and intelligent load scheme in case of blackout.

Keywords: Demand Response (DR); Micro-Grid (MG); Power System Reliability; Distributed Generation (DG); Renewable Energy Resources (RER); Power System Contingencies

1. Introduction

Power supply is the one basic requirement to operate the machine or devices. As technology advances, the need of power is increasing drastically. Centralized power systems for generation thus produces fluctuations at load side as demand is not being fulfilled properly and this basic need is increasing rapidly. Thus utilities feel a great pressure on them for an increasing generation [1] and transmission lines. Consumers are not satisfied fully with the present scenario of available power especially in developing countries. So, Distributed Generation is one of the basic means of generating power for a certain locality, which is environmentally friendly as it minimizes the emissions of hazardous gases. MG consists of power generators capable of power generation rang from several KW to a few MW using Renewable Energy Resources such as solar, wind, biomass etc. and can run on fossil fuels such as coal, gas, oil etc. Operation of Micro Grids can be performed in two basic modes. One form of operation is in autonomous mode and other form of operation is in nonautonomous mode.

In power system planning and operation, Demand Responses are one of the most important factors. Currently modern power system smart grids in new emerging technology using Renewable Energy Resources (RER) plays a vital role in fulfilling current energy demand. Currently different demand response distribution schemes are widely implemented and distribution voltage reduction is one of them. In this paper, demand response impact on smart grids is discussed and renewable energy technologies (bio-gas, solar cell, wind energy) is used to propose algorithm fulfill increasing demand of power. Basically micro-grid is a form of smart grids which provides power (normally in KW) to load controlled by the intelligent energy management system.

2. Objectives of Demand Response

- Decrees in the generation need, under this objective

sum of demand and losses are reduced in distribution and transmission in power system. Different factors effect on this objective like cost of generation, availability of energy resources, and overall system reliability.

- Increase the overall system reliability, under this objective maintain a balance between demand and supply of power to load in different time intervals.
- Elimination of overloads from power systems, under this objective of overflow of current is reduced by Advance Energy Management System (EMS).

3. Proposed Demand Response Micro-Grid Network

In this paper proposed demand and response algorithm can be implemented on non-autonomous micro-grid mode of operation. In non-autonomous mode which MG acts as a standby system with electrical utilities. These DG sets are installed with electric utility to provide strength [2,3] to the power plant by supporting its voltage level or by supporting power for reducing the peak demand during certain times. Stand-by power generating units with present electrical infrastructure can be used for micro-grids. MG can prove to be one of the best choices for installing it in a certain locality, provided that it costs low and also provides a certain acceptable level of reliability [4] for a fighting with present electrical utilities in the modern age. The trends of MG are enormously rising as they share some basic load from the utility. This basic trend is growing especially growing in developing countries where RER is sufficient to generate the demands of people alone as well as by sharing with utility. The basic and the foremost specialty of MG is that they can operate in Islanding mode [5] thus provides safety against troubling effects of faults.

3.1. Demand Response Power Flow Network

In proposed power and control Demand and Response power network show in **Figure 1**. Micro-Grid Central Controller (MGCC) is the link between Electrical utility and load through Local controller. Renewable Energy Resource (RER) and other small power production resources are connected to local Controller and supply the power to own connected load. Power is supplied to load from own power generation.

3.2. TCP/IP Control Network

In decentralized approach, DG source controller [6] is mainly responsible for power management. The DG source controller is called local controller and micro-grid central controller is connected to the local controller for power management from main utility. In **Figure 2**, intelligent TCP/IP controller network figure electrical utility is connected through Micro-Grid Central Controller (MGCC) and Local Controller (LC).

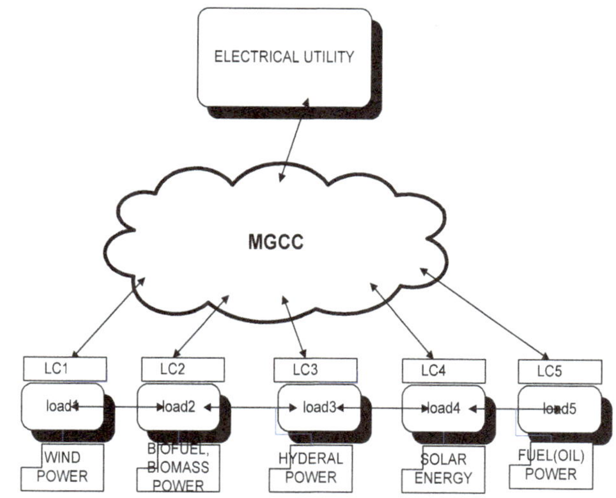

Figure 1. Distributed renewable energy resource micro-grid setup.

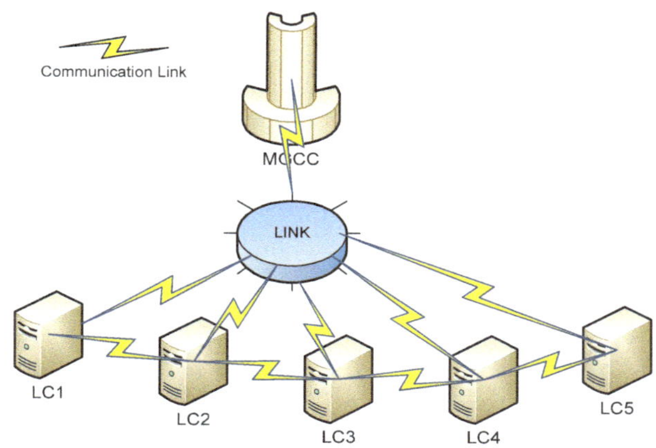

Figure 2. Micro-grid control network.

In peak hour of demand intelligent local controller which is connected by TCP/IP networking cable sends request to micro-grid central controller through TCP/IP communication protocol.

In **Figure 2** TCP/IP controller network intelligent MGCC checks the available power from utility that can supply to load through local controller and local controller control the power supply to the load. If power required to load is available power that can be supplied to load then MGCC send a signal to.

Local controller starts sending power to load through LC. If more power required than local controller send requests to other local controller to send the power. In case of power requirement is not fulfill then intelligent local controller.

Perform intelligent load shedding. The ultimate objective providing the smooth and reliable power supplied to load in different circumstances. Different types of "intelligent load shedding" [7-10] scheme have been discussed in literature for power management. In [11] describe the decentralized software control approach with intelligent agents the on FIPA software where the local controllers compute one another to optimize MGCC operation of the MG.

4. Demand Response Algorithm

The primary components of demand response algorithm show in the flow chart shown in **Figure 3**.

1) Micro-Grid Central Controllers (MGCC) checks the connected Local controller by sending connectivity command using TCP/IP protocol.

2) Micro-Grid Central Controller (MGCC) maintains the database of Local Controller (LC) connected load peak hour and normal hour power demand.

3) Local Controller (LC) compares the current demand and supply of power to load attached to the local controller (LC). If demand is greater than response power supply then local controller send a request to Micro-Grid Central Controller (MGCC) to send the required power and if demand is less than the supply then send an update request of a database of available power for MGCC.

4) If the Micro-Grid Central Controller (MGCC) can not fulfill the current demand of local controller connected load then MGCC performs intelligent load shedding.

5) MGCC performs intelligent load shedding by shed down the low priority local controller loads then 2nd least priority load and so on until the demand and supply is not balanced.

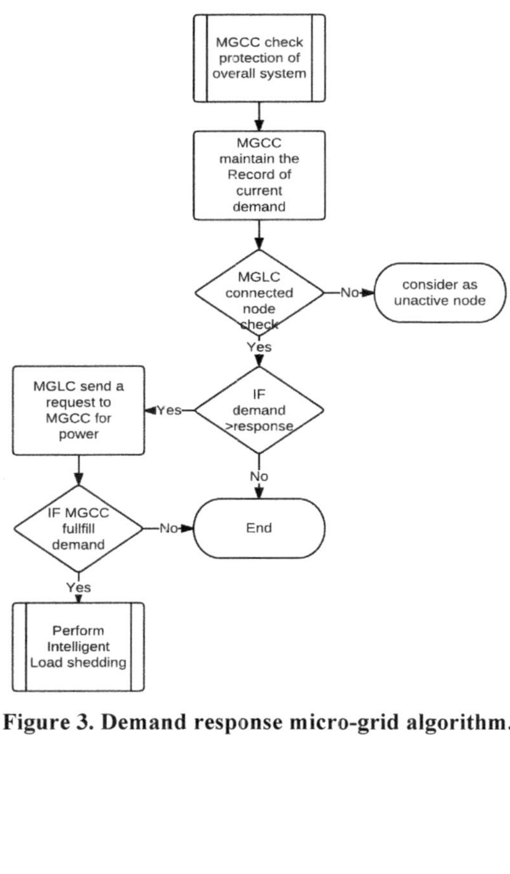

Figure 3. Demand response micro-grid algorithm.

System States

Here we use the notation P_{MGCC} is the power of micro-grid central controller power from connected from main utility and P_{LC} is power generated by the local controller from own connected renewable energy resources. P_{Demand} represents the power demand of connected load.

Case I:

$$P_{LC} > P_{\text{Demand}}$$

In this case local controller sends a request to main utility by using Micro grid central controller through the TCP/IP protocol for power supply.

Case II:

$$P_{LC} < P_{\text{Demand}}$$

If the power generated by the local controller than the power demand then Local controller sends a request to Micro grid central controller for required power. Then system equations as fallows

Case III:

$$P_{LC} + P_{MGCC} \geq P_{\text{Demand}}$$

In this case the micro-gird central controller will perform the intelligent load shedding with the help of Local controller.

5. Benefits of Demand Response in Micro-Grids

- The most promising feature of MG is the establishment of business industry that can worth millions and billions of dollars for providing backup power and support to the masses. With the well equipped setup it is capable of exporting the power to utilize and to some required residential areas as well as to some private sector. Thus, it provokes thirsty businessmen to invest money in this sector and can earn healthy profit. For example in a certain community of 10 - 15 houses there exists an MG setup and each owner of the house has to buy a membership of MG for backup power. So, in case of failure of delivery from utility MG provides its members quality power for certain time. Each member has to pay to MG owner who gains profit through exporting power to residential sector and in utility in times of need. In economics module of formulating the participation of Customer in Demand Response Program shows customer demand varies according to tariff rates and customer is benefited [12].
- The concept of demand response in micro-grids minimizes a lot of risks to electrical utilities. These risks

include fault occurring in central plant, in trans- mission lines, in MG. Thus, it adds a security feature to utility while operating in non-autonomous mode as it is capable of operating in an island mode. During heavy loading conditions MG can be subjected to intentional or emergency islanding. Other types of risks include load forecasting, prices of the fuel regularity issues etc. MG smaller units requires less project capital and lead time compared to large and heavy utility units. Demand Response is a useful tool at a time of high wholesale market price less power consumption [13].

- Rapid increment of power demand has made the need of Demand response more attractive for system operators and customer [14].

MG provides supporting and control services to utilize such as:

1) By the injection/absorption of reactive power for controlling the voltage of the grid.

2) Can act as spinning reserve.

3) Can provide network stability.

4) Avoid system blackouts.

5) By the control of regulation etc. [15].

6. Conclusion

In modern lifestyle, the demand for energy is increasing rapidly day by day and production is increasing with respect to demand increment. So in modern power systems, we must also emphasize the energy management system to fulfill the rapidly increasing demand of energy to the smooth run of industrial and domestic loads. Demand response in micro-grid is a new emerging technology which can play numerous and important rules in world's energy demand and effective load management technique provides an opportunity to overcome the energy imbalance. The demand response technique also provides opportunity to fulfill the demand of different connected loads in the peak and normal hour of consumption. Demand response in non-autonomous microgrid technique improves power quality which is beneficial to people and ensures the interpretable power supply to load. It fit for crucial places such as hospitals, universities, defense intuitions etc.

REFERENCES

[1] B. Renders, L. Degroote, J. Driesen and L. Vandevelde, "Profits of Power-Quality Improvement by Residential Distributed Generation," *42nd International Universities Power Engineering Conference*, 4-6 September 2007, pp. 377-381.

[2] P. A. Daly and J. Morrison, "Understanding the Potential Benefits of Distributed Generation on Power Delivery Systems," *Rural Electric Power Conference*, Little Rock, 29 April-1 May 2001, pp. 211-213.

[3] V. H. M. Quezada, J. R. Abbad and T. G. S. Roman, "Assessment of Energy Distribution Losses for Increasing Penetration of Distributed Generation," *IEEE Transactions on Power Systems*, Vol. 21, No. 2, 2006, pp. 533-540.

[4] P. M. Costa and M. A. Matos, "Reliability of Distributed Networks with Micro Grids," *IEEE Russia PowerTech*, St. Petersburg, 27-30 June 2005, pp. 1-7.

[5] J. O. Anduaga, S. Macedo and J. Ruela, "MICROGRIDS: Large Scale Integration of Micro-Generation to Low Voltage Grids," *Technical Report*, Micro-Grids Project Consortium, 2005.

[6] N. Hatziargyriou, A. Dimeas and A. Tsikalakis, "Hierarchical and Distributed Control Strategies for Micro-Grids," *International Journal of Distributed Energy Sources*, Vol. 1, No. 3, 2005.

[7] A. Maiorano, R. Sbrizzai, F. Torelli and M. Trovato, "Intelligent Load Shedding Schemes for Industrial Customers with Cogeneration Facilities," *Power Engineering Society* 1999 *Winter Meeting*, Vol. 2, No. 31, 1999, pp. 925-930.

[8] D. Novosel and R. L. King, "Using Artificial Neural Networks for Load Shedding to Alleviaie Overloaded Lines," *IEEE Transactions on Power Delivery*, Vol. 9, No. 1, 1994, pp. 425-433.

[9] L. J. Shih, W. J. Lee, J. C. Gu and Y. H. Moon, "Application of df/dt in Power System Protection and Its Implementation in Microcontroller Based Intelligent Load Shedding Relay," *Industrial and Commercial Power Systems Technical Conference*, Memphis, 6-9 May 1991, pp. 11-17.

[10] W. Lee and J. C. Gu, "A Microcomputer-Based Intelligent Load Shedding Relay," *IEEE Transactions on Power Delivery*, Vol. 4, Nov. 4, 1989, pp. 2018-2024.

[11] R. Lasseter, A. Akhil, C. Marnay, J. Stephens, J. Dagle, R. Guttromson, A. Meliopoulos, R. Yinger and J. Eto, "White Paper on Integration of Distributed Energy Resources. The CERTS Micro Grid Concept," *Consortium for Electric Reliability Technology Solutions* (*CERTS*), 2002, Technology Report LBNL-50829.

[12] H. Aalami, G. R. Yousefi and M. P. Moghadam, "Demand Response Model Considering EDRP and TOU Programs," *IEEE/PES Transmission and Distribution Conference & Exhibition*, Chicago, 21-24 April 2008, pp. 1-6.

[13] M. H. Albadi and E. F. El-Saadany, "Demand Response in Electricity Markets: An Overview," *IEEE Power Engineering Society General Meeting*, Tampa, 24-28 June 2007, pp. 1-5.

[14] US Department of Energy, "Benefits of Demand Response in Electricity Markets and Recommendations for Achieving Them," 2006.

[15] M. Qamaraza, S. M. Ali, A. Rasheed and Azam-ul-Asar, "Major Prospectus of Non-Autonomous Microgrid in Modern Power Systems," *Power Generation and Renewable Energy Technologies*, 2010.

The Interaction between the Large-Scale EVs and the Power Grid*

Yuan Huang, Junyong Liu, Xiaodong Shen, Tingting Dai

School of Electrical Engineering and Information, Sichuan University, Chengdu, China.

ABSTRACT

With the problem of global energy shortage and people's awareness of energy saving, electric vehicles receive worldwide attention from government to business. Then the load of the power grid will rapidly increase in a short term, and a series of effects will bring to the power grid operation, management, production and planning. With the large-scale penetration of electric vehicles and distributed energy gradually increased, if they can be effectively controlled and regulated, they can play the roles of load shifting, stabling intermittent renewable energy sources, providing emergency power supply and so on. Otherwise they may have a negative impact, which calls for a good interaction of electric vehicles and power grid. Analyzed the status of the current study on the interaction between the electric vehicles and the power grid, this paper builds the material basis, information architecture and the corresponding control method for the interaction from the aspect of the energy and information exchanging, and then discusses the key issues, which makes a useful exploration for the further research.

Keywords: Smart Grid; Power System; Electric Vehicles; Interaction; V2G

1. Introduction

With the deepening of the global energy crisis, advances in battery technology and energy saving requirements of the growing importance make electric vehicles worldwide widespread attention and promotion in recent years [1]. Electric vehicles as a terminal energy consumption of products, when their number is less, are not enough to have a significant impact on the grid, but when tens of thousands of vehicles are present these can move loads of randomness, intermittent structure and operation of the grid are bound to have a significant impact. The introduction of electric vehicles will fully reflect the smart grid to a large number of "plug and play" of distributed power generation centralized added features to improve the smart grid to become the embodiment of core values [2].

Research has been carried out electric vehicle charging impact on the grid, the demand charge, the charger, charge pile model, charging station monitoring system, charging system standards, network planning and coordination of resources and utilization [3-6]. Studies have shown that large-scale electric vehicle charging at peak load will lead to "peak plus peak" phenomenon, increasing the difficulty of peaking power and backup power' increase [7,8]; large-scale electric vehicle charging loads will change the load distribution network structure, the existing distribution system planning criteria and rules of operation are facing new problems [9]; electric vehicle charging network operation increases the randomness of the uncertainty factor. Various studies have made on vehicle to grid (V2G). V2G is used in the distributed large-scale electric vehicle as an energy storage system. It makes charging stations for electric vehicles or energy storage unit as the interaction with the grid, considering the access and participation in other clean energy sources such as FM [10].

The development of electric vehicles is closely related to electricity, charging facilities penetration, and large scale electric vehicle charging is to bring new load growth, transmission, distribution facilities planning and other issues. The interaction of electric vehicles and grid power is through the interaction of supply and demand and adjust to achieve the rational use of resources. The study of electric cars and charging stations and grid (generation, transmission and distribution, with use) will reveal the interaction kinds of relationships. This includes the use

*The Young Scientific Research Foundation of Sichuan University (No. 2011SCU11062).

of interactive communication, the interface system integration technology, for electric vehicles end-user devices and information integration to achieve the optimal grid resources rational use of the whole process.

This paper analyzes the energy and information in electric vehicles and grid interaction between the process of interaction, and it also summarizes the present situation of two-way interaction of the structural framework for the realization of the process proposed research questions and analysis of relevant technologies to achieve the reduced mass electric vehicle charging and discharging of the power grids, reducing the power grid operation and investment costs, resulting in good economic and social benefits.

2. Interact among the Electric Vehicle, the Charging Stations and the Smart Grid

[11] points out that the interaction is the essence of the grid alternative energy, compatible use and economic interaction. The interaction among the electric vehicles, charging stations and smart grid is based on smart grid technology, by data through the network to read the real-time, high-speed, bi-directional transmission. They make end-users of energy and information with the grid interaction. Interaction of electric vehicles and smart grid is shown in **Figure 1**.

Interaction with the electric car is the essence of smart grid energy to achieve the optimal use of the optimal allocation of social resources; it can be shown in **Figure 2** of its three levels of interaction analysis.

2.1. The Physical Architecture of Interact among the Electric Vehicles, Charging Stations and Smart Grid

The application of the Internet of Things using in electric vehicles and charging stations is the electric vehicles and charging stations to interact with the physical basis of the smart grid, which is an extension and expansion of the Internet [12], it will extend and expand the client to electric cars, charging facilities exchange of information between the grid and communication, and it covers things electric vehicles and battery charging facilities, things, information center of things. In the Internet of Things, Radio Frequency Identification (RFID) tags stores on electric vehicles with the interoperability specifications and information, through wireless data communication network automatically collecting them charging facilities management information systems for electric vehicle identification. Then through open computer network to a dispatching production management system, information exchange and sharing of consumer electric vehicles have achieved as a "transparent" administration.

Internet of things electric vehicles and batteries use the Global Positioning System (GPS) and Internet to do electric vehicles in real time and non real-time information collection, collation and transmission. It is shown in **Figure 3**. Car Smart electric vehicle information terminal is the harmonization of export, which covers the battery-related information, such as the current remaining battery electric vehicles and so on, through the wireless network

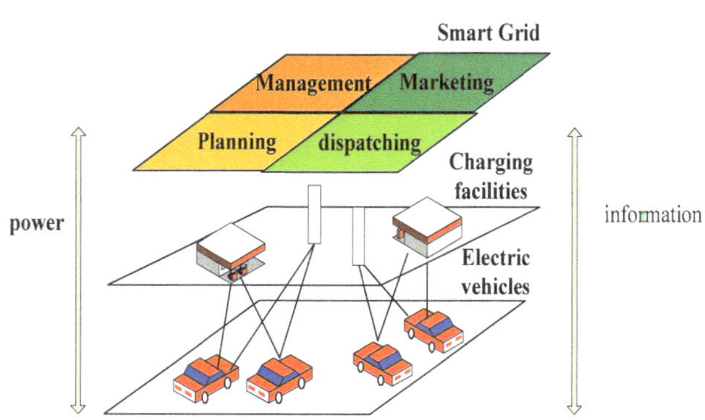

Figure 1. The sketch map of the interaction between the EVs and the smart grid.

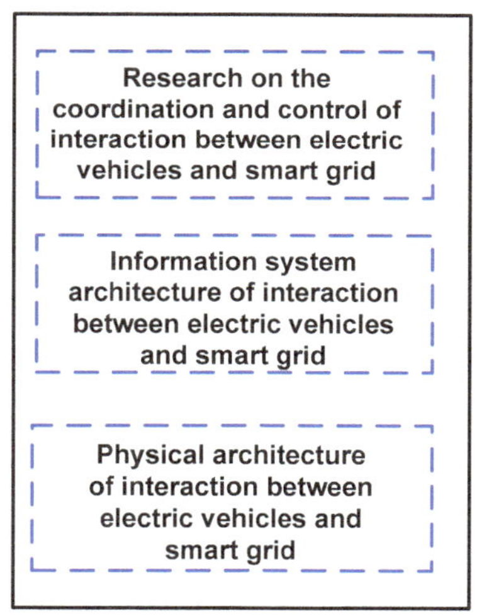

Figure 2. The hierarchical structure of the interaction between the EVs and the smart grid.

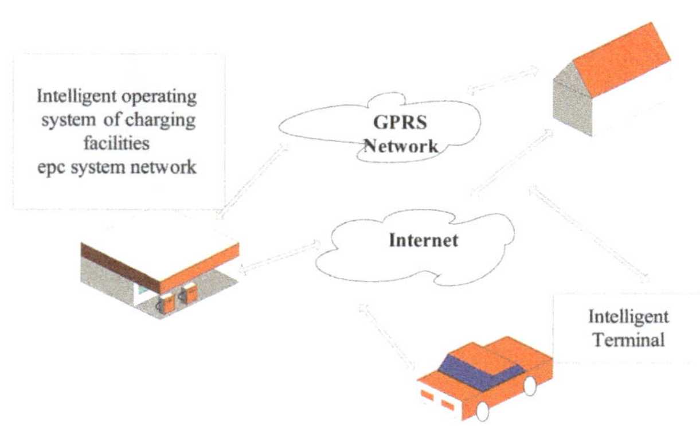

Figure 3. The internet of things framework of EVs and charging station.

and exchange data management center running in the background, through the RFID identification of electric vehicles (battery) and the state location for settlement, management and control for advanced data services, computing support. Things Smart charging facilities operating system is on the one hand the work of running state-aware devices, such as battery charge time remaining, etc., on the other record electricity consumption information and security monitoring and other information.

2.2. Electric Vehicles and Smart Grid Interactive Information System Architecture

Smart Grid will connect more devices, including a variety of intelligent sensors and control components, power equipment, which need to improve the grid digitization, and deepen the generation, transmission, substation, distribution, consumption and scheduling all aspects of data collection, transmission, storage and use of [13]. It achieved through data center centralized collection of information throughout the network, and the use of high-speed communication networks operating equipment online state detection and transmission of information, which constitutes the interaction of electric vehicles and smart grid foundation. In the implementation of the charging stations on the main dispatch operations through information management system (**Figure 4**) and connection dispatching communication and information systems, they must follow the existing IEC 61970, IEC 61968, IEC 61850 and Common Information Model (CIM) standard series, unified naming convention, the public power grid and the data model standard unified modeling [14].

The charging station side, on the one hand response is for the electric vehicle charging management services, on the other hand need to assume management responsibility for electric vehicles. It needs to interact with power systems control centers all state variables, measurement, electric energy, etc., for the operation of the data should be scheduled through the charging station in the background and operational information management system for data collection, sorting, statistics, classification, analysis of the formation of the back data sources.

2.3. Coordination Controls for the Interaction of Electric Vehicles and Smart Grid

Electric vehicles and smart grid interactive coordination control are aimed at considering a large number of random, intermittent load cases, improving the power FM peaking, load capacity of peak load shifting, and promote the economical use of energy, to improve the overall system energy utilization efficiency. This will involve charging stations and grid load characteristics to optimize the coordination of a new model to run, electric vehicle battery energy density, fast and slow charging and the power of technology and tie; plug-in electric vehicles and coordinated operation of the control grid technology and technical standards implementation; charging facilities and their optimal allocation of the interaction between the grid.

3. The Key Issues of Electric Vehicles and Smart Grid Interaction

3.1. Electric Vehicle Charging and Discharging Characteristics of the Loads Analysis

Studies [15,16] from the load management discuss the use of low power for electric vehicle charging time, lower peak-valley difference, change load patterns, which will directly affect the production of grid operation and management. In [17,18], the studies can be seen when hun-

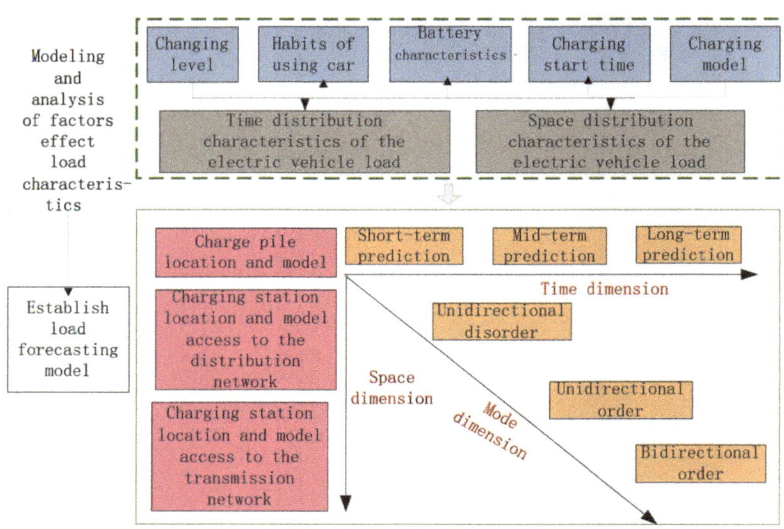

Figure 4. The load forecasting model of electric vehicles.

dreds of millions of electric vehicle charging will increase the disorder in the peak power load, need to increase the system's peak power and peak-shaving backup. They may to existing power equipment operation bring overload, voltage the limit, power quality and other problems that may affect safe and reliable power system operation. [19] studies have shown that electric vehicle charging will bring net loss of disorder increases, impact grid dispatch and economy.

Characteristics of the electric vehicle charging and discharging are the foundation research on the load electric vehicles and grid interaction. Load of electric vehicles is the same penetration of electric vehicles, charging; charge level, battery capacity, electric vehicles the user's habits, such as charging start time closely related. This need based on deterministic and non-deterministic analysis methods combined with artificial intelligence and predictive tools in order to form accurate forecasts of electric vehicle load.

3.2. Electric Vehicle Charging Facilities Planning

The construction of electric vehicle charging stations is the popular premise and foundation for electric vehicles, and in expanding electricity market demand, improve the efficiency of energy supply network is widely used in electric vehicles, one of the necessary conditions. Electric vehicle charging station planning must take full account of the grid structure, transportation structure, and behavioral characteristics of electric vehicle charging, charging infrastructure investment and other aspects, so it is a multi-objective, multi-variable nonlinear optimization problem.

[6] has been proposed based on the amount of electric car ownership, user type, driving range, charging demand planning stages of charging facilities for the program; [20] considers the promotion of electric vehicles in the early stages, through the convenient charging method for electric vehicle development, and establishes of considering the cost of different locations, charge by charge investors the minimum investment and charging stations to minimize the cost of the multi-objective optimization model, the dynamic constraints of time with the heuristic algorithm; the literature [21] analysis electric vehicles charging behavior of stochastic dynamic characteristics. A line of electric vehicle charging service system mathematical model, optimal charging facilities is established using cost model to guide the user at the right time to recharge, combing with the role of TOU pricing system, effective to reduce the peak power difference and improve the power grid load factor; [22] consider the flow due to the load and charge the uneven distribution. The weighted Voronoi diagram method is proposed to determine the priority of charging stations planning to form a charging station network planning combination of traffic.

Therefore, the combination of electric vehicles and electric vehicle development trend of various spatial and temporal characteristics of the load, use of optimization algorithms to achieve the parties to maximize return on investment, to meet the demand to maximize the way to achieve charging facilities planning stages of addressing still has a relatively constant volume high research value.

3.3. V2G

The concept of V2G is first put forward by the Kempton and Letendre [23], and it means the electrical energy storage onboard Electric Vehicles (EV). It refers to the smart grid technology support. It will be in a suspended state for large-scale electric vehicles focused on the latter distributed as a removable storage unit, in a controlled state to achieve the energy grid, two-way information exchange. Electric vehicles can not only gain energy from the grid, but also in the running to meet the needs of users of electric vehicles under the premise of the remaining power control back to the grid, load characteristics for grid regulation, frequency response, standby power, home energy supply, can provide support services for the system to improve power grid reliability [24]. It is the interaction with the grid of electric vehicles as a typical manifestation of the way.

[25,26] predict the scale of development of electric vehicles. Electric vehicles are calculated as energy storage devices to participate in the potential of V2G active output; its findings caused widespread concern at home and abroad. there are studies to V2G as a special business expanding installation [27], and made charging station as close to the 110 kV substation standby power, access to the ring to act as focal point for short-term backup power, and especially important as the load parallel operation and emergency power and other types of more typical electric vehicle charging facilities connected to the grid of the program. [28] considers an unpredictability of wind power access to the Danish, puts forward the set electric car battery energy storage agency V2G balance system energy supply and demand. It is used to simulate the long-term power system dynamic simulation of the process, resulting in that V2G than traditional generators have fast response characteristics. [29] analyzes that in the autonomous micro-network systems, the contact of electric vehicles automatically charge and discharge and the system frequency, and large-scale simulation of the electric vehicle focus on the effect of load frequency control. [30] proposes a centralized frequency adjustment V2G electric vehicle optimization strategy. There is a consumer of electricity and electric vehicles to provide adjustment services to two aspects of the role of market prices, taking into account the factors and the battery capacity constraints in the case established the largest gains the

model.

Existing research usually will put electric vehicles as a constant power load or use electric automobile storage characteristic its polymerization into distributed power supply. And thus it provide for the FM and standby power systems and other auxiliary services, but due to technical limitations of the grid, there no large-scale implementation of V2G, it needs from the electricity demand, supply, energy structure, price, market trading mechanism, load characteristics and other aspects of the potential impact of V2G on the grid. At present, domestic and international research in this area is still in its in- fancy.

3.4. Advanced Interactive between Electric Vehicles and Clean Energy and Grid Smart Control

[18] based on the traditional model of economic dispatch of electric vehicles considers the uncertainty of wind power, with the probability analysis method taking into account their expectations of electricity and contribute to build a power system model of stochastic economic dispatch. The literature [31] proposes that electric vehicle battery as distributed energy storage devices to guide makes it match with the wind power output in order to eliminate excess wind power output, and form electric vehicle charging and coordination scheduling wind power program. [7] establishes a distribution network to minimize network losses and voltage offset for the best goal of the electric vehicle charging policy model, in the model taking into account the uncertainty of load forecasting and trend of constraints. [32] considers in the current model of electric vehicles and uncertainty of wind power. The literature [33] analyzes the distribution of electric cars and solar energy features, complementary analysis.

Electric vehicles and clean energy and smart grid control and high-level interaction in the entire power system, transmission and distribution are reflected on the use side, it should analysis the impace of the grid with electric vehicles and clean energy. It is shown in **Figure 5**. This process is and prices are intertwined to form complex technical and economic issues. Electric vehicle charging and discharging facilities in coordination with the power is needed for the operation of mutual inhibition between the adverse effects, to achieve the optimal use of resources and demand the greatest satisfaction. Because the different characteristics of power generation equipment with vastly different electricity load matching and multi-objective to achieve these higher-order nonlinear dynamic control problems, the traditional research questions adds a great deal of randomness, so the problem solving is more difficult and complicated.

4. Conclusion

The development of large-scale electricity electric vehicle development offers new opportunities for power and energy storage in the face of large-scale distributed network of electric vehicles can achieve a new "load" on the control of the new challenges. That presented in this paper for the interaction of large size of the electric cars

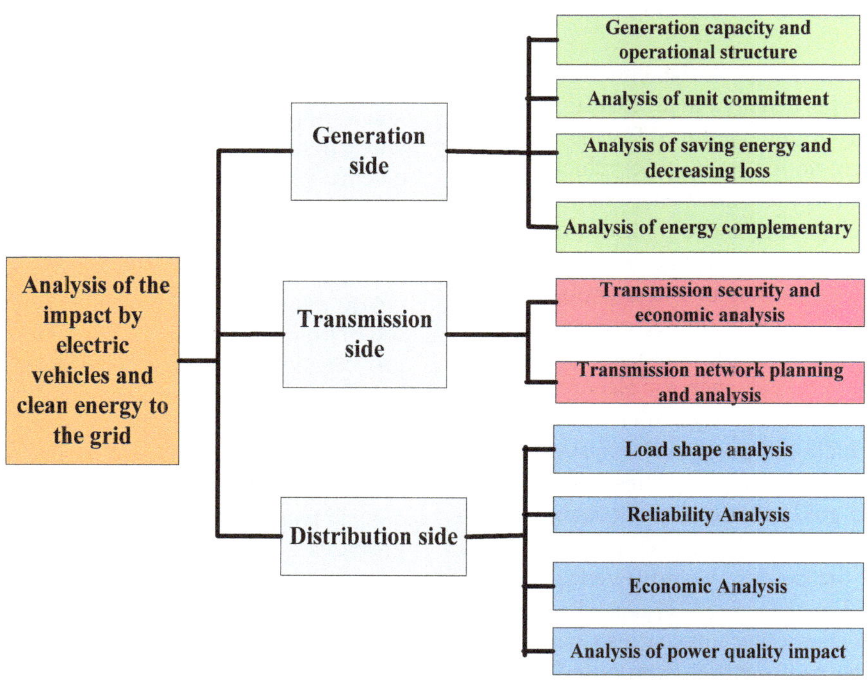

Figure 5. Evaluation of the impact of load includes plug-in electric vehicle and clear energy on power system operations.

and grid build the energy and information exchange program, providing for further exploration of new ideas.

REFERENCES

[1] W. Zhang, B. Wu and W. F. Li, "Discussion on Development Trend of Battery Electric Vehicles in China and Its Energy Supply Mode," *Power System Technology*, Vol. 33, No. 4, 2009, pp. 1-5.

[2] Y. X. Yu and W. P. Luan, "Power System and Clean Energy," *Smart Grid*, Vol. 25, No. 1, 2009, pp. 7-11. (in Chinese)

[3] O. Tremblay, L. A. Dessaint and A. I. Dekkiche, "A Generic Battery Model for the Dynamic Simulation of Hybrid Electric Vehicles," *Vehicle Power and Propulsion Conference*, Arlington, 9-12 September 2007, pp. 284-289.

[4] S. D. Jenkins, J. R. Rossmaier and M. Ferdowsi, "Utilization and Effect of Plug-In Hybrid Electric Vehicles in the United States," *Vehicle Power and Propulsion Conference*, Harbin, 3-5 September 2008, pp. 1-5.

[5] J. H. Zhao, F. S. Wen and Y. S. Xue, "Power System Stochastic Economic Dispatch Considering Uncertain Outputs from Plug-In Electric Vehicles and Wind Generators," *Automation of Electric Power System*, Vol. 34, No. 20, 2010, pp. 22-29.

[6] E.-S. Li and W. M. Wu, "Influence and Countermeasure of Electric Vehicle Battery Charging to Power Systems," *East China Electric Power*, Vol. 38, No. 1, 2010, pp. 0109-0113.

[7] K. Clement-Nyns, E. Haesen and J. Driesen, "The Impact of Charging Plug-In Hybrid Electric Vehicles on a Residential Distribution Grid," *IEEE Transactions on Power Systems*, Vol. 25, No. 1, 2010, pp. 371-380.

[8] S. Acha, T. C. Green and N. Shah, "Effects of Optimised Plug—In Hybrid Vehicle Charging Strategies on Electric Distribution Network Losses," *Transmission and Distribution Conference and Exposition*, New Orleans, 19-22 April 2010, pp. 1-6.

[9] L. T. Tian, S. L. Shi and Z. Jia, "A Statistical Model for Charging Power Demand of Electric Vehicles," *Power System Technology*, Vol. 4, No. 11, 2010, pp. 126-130.

[10] X. L. Yang, "The Development Trend and Foreground of the Electric Vehicle," *Auto Mobile Science & Technology*, Vol. 6, 2007, pp. 10-13.

[11] L. Q. Chen and D. S. Cui, "China New Energy Vehicle Policy," *Auto Time*, Vol. 9, 2011, pp. 21-29.

[12] P. C. Sun, "The Current Situation and Development Trend of the Electric Vehicle," *Scientific Chinese*, Vol. 8, 2006, pp. 44-47.

[13] L. Li and Q. B. Liu, "Study on the Influence on the EV's Usage to the Load Curve," *Electrical Machinery Technology*, Vol. 1, 2000, pp. 37-39.

[14] W. Kempton, J. Tomic and S. Letendre, "Vehicle-to-Grid Power: Battery, Hybrid, and Fuel Cell Vehicles as Resources for Distributed Electric Power in California," *Institute of Transportation Studies Report*, 2001.

[15] W. Jason and P. Lincoln, "Impact of Plug-Hybrid Electric Vehicles on California's Electricity Grid," North Carolina: Duke University, 2009.

[16] W. Kempton and J. Tomic, "Vehicle-to-Grid Power Fundamental: Calculating Capacity and Net Revenue," *Journal of Power Sources*, Vol. 144, No. 1, 2005, pp. 268-279.

[17] W. Kempton and J. Tomic, "Vehicle-to-Grid Power Implementation: From Stabilizing the Grid to Supporting Large-Scale Renewable Energy," *Journal of Power Sources*, Vol. 144, No. 1, 2005, pp. 280-294.

[18] G. Christophe and G. George, "A Conceptual Framework for the Vehicle-to-Grid (V2G) Implementation," *Energy Policy*, Vol. 37, No. 11, 2009, pp. 4379-4390.

[19] W. Kempton, V. Udo and K. Huber, "A Test of Vehicle-to-Grid (V2G) for Energy Storage and Frequency Regulation in the PJM System," University of Delaware, Delaware, 2008.

[20] D. Larry and H. Jessica, "A New Car, A New Grid," *IEEE Power & Energy Magazine*, Vol. 8, No. 2, 2010, pp. 55-61.

[21] S. W. Hadley and A. A. Tsvetkova, "A Potential Impacts of Plug-In Hybrid Electric Vehicles on Regional Power Generation," *The Electricity Journal*, Vol. 22, No. 10, 2009, pp. 56-68.

[22] W. H. Stanton, "Impact of Plug-In Hybrid Vehicles on the Electric Grid," Oak Ridge National Laboratory, Tennessee, 2006.

[23] K. Parks, P. Denholm and T. Markel, "Costs and Emissions Associated with Plug-In Hybrid Electric Vehicle Charging in the Xcel Energy Colorado Service Territory," National Renewable Energy Laboratory, Colorado, 2007.

[24] A. W. Rashid, D. G. Matthias and B. Michael, "Plug-in Hybrid Electric Vehicles and Smart Grid: Investigations Based on a Micro-Simulation," *The 12th International Conference of the International Association for Travel Behaviour Research*, Jaipur, 2009.

[25] M. Rowand, "The Electricity Utility-Business Case," *Plug-In Conference*, San Jose, 2009.

[26] J. Taylor, A. Maitra and M. Alexander, "Evaluation of the Impact of Plug-In Electric Vehicle Loading on Distribution System Operations," *Power & Energy Society General Meeting*, Calgary, 26-30 July 2009, pp. 1-6.

[27] E. S. Li and W. M. Wu. "Influence and Countermeasure of Electric Vehicle Battery Charging to Power Systems," *East China Electric Power*, Vol. 38, No. 11, 2010, pp. 109-113.

[28] Y. J. Chen, "Study on the Influence on Power Quality By Electric Vehicle Charging Equipment," *Hubei Electric Power*, Vol. 33, No. 6, 2009, pp. 48-50.

[29] J. G. Kang, Z. L. Wei and D. M. Cheng, "Research on Electric Vehicle Charging Mode and Charging Stations Construction," *Power Demand Side Management*, Vol. 11,

No. 5, 2009, pp. 64-66.

[30] L. Steven, "Plug-In Hybrid Electric Vehicles and the Vermont Grid: A Scoping Analysis," University of Vermont Transportation Center, Vermont, 2007.

[31] M. K. Meyers, K. Schneider and R. Pratt, "Impacts Assessment of Plug-In Hybrid Vehicles on Electric Utilities and Regional US Power Grids Part 1: Technical Analysis," Pacific Northwest National Laboratory, Washington DC, 2007.

[32] P. Denholm and W. Short, "An Evaluation of Utility System Impacts and Benefits of Optimally Dispatched Plug-In Hybrid Electric Vehicles," National Renewable Energy Laboratory, Colorado, 2007.

[33] Y. Gurkaynak and A. Khaligh. "A Control and Power Management of a Grid Connected Residential Photovoltaic System with Plug-In Hybrid Electric Vehicle (PHEV) Load," *Applied Power Electronics Conference and Exposition*, Washington DC, 15-19 February 2009, pp. 2086-2091.

Modeling and Simulation of a Renewable Energy Based-Medical Herb Dryer

Abdel-Shafy A. Nafeh, Emad Ahmed Sweelem, Faten Hosny Fahmy

PV Cells Department, Electronics Research Institute, Cairo, Egypt.

ABSTRACT

The medical herbs should be dried directly after harvesting, otherwise, insects and fungi spoilage them. Conventional drying methods such as open sun drying and conventional fuel dryers are not suitable, since they contaminate the herbs, decrease the drying efficiency and at the same time increase the drying cost. Therefore, the recent trend is toward the harnessing the renewable energy to dry the medical herbs. This paper presents a new controlled drying system, which uses a solar collector and an electric heater to heat the drying air, and a PV-WIND hybrid system to supply the required electric energy to the dryer electric load. This paper also presents the dynamic modeling, simulation, and control of the suggested thermal system (*i.e.*, the solar thermal system and the electrical heater). Moreover, this paper, exhibits the different results of the thermal system, to check the effectiveness of this system that fulfills the requirements of the drying operation. The system results are found to satisfy the ultimate goals of the drying operation.

Keywords: Medical Herb Dryer; Solar Thermal System; Renewable Energy; Dynamic Modeling

1. Introduction

In most countries, enormous of food losses have resulted from spillage, contamination, attack by inserts/rodents/birds and deterioration during the storage period. There is a number of factors, which are responsible for the post harvest losses, such as system of harvesting, processing, storage, handling, and marketing. Drying of product is one of the important post harvest processes and it has enough potential to reduce the post harvest losses, and to prevent spoilage of the product in storage drastically. Moreover, good drying technique can enhance the quality of the product significantly [1]. Therefore, the drying of medical herbs must be accomplished as soon as possible after harvesting, to increase the quality of the herb and to prevent the expected contamination and losses caused by infestation of rodents, birds, insects, and fungi which thrive in moist conditions [1,2].

Traditional open sun drying, where the product to be dried is exposed directly to the sun, has many disadvantages such as degradation by wind-blown debris, rain, insect infestation, human and animal interference which will result in contamination of the product. Also, it extends the drying time significantly. High-speed dryers, on the other hand, use high-cost conventional energy sources for their operation. Moreover, they may not be reliable and environmentally safe. Instead these dryers may be operated by using the renewable energy sources, such as photovoltaic (PV) modules, solar collectors and wind turbines etc.

In this paper a new controlled herb dryer system is developed. This dryer uses a flat plate solar collector to heat the drying air, and a PV-WIND hybrid system to feed the electrical energy to the dryer electric load.

2. The Suggested Herb Drying System

The configuration of the suggested medicinal herb dryer system is shown in **Figure 1**. The diagram consists of three main systems: the solar thermal system, the dryer, and the renewable energy system. The main function of the solar thermal system is to supply the inlet air to the dryer with the required solar thermal energy that is suitable to accomplish the drying process. While, that of the renewable energy system is to feed the electrical load of the dryer with the required electrical energy, to ensure good and continuous drying operation [3].

3. Modeling of the Thermal System

In this section, the energy balance principle is utilized to develop the complete dynamic modeling of the compo-

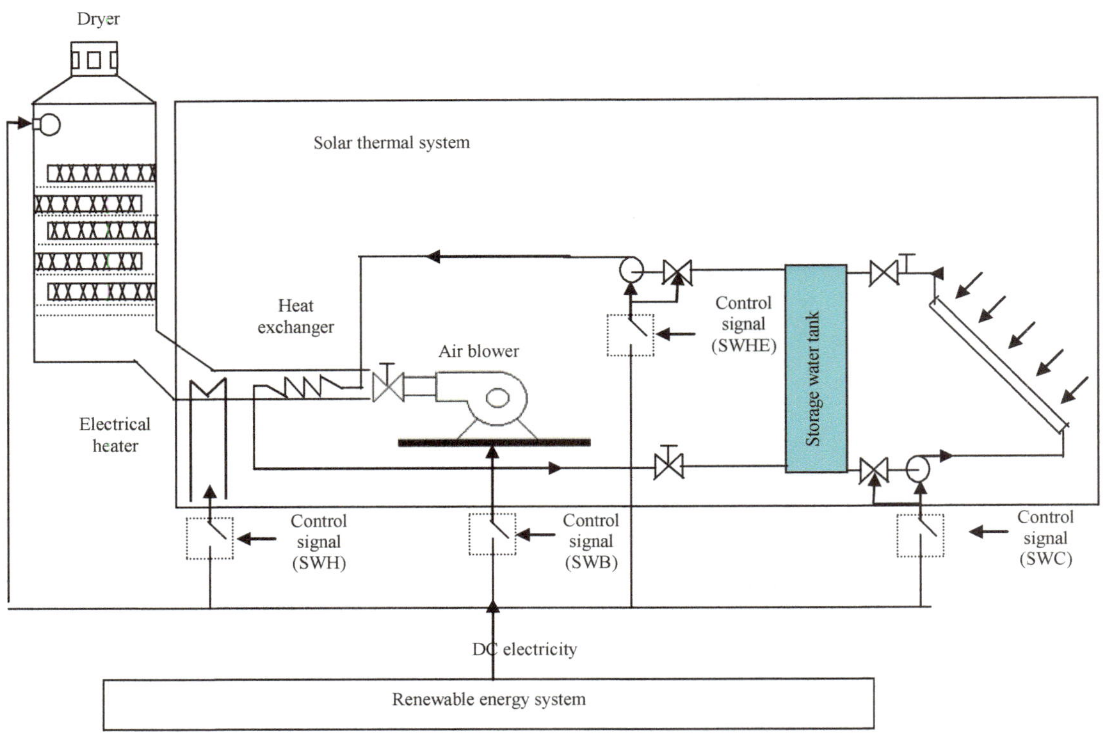

Figure 1. The medical herb dryer system.

nents of the suggested solar thermal system and that of the electrical heater.

3.1. Modeling of the Solar Collector

A flat plate solar collector type is used, to absorb the solar energy from the sun and transfers it into the water that flows inside its parallel tubes, which are attached to a black metal plate. This type of collectors works on both the direct and diffuse components of the solar radiation. Therefore, it stills function when the direct component is cut off by the effect of clouds [4,5].

The water in the collector will absorb the solar energy with the rate

$$\dot{Q}_{abs} = (\tau\alpha)_{eff} A_c \text{Rad} \tag{1}$$

and at the same time will loss a certain amount of heat energy to the environment, $\dot{Q}_{amb}$, given by

$$\dot{Q}_{amb} = \bar{U}_c A_c (T_{cout} - T_{air}) \tag{2}$$

where

τ transmissivity of the collector (fraction of incident energy transmitted);

α absorbitivity of the collector (fraction of insolation);

$(\tau\alpha)_{eff}$ effective absorbance of collector's cover (=0.8);

A_c collector's area (=2 m^2);

Rad solar illumination

$\bar{U}_c$ overall thermal loss coefficient (=8.38 W/m^2 °C);

T_{cout} water temperature at the collector outlet;

T_{air} ambient temperature.

Thus, an energy balance [2,4] on the collector (**Figure 2**) will lead to the following differential equation

$$\frac{dT_{cout}}{dt} = -\frac{\dot{m}_{cw}}{\rho_w V_{cw}}(T_{cout} - T_{cin}) + \frac{\dot{Q}_{abs} - \dot{Q}_{amb}}{\rho_w V_{cw} C_{pw}} \tag{3}$$

where T_{cin} water temperature at the collector inlet;

$\dot{m}_{cw}$ mass flow rate of water in the collector;

C_{pw} specific heat of water;

ρ_w density of water;

V_{cw} volume of water in the collector.

3.2. Modeling of the Storage Tank

Due to the stratified nature of the storage water tanks, the used tank is considered to be composed of five identical layers, as shown in **Figure 3**. The top layer (*i.e.*, layer #1) is the highest temperature layer, while the bottom layer (*i.e.*, layer #5) is the lowest temperature layer. The outlet water from the collector enters the tank at the top layer

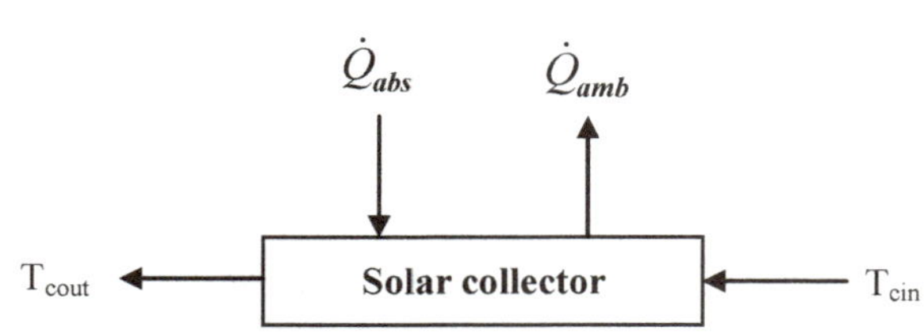

Figure 2. Energy balance on the collector.

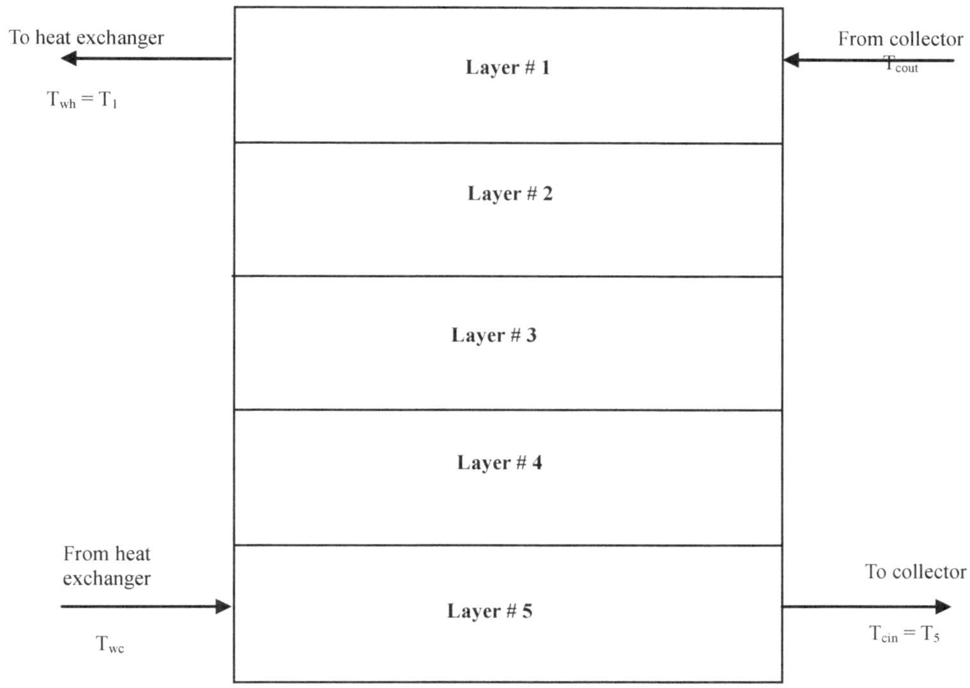

Figure 3. Stratified water storage tank.

with a temperature T_{cout}, and the water leaves the tank to the collector from the bottom layer (*i.e.*, the water temperature at the collector inlet T_{cin} is nearly the same as that of the bottom layer T_5). At the same time, the water outputs from the top layer of the tank at a temperature $T_{wh} = T_1$, to feed the heat exchanger side. Also, the water returns from the load side to the bottom layer of the tank with temperature T_{wc}.

Generally, the circulation of water in the collector side increases the thermal energy of the tank, while the circulation of water in the load side may increase or decrease the stored energy of the tank. Therefore, an energy balance [2,3] on a general layer (m) of the tank (**Figure 4**) yields

$$\frac{dT_m}{dt} = \frac{\dot{Q}_{cm}}{m_m C_{pw}} + \frac{\dot{Q}_{HEm}}{m_m C_{pw}} - \frac{\dot{Q}_{Loss}}{m_m C_{pw}} \quad (4)$$

where

$$\dot{Q}_{cm} = \begin{cases} \dot{m}_{cw} C_{pw} \left(T_{m-1} - T_m\right), & T_{cout} > T_1 \\ \dot{m}_{cw} C_{pw} \left(T_{cout} - T_m\right), & T_{m-1} > T_{cout} > T_m \end{cases}$$

$$\dot{Q}_{HEm} = \begin{cases} \dot{m}_{EHw} C_{pw} \left(T_{m+1} - T_m\right), & T_{wc} < T_5 \\ \dot{m}_{HEw} C_{pw} \left(T_{wc} - T_m\right), & T_{m+1} < T_{wc} < T_m \end{cases}$$

and

$$\dot{Q}_{Loss} = \bar{U}_{st} A_m \left(T_m - T_{air}\right)$$

The symbols of Equation (4) are defined as

T_m temperature of layer m $\left(m = 1, 2, \cdots, 5\right)$;

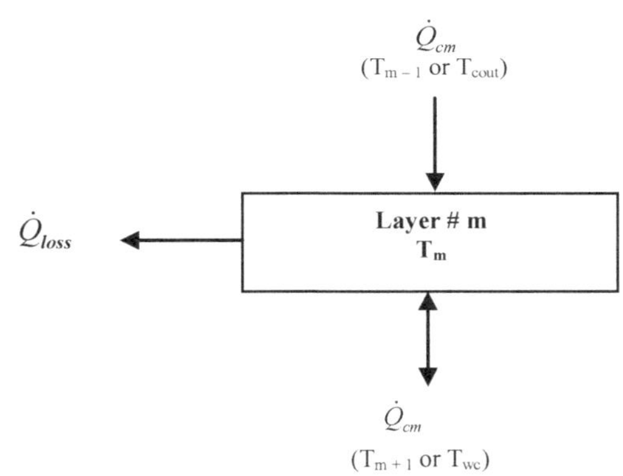

Figure 4. Energy balance on layer *#m*.

T_{wc} outlet temperature of water from the heat exchanger side;

m_m mass of water in layer m;

$\dot{m}_{HEw}$ mass flow rate of water in the heat exchanger side;

A_m surface area of layer m;

$\bar{U}_{st}$ overall loss coefficient for the exposed storage surface (=10.08 W/m^2 °C).

3.3. Modeling of the Heat Exchanger

The used heat exchanger is parallel flow water to air type, which consists of two concentric pipes as shown in **Figure 5(a)**. The hot fluid (*i.e.*, the water) passes inside the smaller diameter pipe, while the cold fluid (the air) passes through the space between the two pipes. It is

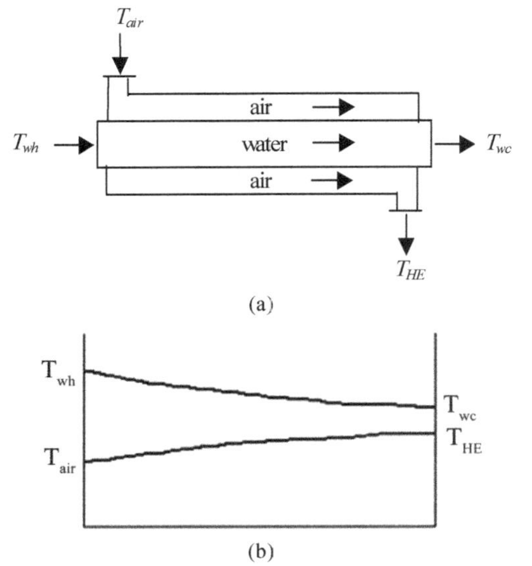

Figure 5. Parallel-flow heat exchanger: (a) Schematic diagram; (b) Temperature distribution along tube axis.

necessary to note here that the two fluids are co-flow. Where, during the co-flow of the two fluids along the heat exchanger length, the hot fluid transfers heat through the shell of the smaller diameter pipe to the cold fluid. Therefore, a temperature drop will occur in the hot fluid continuously and at the same time a temperature rise will occur in the cold fluid along the length of the heat exchanger (**Figure 5(b)**). Where, the hot water and the ambient air enter the heat exchanger with temperatures T_{wh} and T_{air}, respectively; whereas their outlet tempe- ratures are T_{wc} and T_{HE}, respectively.

Knowing the heat exchanger efficiency η_{HE}, the rate of transferring the heat energy from the water to the air $Q_{w\text{-}a}$, will be

$$\dot{Q}_{w\text{-}a} = \dot{m}_{HEw} C_{pw} \left(T_{wh} - T_{wc}\right) \eta_{HE} \tag{5}$$

Therefore, applying the energy balance equation [2,4] on the heat exchanger (**Figure 6**) will lead to the differential equation

$$\frac{\mathrm{d}T_{HE}}{\mathrm{d}t} = -\frac{\dot{m}_{air}}{\rho_{air} V_{HE}} \left(T_{HE} - T_{air}\right) + \frac{\dot{Q}_{w-a}}{\rho_{air} V_{HE} C_{pair}} \tag{6}$$

where $\dot{m}_{air}$ mass flow rate of the air;

ρ_{air} density of the air;

V_{HE} volume of the heat exchanger;

C_{pair} specific heat of the air.

The relation between the inlet and outlet temperatures of the heat exchanger, can be outlined using the heat exchanger effectiveness ε_{HE}, as [6]

$$\varepsilon_{HE} = \begin{cases} \dfrac{T_{HE} - T_{air}}{T_{wh} - T_{air}}, & \dot{m}_{HEw} C_{pw} > \dot{m}_{air} C_{pair} \\ \dfrac{T_{wh} - T_{wc}}{T_{wh} - T_{air}}, & \dot{m}_{HEw} C_{pw} < \dot{m}_{air} C_{pair} \end{cases} \tag{7}$$

3.4. Modeling of the Electrical Heater

The main function of the electrical heater is to make the auxiliary heating of drying air. The auxiliary heating means that the electrical heater will operate only if the temperature of the output air T_{HE} from the heat exchanger is lower than the required value to adjust the air drying temperature T_{dryer} at 40˚C.

If the rated power of the electrical heater is P_{heater} and the heater efficiency is η_{heater}, then the rate of heat added to the drying air (in the electrical heating chamber) [7] will be

$$\dot{Q}_{heater} = P_{heater} \times \eta_{heater} \tag{8}$$

Thus, by applying the energy balance principle on the electrical heating chamber (**Figure 7**) will yield [8]

$$\frac{\mathrm{d}T_{dryer}}{\mathrm{d}t} = -\frac{\dot{m}_{air}}{\rho_{air} V_{HC}} \left(T_{dryer} - T_{HE}\right) + \frac{\dot{Q}_{heater}}{\rho_{air} V_{HC} C_{pair}} \tag{9}$$

where V_{HC} volume of the heating chamber of the electrical heater.

4. Control of the Thermal System Including the Air Blower

The control system of the thermal system including the air blower performs control and protection functions for the components of the thermal system and the air blower in the dryer system of **Figure 1**. Where, it optimizes the operation of the following electrical loads:

1) The air blower;

2) The water circulation valve and pump of the collector side;

3) The water circulation valve and pump of the heat exchanger side;

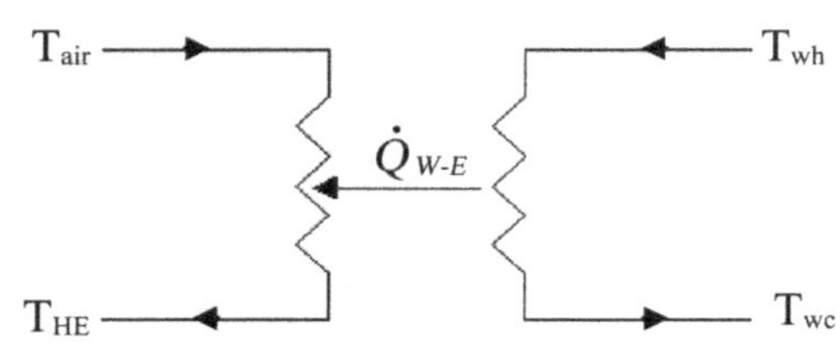

Figure 6. Energy balance on the heat exchanger.

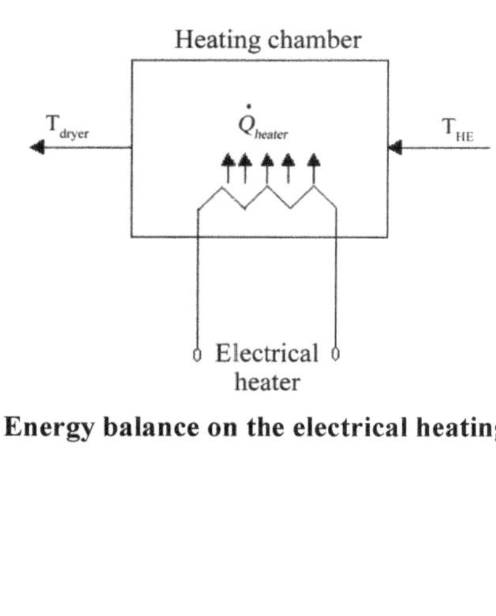

Figure 7. Energy balance on the electrical heating chamber.

4) The electrical heater.

Therefore, there will be four control signals in the system (**Table 1**). The first three control signals are switched control signals based on logic control decision and the fourth is a switched control signal based on/off controller (or thermostat).

Where, T_{ref} is the desired drying temperature of the air and SWB is the switching control signal of the air blower, SWC is the switching control signal of the valve and pump in the collector side, SWHE is the switching control signal of the valve and pump in the heat exchanger side.

Simulation of the Thermal System

In this simulation the mathematical models of the different components of both the solar thermal system and the electrical heater are used to simulate the complete suggested thermal system, by using the MATLAB-SIMULINK. **Figure 8** shows the designed SIMULINK model of the dryer thermal system. Each component, in the simulation model, has its own parameters and input and output variables.

5. Simulation Results of the Thermal System

The performance of the complete components in both the solar thermal system and the electrical heater with the corresponding control signals are necessary to evaluate the control capability of the proposed system. The performance of the system is preferably evaluated by compar-

Table 1. Thermal system control.

Component	Control signal	Control rule
Air blower	SWB	If ($T_{ref} \neq 0$) then (SWB will be on and vice versa)
Valve and pump in collector side	SWC	If (($Rad > 0$) & ($T_{cout} > T_{cin}$)) then (SWC will be on and vice versa)
Valve and pump in HE side	SWHE	If ($T_{air} < T_{ref}$) then (SWHE will be on and vice versa)
Electrical heater	SWheater	On/off control (thermostat)

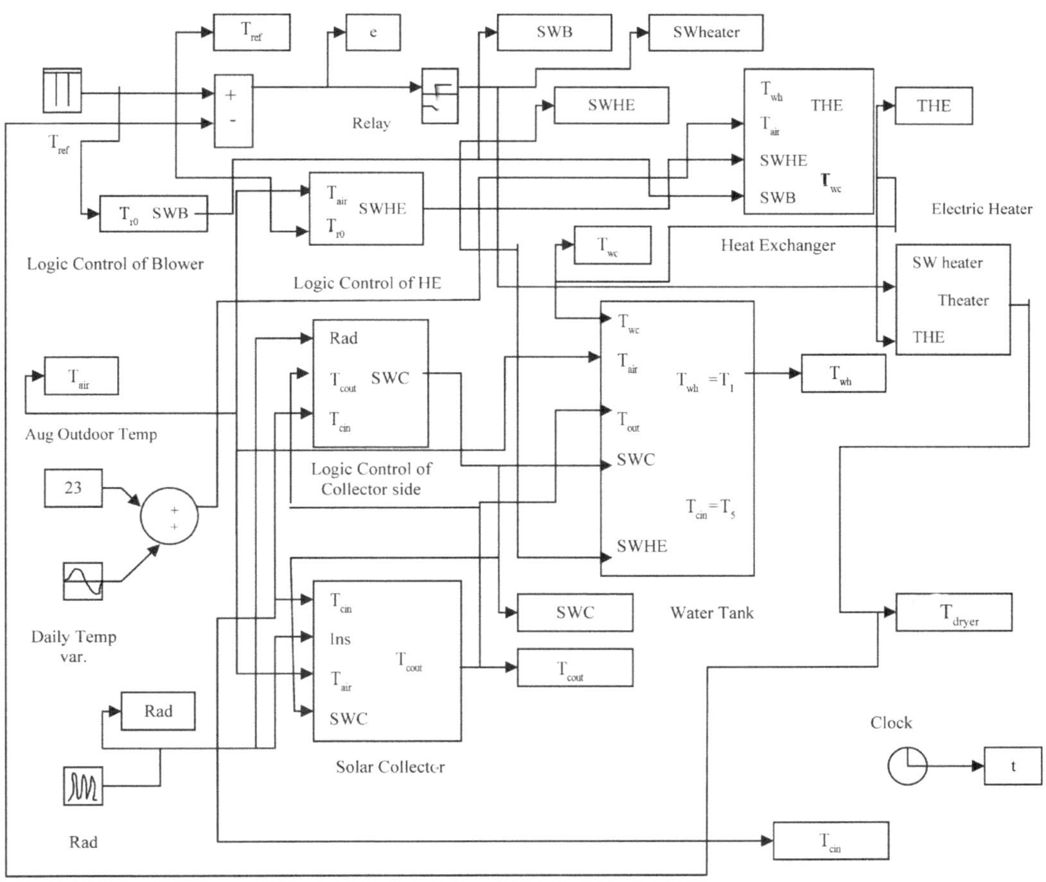

Figure 8. SIMULINK model of the thermal system.

ing the performance of the system in a summer day with that of a winter day.

The meteorological data in this work are obtained at two different days with different atmospheric conditions. **Figures 9(a)** and **(b)** show the variations of the solar insulation incident on the surface of the solar collector on summer and winter days, respectively (in Sinai Peninsula (Abu Rudies) [9]). While, **Figure 10** shows variation of the corresponding ambient temperature.

The performance of the solar collector during the summer and winter days is shown in **Figures 11** and **12**, respectively. **Figure 11**, as shown, includes the switching control signal of the collector-side path, while **Figure 12** shows the temperatures of the water at the inlet and outlet sides of the collector. It is shown from **Figures 11** and **12** that the used collector is able to heat the water that flows through it, whenever there is solar illumination;

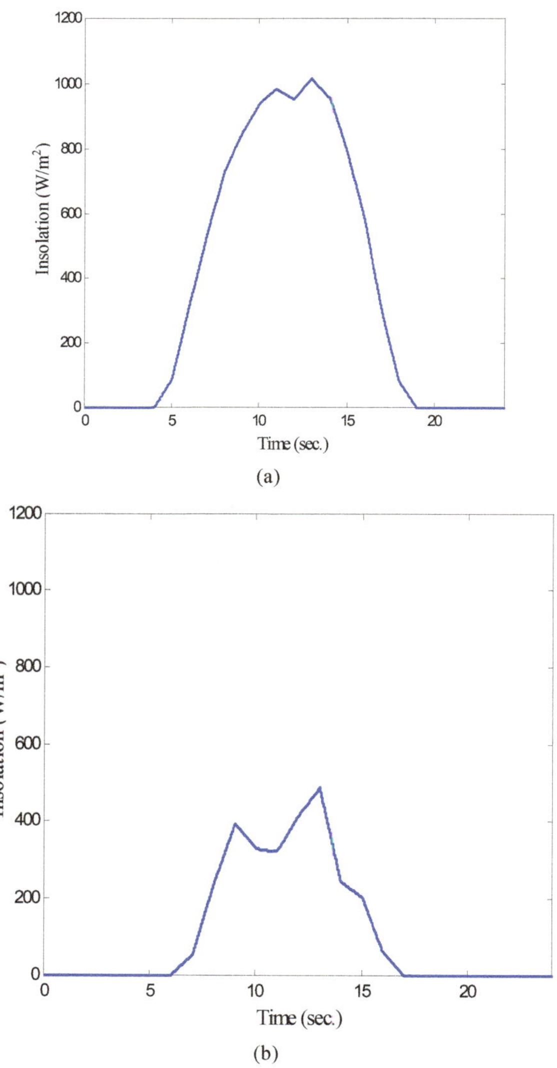

Figure 9. Variation of solar insolation: (a) Insolation of the summer day; (b) Insolation of the winter day.

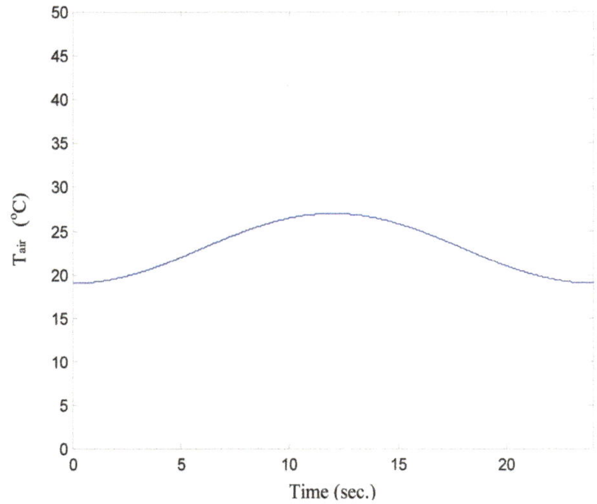

Figure 10. Variation of ambient temperature.

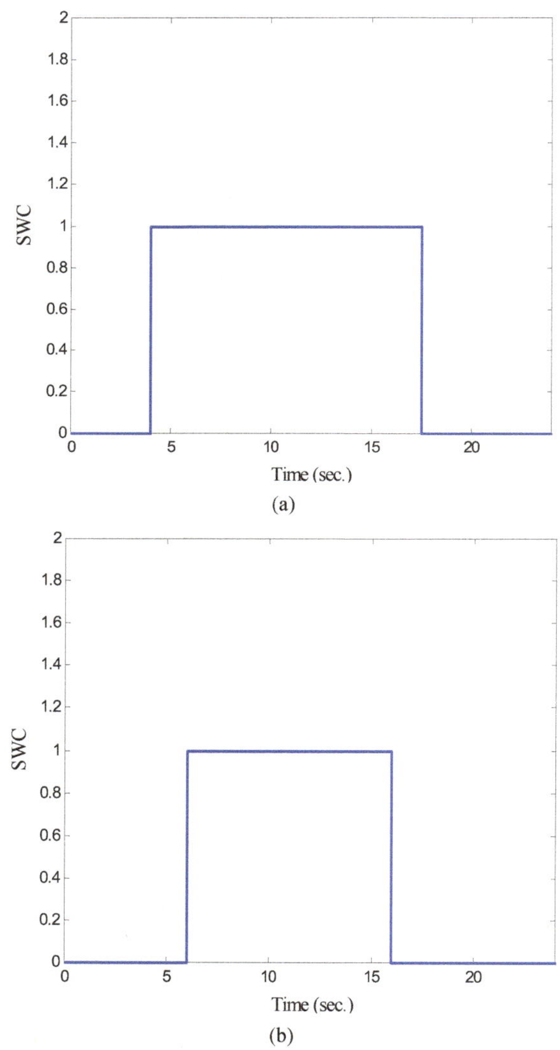

Figure 11. Switching signal of the collector side: (a) During summer day; (b) During winter day.

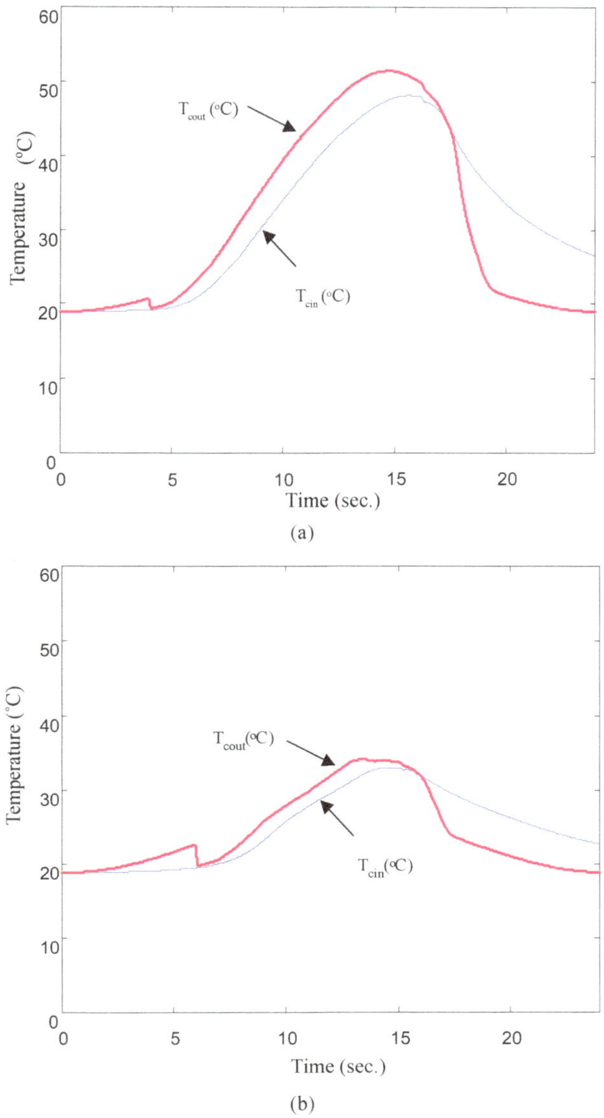

Figure 12. Inlet and outlet temperatures of the collector: (a) During summer day; (b) During winter day.

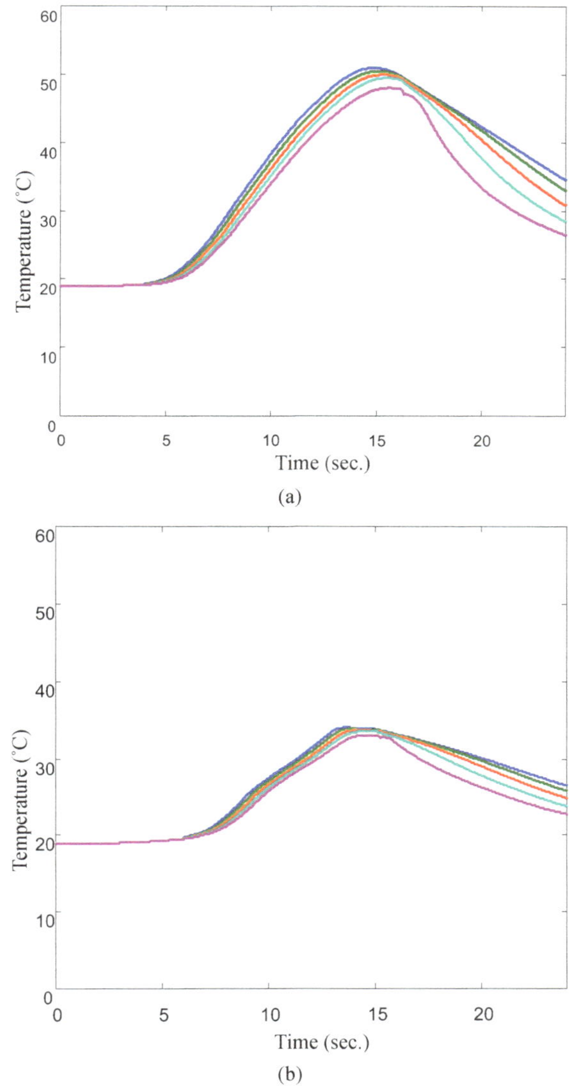

Figure 13. Temperature of the tank layers: (a) During summer day; (b) During winter day.

where the heating of the water is proportional to the incident insolation level. Also, it is cleared that the water continues to flow in the collector as long as the temperature of the water at the collector outlet is greater than that at the collector inlet. The initial apparent-rise of the outlet temperature compared with the inlet one, for the two cases, is due to: using the same value of the initial conditions for the two temperatures, the water does not flow in the collector during the night period, and the corresponding variation in the ambient temperature.

Figures 13(a) and **(b)** show the performance of the storage water tank during the summer and winter days, respectively. Thus, the effect of stratification on the tank is shown in this figure, where the bottom layer exhibits the lowest temperature and the top layer exhibits the highest temperature of all the tank layers. It is indicated that the temperatures of the tank layers increase during the daylight hours and decrease during the night hours, such that the variation of the temperatures is proportional to the variation in the solar insolation.

The performance of the heat exchanger during the summer and winter days is shown in **Figures 14-17**, respectively. **Figures 14** and **15** include the switching control signals of the air blower and the heat exchanger side, respectively. While the performance of the inlet and outlet temperatures of the heat exchanger water and the performance of the inlet and outlet temperatures of the heat exchanger air are shown in **Figures 16** and **17**, respectively. Thus, it is clear from these figures that the switching control signal of the air blower exists as long as there are herbs in the dryer, and that of the heat exchanger

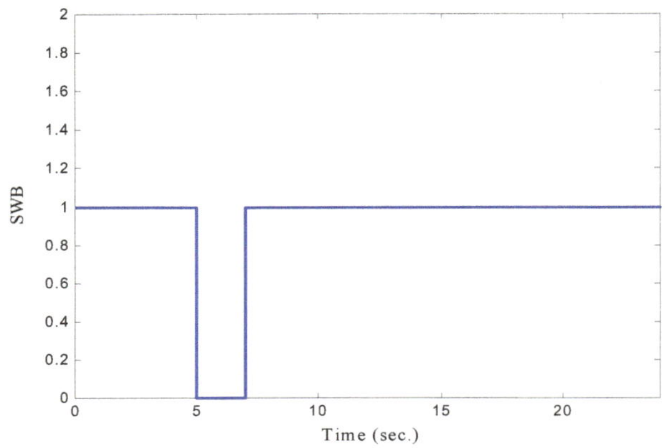

(a)

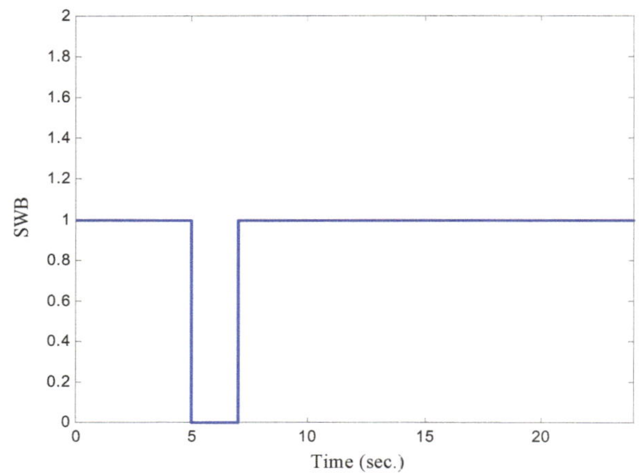

(b)

Figure 14. Switching signal of the air blower: (a) During summer day; (b) During winter day.

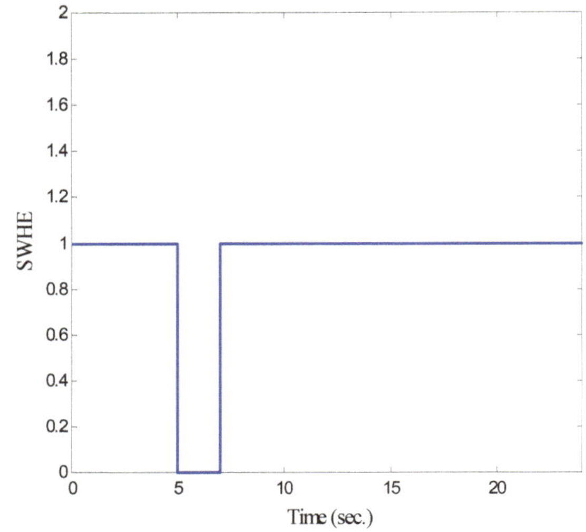

(a)

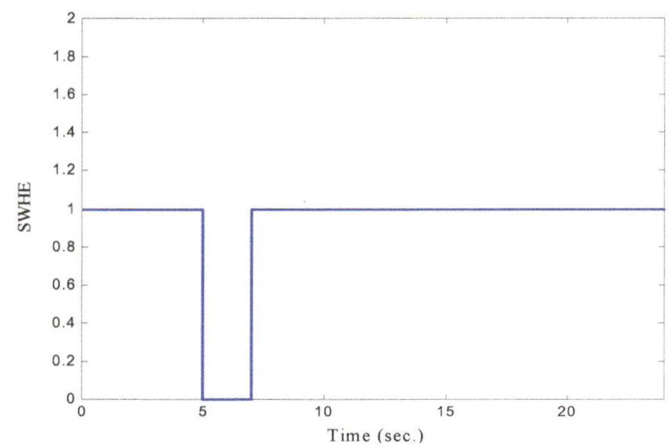

(b)

Figure 15. Switching signal of the heat exchanger side: (a) During summer day; (b) During winter day.

side exists as long as the ambient temperature is lower than the desired drying temperature T_{ref}. Also, the inlet temperature of the heat-exchanger water is the temperature of the tank top layer (**Figure 13**) and the shown difference in temperature between the inlet and the outlet water (which depends on the solar insolation incident on the collector) expresses the rate of the thermal energy transferred to the heat exchanger air. Moreover, the outlet temperature of the heat-exchanger air is increased with respect to the ambient temperature according to the quantity of heat transferred from the water.

6. Conclusion

This paper suggests and develops a new controlled drying system for the medical herbs. The suggested drying system includes the following three systems: the solar thermal system, the dryer, and the renewable energy system. The main function of the solar thermal system is to supply the dryer inlet air with the required thermal energy through utilizing the flat plate solar collector. While, that of the renewable energy system, which comprises hybrid photovoltaic and wind energy sources, is to feed the electrical load of the dryer with the required electrical energy. This paper presents, also, the dynamic modeling and simulation of the suggested drying system. The system results are indicated that the used collector is able to heat the water that flows through it, whenever there is solar illumination; where the heating of the water is found to be proportional to the incident insolation level. Also, the water flows in the collector as long as the temperature of the water at the collector outlet is greater than that at the collector inlet and there is illumination. The temperatures of the layers of the storage water tank increase during the daylight hours and decrease during the night hours; such that the variation of the temperatures is proportional to the variation in the solar insolation. The air blower operates as long as there are herbs in the dryer, and the water flows in the heat exchanger as long as the ambient temperature is lower than the desired drying temperature. Also, the drop in water temperature between the inlet and the outlet sides of the heat exchanger is used

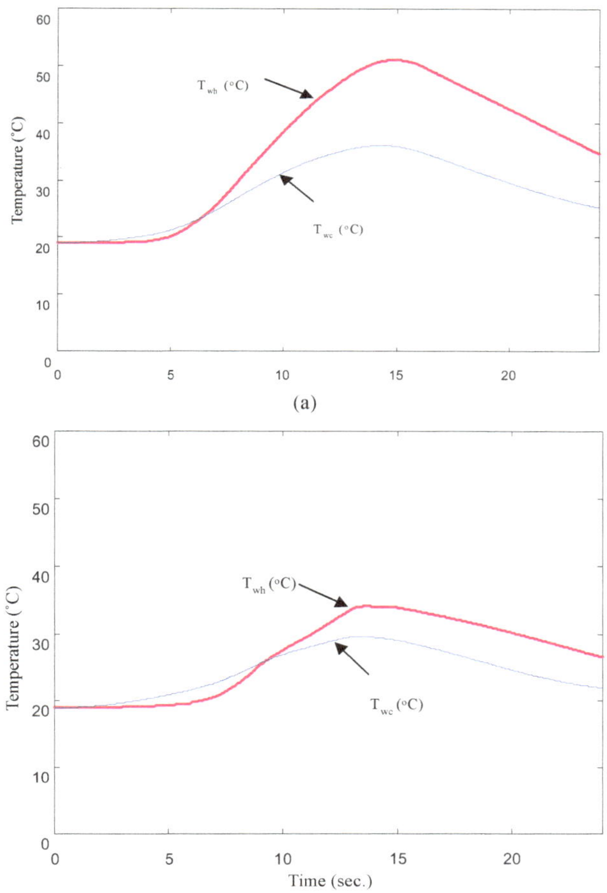

(a)

(b)

Figure 16. Inlet and outlet temperature of the heat exchanger water: (a) During summer day; (b) During winter day.

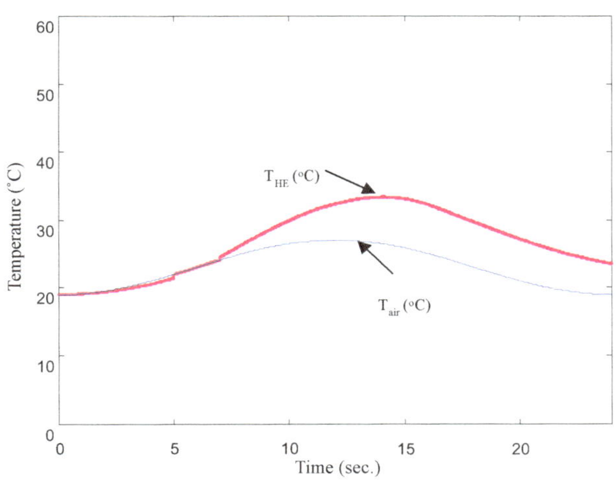

(a)

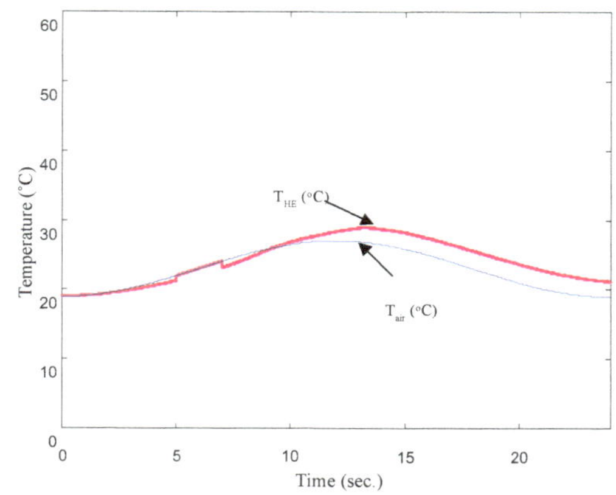

(b)

Figure 17. Inlet and outlet temperature of the heat exchanger air: (a) During summer day; (b) During winter day.

to rise the temperature of the outlet air from heat exchanger.

REFERENCES

[1] G. Grang and H. P. Kumar, "Developments in Solar Drying," *Proceedings of the Second Asian-Oceania Drying Conference* (*ADC* 2001), Pulau Pinang, 2001, pp. 297-319.

[2] J. Howell, R. Bannerot and G. Vliet, "Solar-Thermal Energy Systems: Analysis and Design," McGraw-Hill, Inc., New York, 1982.

[3] A. Vieira da Rosa, "Fundamentals of Renewable Energy Processes," 3rd Edition, Academic Press, Waltham, 2013.

[4] J. W. Twidel and A. D. Weir, "Renewable Energy Resources," Cambridge, London, 1986.

[5] X. Xu, M. M. Meyers, B. G. Sammakiak and B. T. Murray, "Thermal Modeling and Life Prediction of Water-Cooled Hybrid Concentrating Photovoltaic/Thermal Collectors," *Journal of Solar Energy Engineering*, *Transactions of the ASME*, Vol. 135, No. 1, 2013.

[6] T. A. Stuetzle, "Automatic Control of the 30 MWe SEG VI Parabolic Trough Plant," M.Sc. Thesis, Wisconsin-Madison University, Madison, 2002.

[7] T. A. Stuetzle, "Automatic Control of the 30 MWe SEG VI Parabolic Trough Plant," M.Sc. Thesis, Wisconsin-Madison University, Madison, 2002.

[8] P. M. Doran, "Bioprocess Engineering Principles," 2nd Edition, Academic Press, Waltham, 2013.

[9] N. Elshahat and A. Zeit, "Aromatic Plants & Its Agricultural-Medical Products," Arabic, Inc., Egypt, 2003.

13

Temperature Control in a PV-WIND Medical Herb Dryer System

Abdel-Shafy A. Nafeh, Emad Ahmed Sweelem, Faten Hosny Fahmy

PV Cells Department, Electronics Research Institute, Cairo, Egypt.

ABSTRACT

This paper introduces a temperature control technique for a medical herb dryer system. The technique fixes the drying temperature of the medical herbs at 40˚C even in cases of rapidly changing atmospheric conditions. The control of the dryer temperature is achieved through using on/off controller. The designed dryer contains two systems, which are the thermal and electrical systems. The thermal system is designed to heat the drying air by using solar water collector and electric heater. While, the electrical system which contains a photovoltaic (PV)-WIND hybrid system is used to feed the different electrical loads of the dryer system. The control technique is investigated through simulation work by using MATLAB-SIMULINK. The simulation results indicate the high capability of the proposed technique in controlling the drying temperature, even in case of rapidly changing atmospheric conditions.

Keywords: Medical Herb Dryer; Solar Thermal System; PV-WIND System; On/Off Controller

1. Introduction

Most agricultural crops, which are intended to be stored before use, have to be dried first. Otherwise insects and fungi, which thrive in moist conditions, render them unusable. Examples include wheat, rice, coffee, herbs, etc. All crops drying involve a transfer of water from the crop to the air around it, and this process can be accomplished by passing a hot and unsaturated air over the wet crop. It is also important to realize that there are limiting temperatures for drying corps for storage, so the product does not crack and allow bacterial attack [1]. If the product to be dried is left in the dryer for the exact drying time, then a moist crop or plant will give water to the surrounding air until the product reaches its equilibrium content (*i.e.*, its final moisture content). The value of the equilibrium moisture content depends on the crop, and the speed to reach this value depends on the temperature and humidity of the surrounding air. Thus, for unsaturated air, the drying temperature is an important factor to accomplish the drying process. Because the plants (especially the medical herbs) are sensitive to heat, there is an optimal drying temperature for each plant or herb. The optimal drying temperature for most of the medical herbs is about 40˚C [2]. Therefore, conventional direct or indirect sun-drying methods [3], where the product to be dried is exposed to temperature variation during the drying time, are not suitable. These methods produce poor quality products. In this paper, a temperature control technique is proposed to fix the drying temperature of the medical herbs at 40˚C.

2. The Suggested Thermal System of the Dryer

The block diagram of the complete thermal system is illustrated in **Figure 1**, which includes the solar thermal system and the electrical heater. The suggested solar thermal system of the herb dryer consists mainly of three components that are:

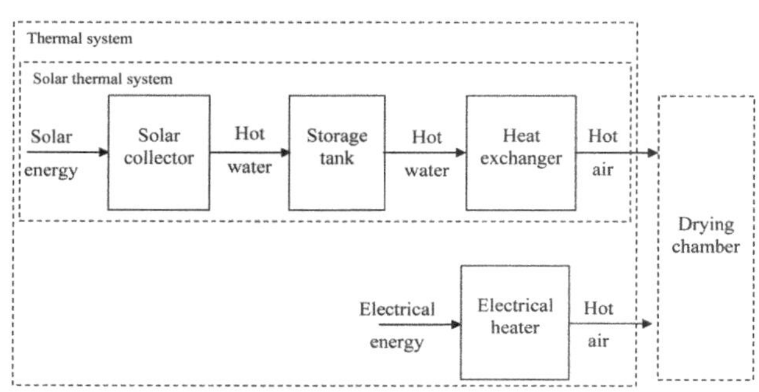

Figure 1. Block diagram of the thermal system.

1) The solar collector;
2) The storage tank;
3) The heat exchanger.

The solar collector is considered to be the main source of the thermal energy or heating in the system. It is used to directly utilize the incident solar energy from the sun and transfers it into the heat transfer fluid, which is chosen to be the water in this work. Also, the storage tank is used, in this work, to store the thermal energy of the water gained in the collector stage. The output water from the tank, which carries certain amount of thermal energy and at specified temperature, enters the heat exchanger stage. This heat exchanger is utilized to transfer a certain amount of the thermal energy of the hot water to the dryer's inlet air. In this way, the temperature of the dryer's inlet air can raise to certain temperature. If this temperature is lower than the required drying temperature, then an auxiliary heating source (the electrical heater) will be utilized to supply the deficit in temperature.

3. Control of the Drying Temperature

It is known from the previous discussion that the main function of the electrical heater is to supply the auxiliary heating required by the drying air that supplements the thermal energy deficit of the solar thermal system; to adjust the drying air temperature at the desired optimal value (*i.e.*, 40˚C). Therefore, the suggested block that controls the air drying temperature is shown in **Figure 2**. Where, in this case, the on/off controller is utilized to provide the electrical heater with the suitable control action. The heater, in turn, can adjust the drying temperature T_{dryer} to the desired value, by providing additional air temperature ΔT to the available T_{HE}. Note that the auxiliary heating, in this case, means that the electrical heater will operate only if the temperature of the output air from the heat exchanger T_{HE} is lower than the required value to adjust the drying temperature T_{dryer} at 40˚C.

3.1. The On/Off Controller

On-off control is perhaps the simplest type of feedback control strategy. The strategy is similar to that of a relay. The following equations define the output of on-off controller [4]:

Output = *ON* if the controller is in the ON state;

Output = *OFF* if the controller is in the OFF state.

where e is a error and *output when ON* and *output when OFF* are values determined based on the application.

When the ON-OFF controller is in the ON state, the controller remains in the ON state until the input to the controller falls below the value of the switch OFF point state until the input exceeds the value of the SWITCH ON point.

The ON-OFF controller is initially in the OFF state. In short, the On-OFF controller output one of two values ("output when ON" or "output when OFF") based on the value of the input signal.

A system under ON-OFF control will always oscillate. To avoid rapid switching between the ON and OFF states, it is common to introduce a hysteresis (ζ) in the switch. The input-output relation for an ON-OFF controller with hysteresis is shown in below **Figure 3**.

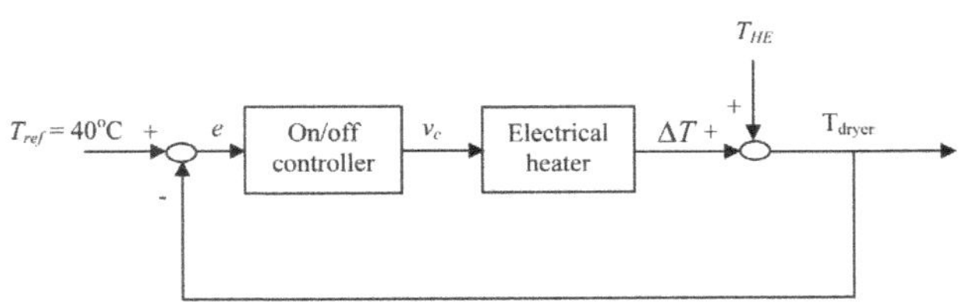

Figure 2. Block diagram of the temperature control system.

3.2. The Electrical Heater

The main function of the electrical heater is to make the auxiliary heating of drying air. The auxiliary heating means that the electrical heater will operate only if the temperature of the output air T_{HE} from the heat exchanger is lower than the required value to adjust the air drying temperature T_{dryer} at 40˚C.

If the rated power of the electrical heater [5] is P_{heater} and the heater efficiency is η_{heater}, then the rate of heat added to the drying air (in the electrical heating chamber) will be

$$\dot{Q}_{heater} = P_{heater} \times \eta_{heater} \quad (1)$$

Thus, by applying the energy balance principle on the electrical heating chamber will yield [6]

$$\frac{dT_{dryer}}{dt} = -\frac{\dot{m}_{air}}{\rho_{air} V_{HC}}\left(T_{dryer} - T_{HE}\right) + \frac{\dot{Q}_{heater}}{\rho_{air} V_{HC} C_{pair}} \quad (2)$$

where V_{HC} volume of the heating chamber of the electrical heater

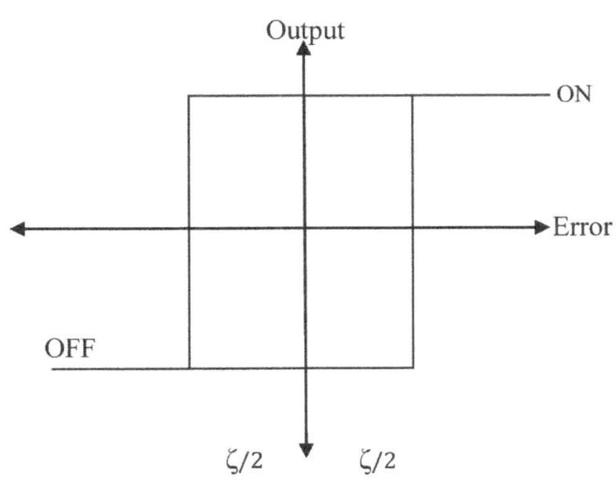

Figure 3. The input-output relation for an ON-OFF controller with hysteresis.

4. Simulation Results of the Thermal System

The complete components of the considered thermal system are simulated in this work by using MATLAB-SIMU-LINK [7].

The purpose of the temperature control system is to adjust the dryer temperature T_{dryer} at 40˚C, which is a necessary value for drying the medical herbs. Therefore, to investigate the performance of the proposed temperature control system the response of the drying temperature must be indicated at different cases. **Figure 4** illustrates the response of T_{dryer} in case of using the on/off controller. Where, in this case, the deadband of the controller ε is set to be $\varepsilon/2 = 1$ (since the drying temperature of medical herbs ranges from 35˚C - 45˚C with optimal value 40˚C).

The robustness of the used controller in rejecting the sudden variations in the insolation level or in the ambient temperature is shown in **Figures 5** and **6**, respectively. Where, **Figures 5(a)** and **(b)** illustrate the sudden variation in the insolation level at constant ambient temperature of 23˚C and the corresponding response of T_{dryer}, respectively. Also, **Figures 6(a)** and **(b)** indicate the sudden variation in the ambient temperature at constant insolation of 1000 W/m^2 and the corresponding response of T_{dryer}, respectively. Therefore, the used on/off controller is sufficient for controlling this system and also has a good capability in rejecting the imposed sudden variations in the solar insolation or in the ambient temperature.

The meteorological data in this work are obtained at two different days with different atmospheric conditions. **Figures 7(a)** and **(b)** show the variations of the solar insolation incident on the surface of the solar collector on summer and winter days (in Sinai Peninsula (Abu Rudies) [8,9]), respectively. While, **Figure 8** shows variation of the corresponding ambient temperature.

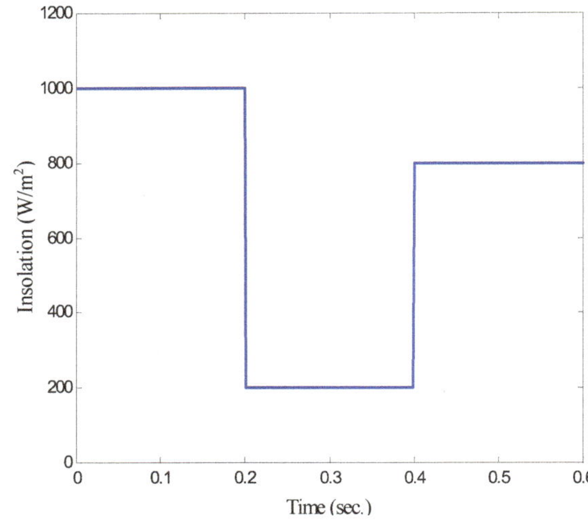

(a)

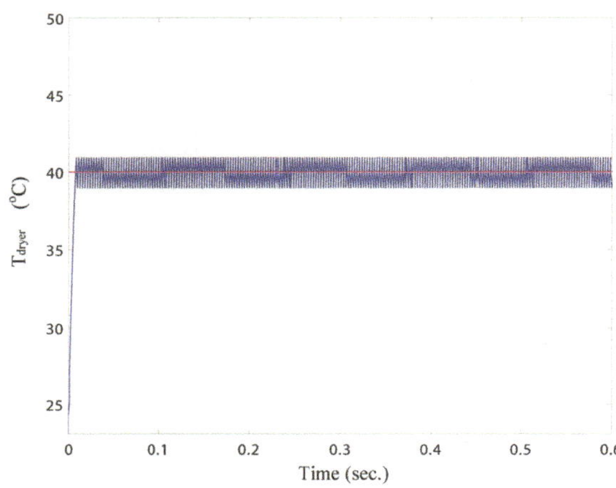

(b)

Figure 5. Response of T_{dryer} with insolation change: (a) Insolation versus time; (b) T_{dryer} versus time.

The performance of the solar collector during the summer and winter days is shown in **Figures 9** and **10**, respectively. **Figure 10**, as shown, includes the switching control signal of the collector-side path, while **Figure 11** shows the temperatures of the water at the inlet and outlet sides of the collector. It is shown from **Figures 9** and **10** that the used collector is able to heat the water that flows through it, whenever there is solar illumination; where the heating of the water is proportional to the incident insolation level. Also, it is cleared that the water continues to flow in the collector as long as the temperature of the water at the collector outlet is greater than that at the collector inlet. The initial apparent-rise of the outlet temperature compared with the inlet one, for the two cases, is due to: using the same value of the initial conditions for the two temperatures, the water does not flow in the collector during the night period, and the corresponding variation in the ambient temperature.

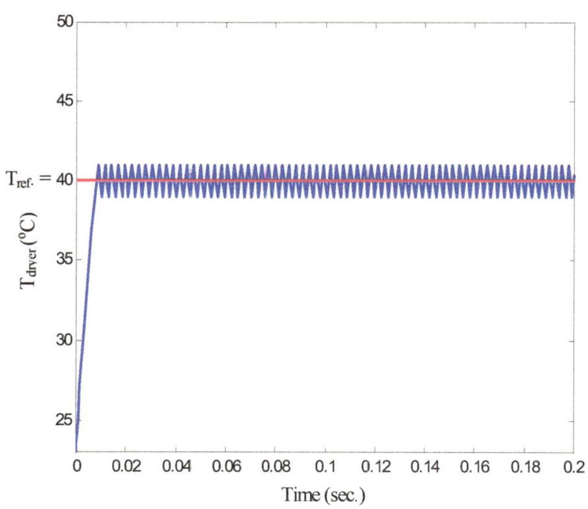

Figure 4. Response of T_{dryer} with the on/off controller.

Figures 11(a) and **(b)** show the performance of the storage water tank during the summer and winter days,

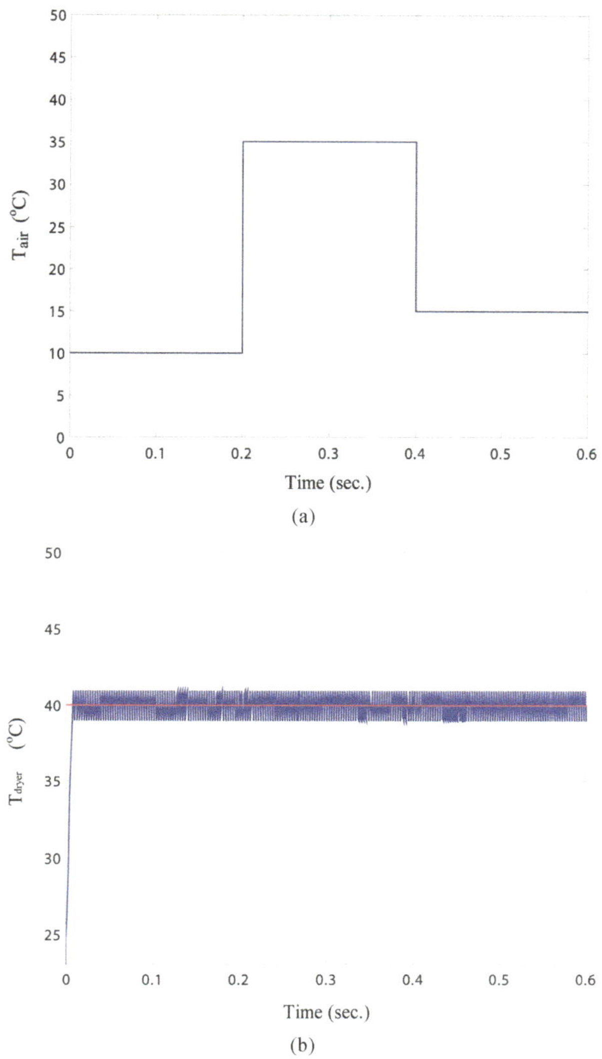

Figure 6. Response of T_{dryer} with ambient temperature change: (a) Ambient temperature versus time; (b) T_{dryer} versus time.

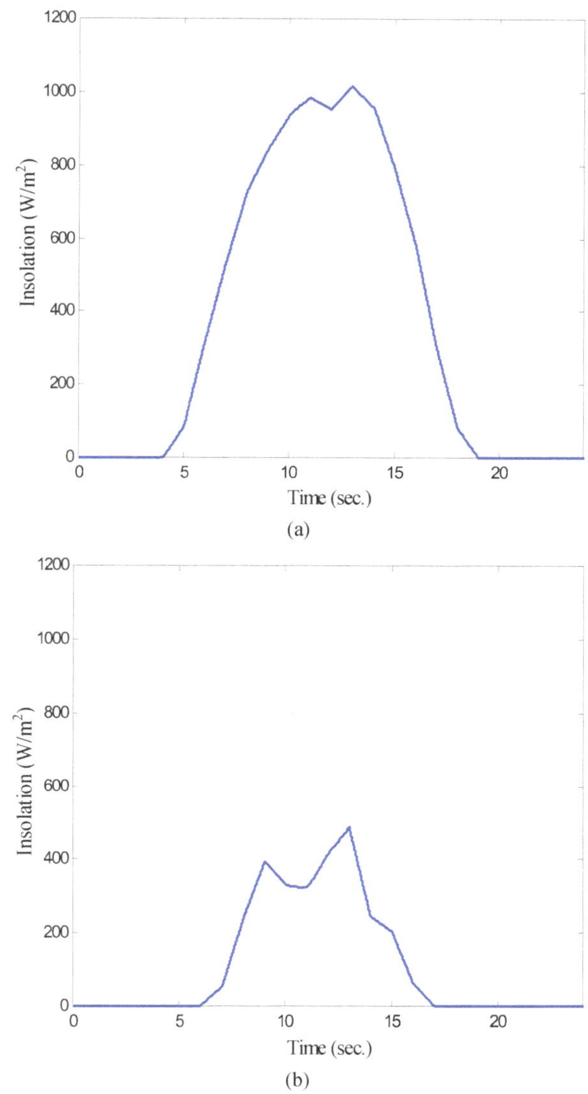

Figure 7. Variation of solar insolation: (a) Insolation of the summer day; (b) Insolation of the winter day.

respectively. Thus, the effect of stratification on the tank is shown in this figure, where the bottom layer exhibits the lowest temperature and the top layer exhibits the highest temperature of all the tank layers. It is indicated that the temperatures of the tank layers increase during the daylight hours and decrease during the night hours, such that the variation of the temperatures is proportional to the variation in the solar insolation.

The performance of the heat exchanger during the summer and winter days is shown in **Figures 12-15**, respectively. **Figures 12** and **13** include the switching control signals of the air blower and the heat exchanger side, respectively. While the performance of the inlet and outlet temperatures of the heat exchanger water and the performance of the inlet and outlet temperatures of the heat exchanger air are shown in **Figures 14** and **15**, respectively. Thus, it is clear from these figures that the switching control signal of the air blower exists as long as there are herbs in the dryer, and that of the heat exchanger side exists as long as the ambient temperature is lower than the desired drying temperature T_{ref}. Also, the inlet temperature of the heat-exchanger water is the temperature of the tank top layer (**Figure 11**) and the shown difference in temperature between the inlet and the outlet water (which depends on the solar insolation incident on the collector) expresses the rate of the thermal energy transferred to the heat exchanger air. Moreover, the outlet temperature of the heat-exchanger air is increased with respect to the ambient temperature according to the quantity of heat transferred from the water.

Finally, the performance of the electrical heater is shown in **Figures 16** and **17**, respectively for the summer

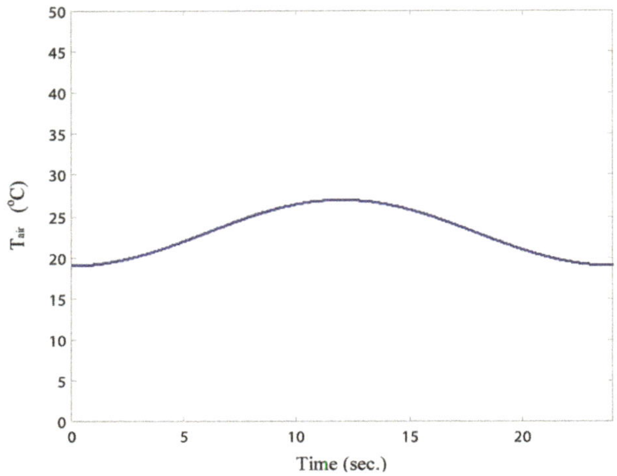

Figure 8. Variation of ambient temperature.

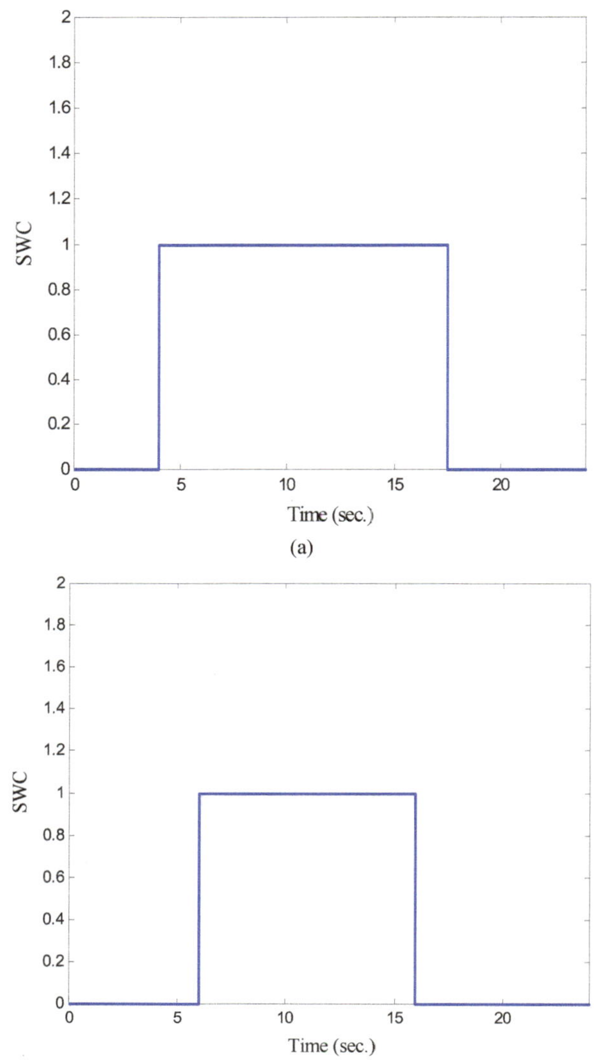

Figure 9. Switching signal of the collector side: (a) During summer day; (b) During winter day.

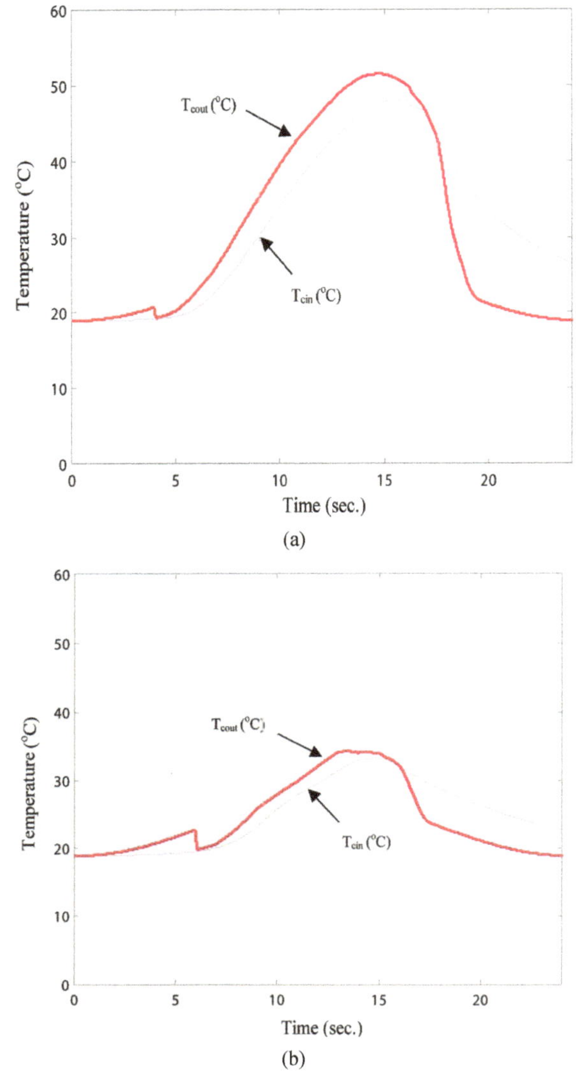

Figure 10. Inlet and outlet temperatures of the collector: (a) During summer day; (b) During winter day.

and winter days. **Figure 16** shows the output control signal of the on/off controller, while **Figure 17** shows the corresponding air temperature at the inlet and outlet sides of the electrical heater chamber. It is shown from these figures that the electrical heater will be active only if the temperature of the outlet air from the heat exchanger is lower than the desired drying temperature of the medical herbs. Also, these figures and **Figure 15** indicate that the temperature of the inlet air to the heating chamber has the same temperature as the outlet air from the heat exchanger, and the temperature of the outlet air from the heating chamber (*i.e.*, the drying temperature) tracks the desired drying temperature.

5. Conclusion

The optimal drying temperature of the medical herbs

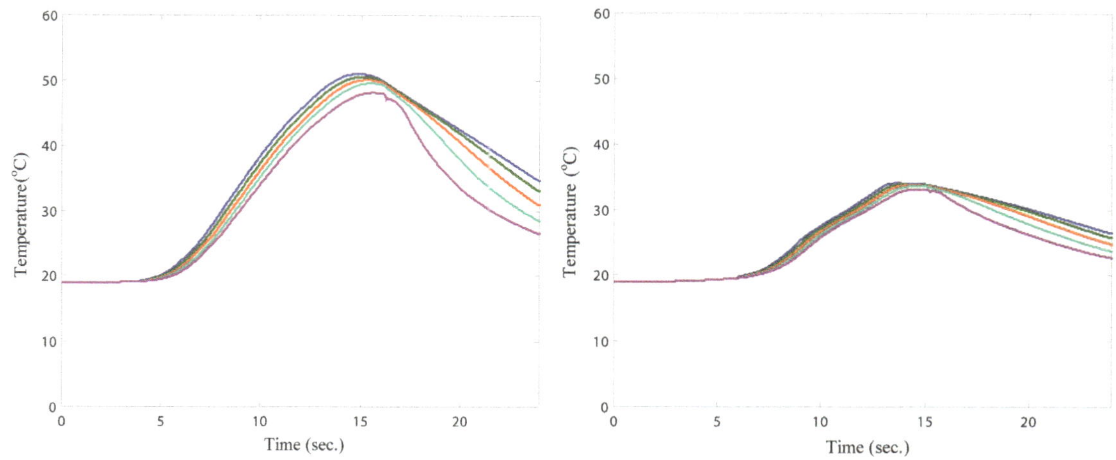

(a) (b)

Figure 11. Temperature of the tank layers: (a) During summer day; (b) During winter day.

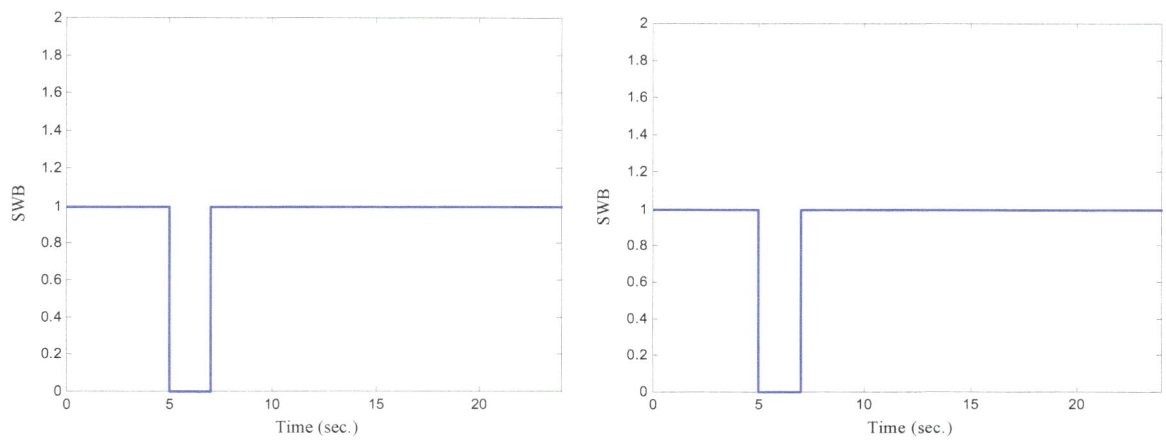

(a) (b)

Figure 12. Switching signal of the air blower: (a) During summer day; (b) During winter day.

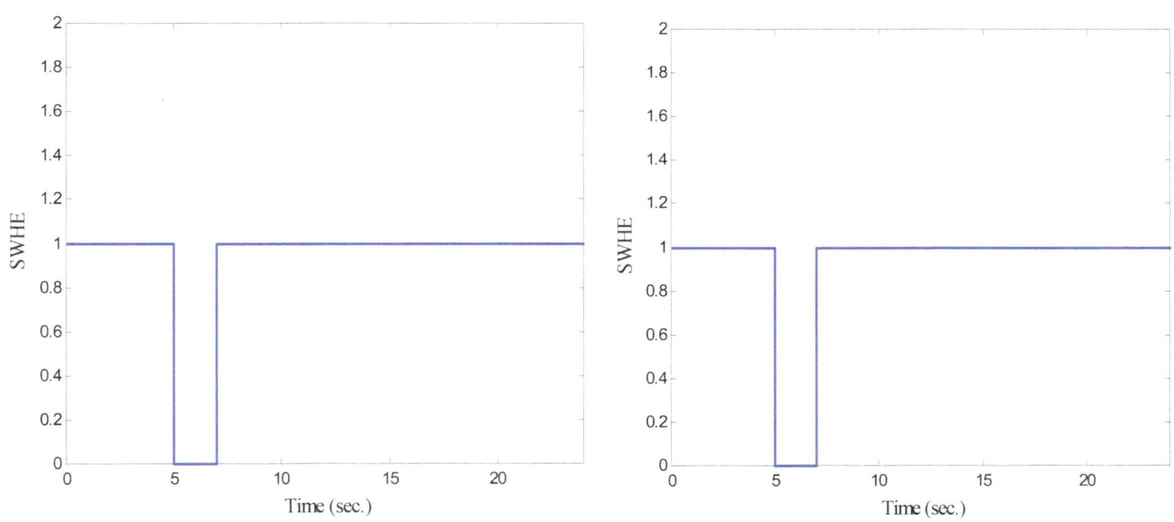

(a) (b)

Figures 13. Switching signal of the heat exchanger side: (a) During summer day; (b) During winter day.

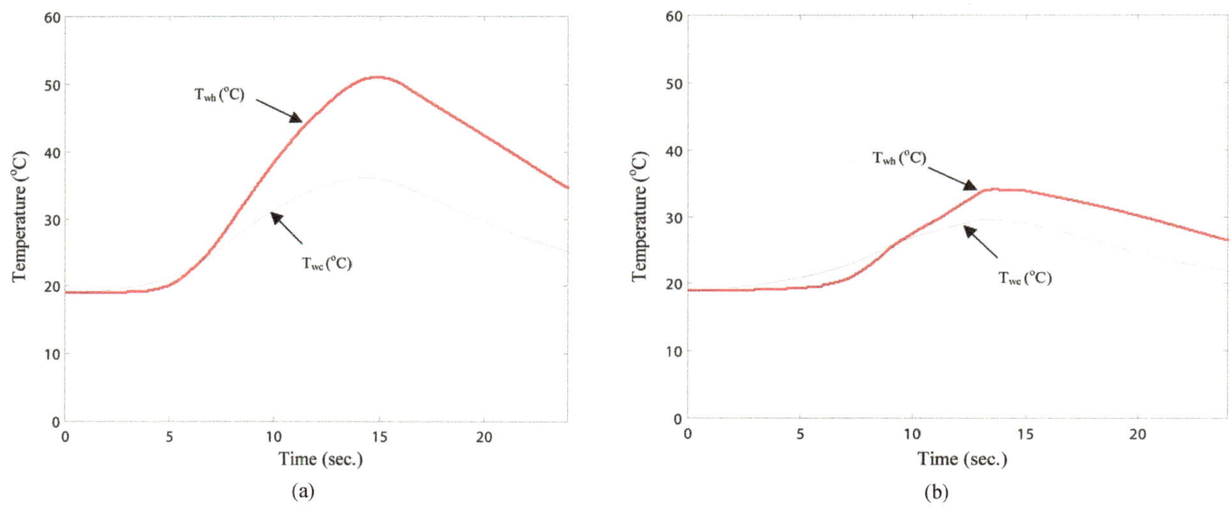

Figure 14. Inlet and outlet temperature of the heat exchanger water: (a) During summer day; (b) During winter day.

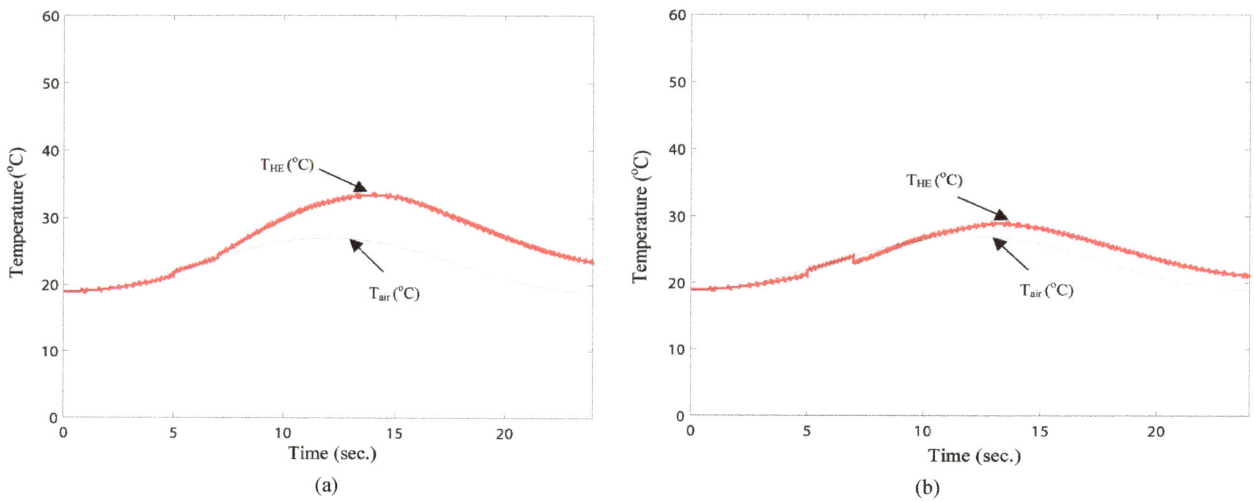

Figure 15. Inlet and outlet temperature of the heat exchanger air: (a) During summer day; (b) During winter day.

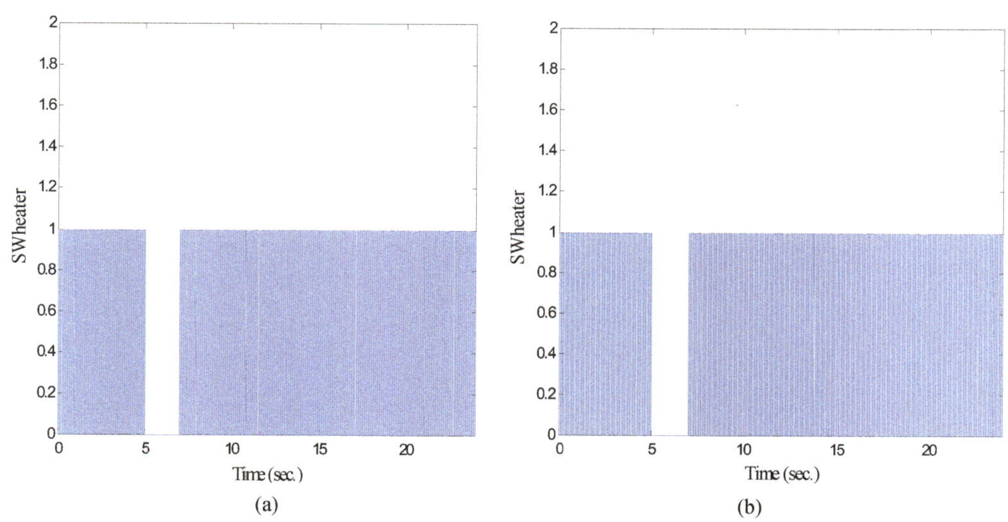

Figure 16. Output signal of the on/off controller: (a) During summer day; (b) During winter day.

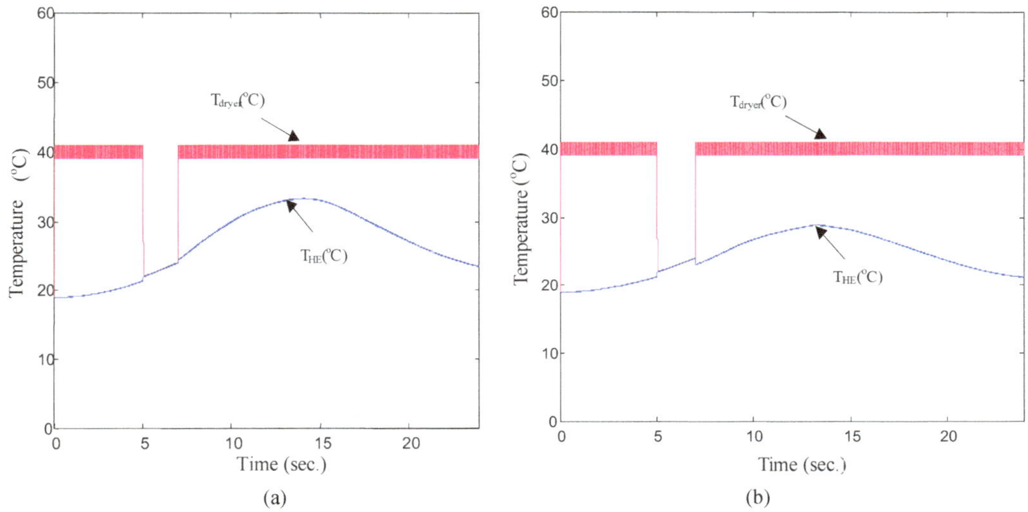

Figure 17. Inlet and outlet temperatures of the heating chamber air: (a) During summer day; (b) During winter day.

must be adjusted to about 40°C. A control technique is proposed to adjust the dryer inlet temperature at the required optimal value. The proposed control technique is based on using the on/off controller, to control the operation of the electric heater whenever necessary. The electric heater will be active only when the temperature of the outlet hot air from the heat exchanger T_{HE} is less than 40°C. The simulation results indicate that the used on/off controller with the tuning of the deadband gives a good capability in rejection, the imposed sudden variations in the solar insolation or in the ambient temperature. Therefore, the developed herb dryer system will be incorrect and continuous operation during the summer/winter day and night hours.

REFERENCES

[1] G. Grang and H. P. Kumar, "Developments in Solar Drying," *Proceedings of the Second Asian-Oceania Drying Conference* (*ADC* 2001), Pulau Pinang, 2001, pp. 297-319.

[2] A. Shafy, F. Famhy and M. Hassan, "Naval Temperature Control Technique for a Medical Herb system Powered by a Photovoltaic Array," *International Journal of Green Energy*, Vol. 1, No. 2, 2005, pp. 147-156.

[3] J. R. Howell, R. B. Bannerot and G. C. Vliet, "Solar-Thermal Energy Systems: Analysis and Design," McGraw-Hill, Inc., New York, 1982.

[4] T. Markvart, "Solar Electricity," John Wiley & Sons, Inc., Hoboken, 1994.

[5] X. Jiang and J. Li, "Design of the Batch Predictive-PID Controller Based on the Nonlinear Electric Heating Model," *Lecture Notes in Electrical Engineering*, Vol. 129, No. 6, 2012, pp. 293-298.

[6] B. Hahn and K. Oldham, "On-Off Iterative Adaptive Controller for Low-Power Micro-Robotic Step Regulation," *Asian Journal of Control*, Vol. 14, No. 3, 2012, pp. 624-640.

[7] X. Xu, M. M. Meyers, B. G. Sammakia and B. T. Murray, "Thermal Modeling and Life Prediction of Water-Cooled Hybrid Concentrating Photovoltaic/Thermal Collectors," *Journal of Solar Energy Engineering*, *Transactions of the ASME*, Vol. 135, No. 1, 2013.

[8] Math Works, "Dynamic System Simulation for MatLab," Software, Version 10, 2010.

[9] N. Elshahat and A. Zeit, "Aromatic Plants & Its Agricultural-Medical Products," Arabic, Inc., Egypt, 2003.

Uniformly Illuminated Efficient Daylighting System

Irfan Ullah, Seoyong Shin

Department of Information and Communication Engineering, Myongji University, Yongin 449-728, South Korea.

ABSTRACT

Different approaches have been introduced for the daylighting system to reduce energy consumption, but they were not populated due to complex, high cost, and insufficient designs. There has been a recent problem in achieving uniform distribution of sunlight at a destination deep inside the building. Therefore, we propose a system to achieve high illumination by illuminating the surface of the absorber (optical fibers) uniformly. To capture sunlight, parabolic reflector and Fresnel lens are used. Different reflectors and lenses are considered, compared, and investigated to direct uniform light, which solves the heat problem, into the optical fibers. In addition, this study includes comparison between fiber-based daylighting systems. The proposed system has been verified through simulation and experimental results.

Keywords: Uniform Illumination; Daylighting System; Optical Fiber; Solar Concentrator

1. Introduction

By developing a suitable and more efficient daylight collecting and distributing system will lead us to a solution of the energy problem which is becoming more serious. Natural light is the requirement of our needs. There has been a research that sunlight has positive effects on the human body. It can be used to reduce seasonal affective disorder and other illnesses [1]. Daylighting system gives the reduction in power consumption and makes the cost at one time. It was not progressed due to less efficiency and poor design. In the sense of implementing the system at low cost, different methods have been introduced, but since the system was inefficient, it was difficult for it to be applied to all different kinds of building structures. Mostly, daylighting systems have some problems like: low accuracy in hardware designing, installing, and routing. There is a need of the best design that has high efficiency and simple architecture. Different solar light transport systems have been demonstrated using light pipes, prisms, and optical fibers [2-4]. Available fiber based system is the Himawari solar lighting system [4]. It has two series of models with the different number of lenses and optical fibers. The system contains quartz-glass optical fibers that make the system very high cost. The recent requirement is to get high illumination that is not present in these models due to the limited number of optical fibers and size of collecting system. It contains the outdoor collecting system, which has small illuminated surface area. Therefore, a low illumination is achieved at the indoor unit, which has small surface area. Furthermore, the distribution of uniform light is becoming a very critical issue in the solid-state lighting. On the other hand, this issue is more serious in the daylighting system. In [5], a solar lighting and energy system has been presented. They introduced a fiber-based system to use solar light spectrum for lighting and energy. Our proposed system is also better for solar energy systems to direct uniform light on the solar cells to produce the same and large amount of energy from each cell, because only required light will be directed on the solar cells. Thus, our proposed system can help to produce an efficient solar energy system. Our current research and development show an improvement in the illumination of the optical fiber-based system. This study will focus on uniform distribution of the light. Therefore, transmission of uniform sunlight is possible deep inside the building areas like: store rooms, basement, room without any window, and different floor areas. Furthermore, different lenses and reflectors are investigated and implemented to achieve accuracy.

2. System Architecture

Size of the system is a very critical issue in the daylighting system. In order to install it over different free regions (e.g. roof) of the building, it should occupy a small area with effective output and adopt surrounding environment according to the required size. The size of indoor and outdoor modules is selected according to the required range of illumination. Normally, a large surface area is preferred to capture more light. It is tried to make the size small of the indoor apparatus as compared to the outdoor apparatus to deliver light at a deep inside the building. The light collecting system must capture a large amount of sunlight and focus it over the small area. To achieve this goal, the system is presented using optical fibers, lenses, and reflectors. A sun tracking module is installed to achieve direct light all times of the day, and the light collecting module is interconnected with it. There has been a research using Fresnel lens, but effective results were not achieved due to less efficient hardware designing. Fresnel lens has various advantages like low-cost, less affective from dust, easiness to install, and fewer hardware components. If we compare the surface area of the Fresnel lens and parabolic reflector, it will be concluded that parabolic reflector has more surface area than Fresnel lens with the same diameter. Therefore, more light can be captured using parabolic reflector. Parabolic reflector is used due to some features (e.g., easiness in manufacturing) as compared to other reflectors. In both cases, a second component is necessary in order to get the collimated light. It can also be easily converged or diverged at a certain angle. It was not under discussion in all previous systems to distribute sunlight uniformly on the surface of the absorber. They also tried to make their systems using different modules, but they could not achieve better performance due to lake of the simulation. Therefore, the purpose of this study is to achieve uniform daylight at the destination. In case of non-uniform, each optical fiber will have different amounts of light. Therefore, light will not be uniformly distributed over the different regions of the building, and the heat issue will be produced in case of high illumination. To solve this problem, accurate size and placement of components are mandatory. Previously, sunlight collecting systems were developed using different types of primary and secondary reflectors. In some cases, more than two reflectors were utilized to illuminate the absorber; however, the system becomes costly by increasing the amount of reflectors. A system was also presented that had similar reflectors, and they were moved according to the position of sun. Generally, it is tried to make the system with fewer numbers of modules for manufacturing and installing purposes.

In case of reflectors, diameter of the secondary reflector must be same as the diameter of the absorber to achieve uniform and collimated light. If we use secondary circular reflector, it does not produce collimated light and surface of the absorber remains non-uniform. Results show that only secondary parabolic reflector gives collimated and uniform light. Therefore, it is concluded that primary parabolic reflector with secondary parabolic reflector is a good combination for better performance. In case of Fresnel lens, plano-concave lens is used to produce collimated light and to illuminate the surface of the absorber uniformly. If only a Fresnel lens is used in the system, collimated light can not be produced. The diameter of the concave region, in the plano-concave lens, should be same as the diameter of the absorber to illuminate the absorber equally. Therefore, it is concluded that Fresnel lens as a concentrator with plano-concave lens is a good combination. Furthermore, each material has its own properties of reflectance and transmittance. Surface reflectivity of aluminum and silver are 92% and 96% for solar light, respectively [6], and a lens can transmit light up-to 92% [7].

Optical fibers have many advantages in the daylighting system, including highly acceptable for wiring and routing purposes in the building and the best source for the transmission of light from one place to another. The cost problem of using optical fibers is eliminated by introducing plastic optical fiber (POF) for most of the transmitting part. POF has a big core diameter to transmit a large amount of light by efficient total internal reflection. Distribution of light is an important issue in the solid-state lighting. In our case, we have a light source (optical fibers) of small size with high beam of light. Different lenses and reflectors can be used to distribute uniform light over a certain region at the destination. We can divide optical fibers in every region by considering the required light in that region. The amount of optical fibers depends on the size of the concentrator. If we want to illuminate a big area, we shall use a big concentrator. In order to investigate the efficiency of the system, we have done simulation. There is a compromise between size and efficiency of the Fresnel lens. If we increase the size of the Fresnel lens having a constant pitch, we get less efficiency due to blockage of the light. The Fresnel lens with a constant pitch of 0.5 mm is used. If we increase the diameter of the parabolic reflector, we shall achieve high efficiency due to the large surface area. Finally, it is concluded that parabolic reflector gives better efficiency, and Fresnel lens gives good performance with low cost.

3. Solar Concentrator Design

A concave parabolic reflector is used to concentrate light on the small area, and a convex parabolic reflector is used to direct collimated light on the surface of the absorber. Reflectors having less f-ratios and large aperture areas are used to collect more light. To achieve collimated light, both reflectors have the same f-ratio. A bundle of fifty four optical fibers is utilized to get high illumination and to reduce heat effect. Silica optical fiber (SOF) is mounted before POF. In order to reduce losses due to the air gap between them, index matching jell is applied. A parabolic reflector is designed by

$$y = \frac{x^2}{4 \times f} \quad (1)$$

$$d = \frac{D^2}{16 \times f} \quad (2)$$

$$A = \left(\frac{f}{2 \times N}\right)^2 \quad (3)$$

$$S = \frac{8\pi f^2}{3} \times \left(\left(\left(\frac{D}{4f}\right)^2 + 1\right)^{\frac{3}{2}} - 1\right) \quad (4)$$

In Equation (1), y is value on vertical axis, x is value on horizontal axis, and f is the focal length. In Equation (2), D is the diameter, and d is the depth. In Equation (3), A is the area of aperture, and N is the focal number (f/D). In Equation (4), S is the surface area of the parabolic reflector. Primary and secondary parabolic reflectors have surface areas of 87830.37 mm^2 and 162.16 mm^2. The diameter of the SOF and POF are 1.8 mm and 2 mm, respectively. The bundle of POFs at the destination has a surface area of 216.42 mm^2. It is noted that the input surface area is about 371 times larger than the output surface area. **Figure 1(a)** shows that light is focused from a large area to a small area using reflectors. In **Figure 1(b)**, we show that direct light is focused using a Fresnel lens and distributed on the surface of the absorber using a plano-concave lens. Similar designs for real experiment and software simulation were developed to achieve high accuracy. Therefore, losses in the simulation and real experiments were considered similar.

Here, we shall discuss three approaches using secondary circular reflector, secondary parabolic reflector, and lenses. Circular reflector is used to illuminate the surface uniformly, as shown in **Figure 2(a)**. Direct ray has an angle α, which is 64 degrees, on the surface of the primary parabolic reflector with the horizontal axis. After that, it is reflected from the surface of the secondary reflector at a certain angle. We tried both patterns like: rays do not have a focal point and when they have a focal point. However, rays are not uniform in both cases. A large amount of rays are reflected at the center, but there are fewer rays at the ends. As a result, heat is maximized at the center. Secondary parabolic reflector gives uniform light at the surface of the absorber.

Secondary parabolic reflector has a radius r, depth a, focal length f_2. Primary parabolic reflector has a focal length f_1 and focal point F. It can be seen that focal points of both parabolic reflectors are at the same point. We can determine diameter and focal point of the convex reflector from the triangle as shown in **Figure 2(b)**. If we know the focal point, the diameter will be calculated. If we know the diameter, the focal point will be calculated. Therefore, placement of the secondary parabolic reflector is based on the focal point of the primary reflector. From these relationships, values of all parameters can be calculated. Previously, Fresnel lens based system has been already presented, but there is a lake of uniform distribution of light. Also, we want to compare three approaches to get the best design. In **Figure 2(c)**, the design using lenses is described to show placement and positions of lenses.

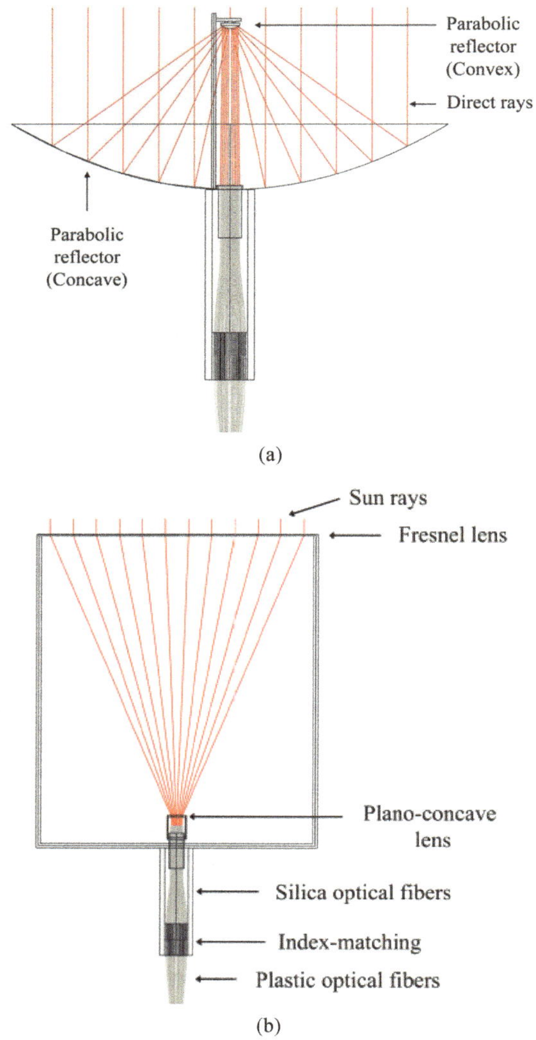

Figure 1. Structure of the system using (a) reflectors and (b) lenses.

4. Simulation Results

LightTools® was used to verify the proposed system. We have generated real time virtual environment in the simulation by designing each part and considering almost each loss. Source, receiver, reflectors, lenses, bending fibers, index-matching, and other hardware components

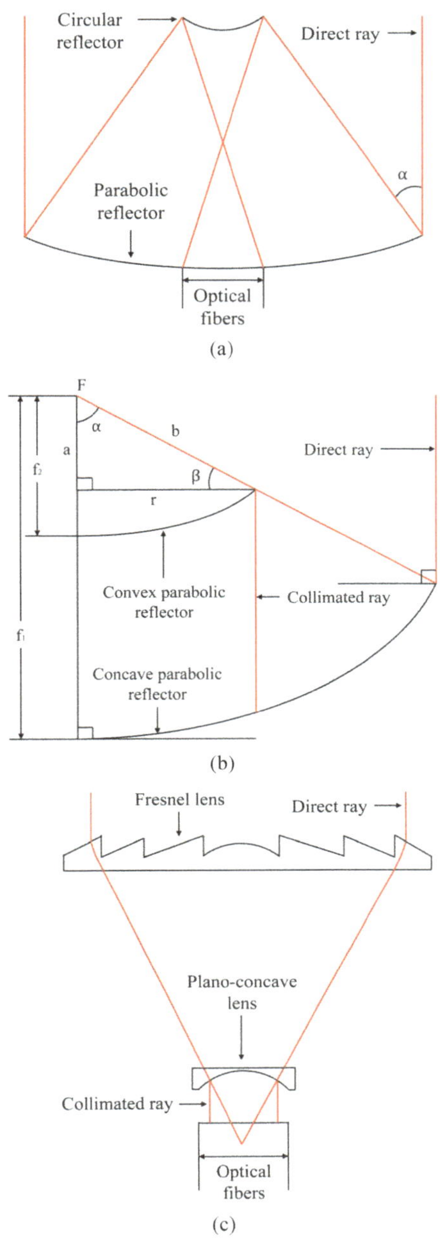

Figure 2. Configuration of the system using (a) secondary circular reflector; (b) parabolic reflectors; and (c) lenses.

are designed in the software. Bending fibers and their losses are considered in the simulation in order to check overall losses. Different refractive indices, reflection coefficients, and transmission coefficients are considered for each material. To see the actual losses, a light source is developed that has a spectrum similar to the sunlight. We also measured the heat effect on the optical fibers that can be estimated from the illuminance charts. Direction cosines of rays have been determined by the ray tracing method. Light rays after reflection from secondary parabolic reflector become perpendicular to the surface of the absorber. For Fresnel lens, rays after passing from the plano-concave lens become perpendicular to the surface of the absorber. Therefore, required lens can be easily designed according to the diameter of the absorber. Simulation is performed step by step from the initial stage to the end by examining the difference between results. Firstly, we considered a circular reflector to get collimated light, but it did not give an equal distribution of the light due to unparalleled rays. Thus, uniform light can't be achieved using circular reflector. In order to achieve collimated light, a convex parabolic reflector is the best choice. The output results, considering the losses due to bending of optical fibers, can be easily seen in **Figure 3**. A light source is created in the simulation that transmits direct rays and produces uniform illuminance on the collecting system. A very large number of rays are transmitted to check the results with high resolution. The length of optical fibers in the simulation is considered same as that of the real time hardware. To reduce spacing between optical fibers, they are arranged in a circular shape. In the bundle of optical fibers, we tried to make the region empty, which has shadow due to any mechanical component.

In **Figure 3(a)**, we show 2D illuminance, 3D illuminance, and rays scattered charts for secondary circular reflector. In this case, the numbers of rays are different in each fiber. Therefore, some optical fibers have high illuminance, but some optical fibers have low illuminance due to unparalleled reflection of rays from the surface of the circular reflector. In **Figure 3(b)**, we illustrate illuminance charts using secondary parabolic reflector, where uniform distribution of light is attained. The surface is uniformly illuminated and high illuminetion is achieved in the bundle of optical fibers. There are two empty regions due to the shadow of the supporting rod and secondary reflector. In order to make it small, we used a very thin rod and a very small secondary reflector. In **Figure 3(c)**, it is shown that all optical fibers have the same amount of rays using lenses. Here, we see that rays are collimated in this case; thus, uniform illumination is achieved. Fresnel lens is designed from the required diameter and focal length. It can be controlled by varying the groove pitch. It can also be designed by varying the groove depth, having a constant draft angle. Fresnel lens has a big advantage of long life time, and it is better for the system to implement, which is the requirement of lighting industry. A Fresnel lens is designed by

$$EFL = \frac{r}{n-1} \tag{5}$$

where EFL is the effective focal length, r is the radius, and n is the refractive index of the material, which has a value of 1.494.

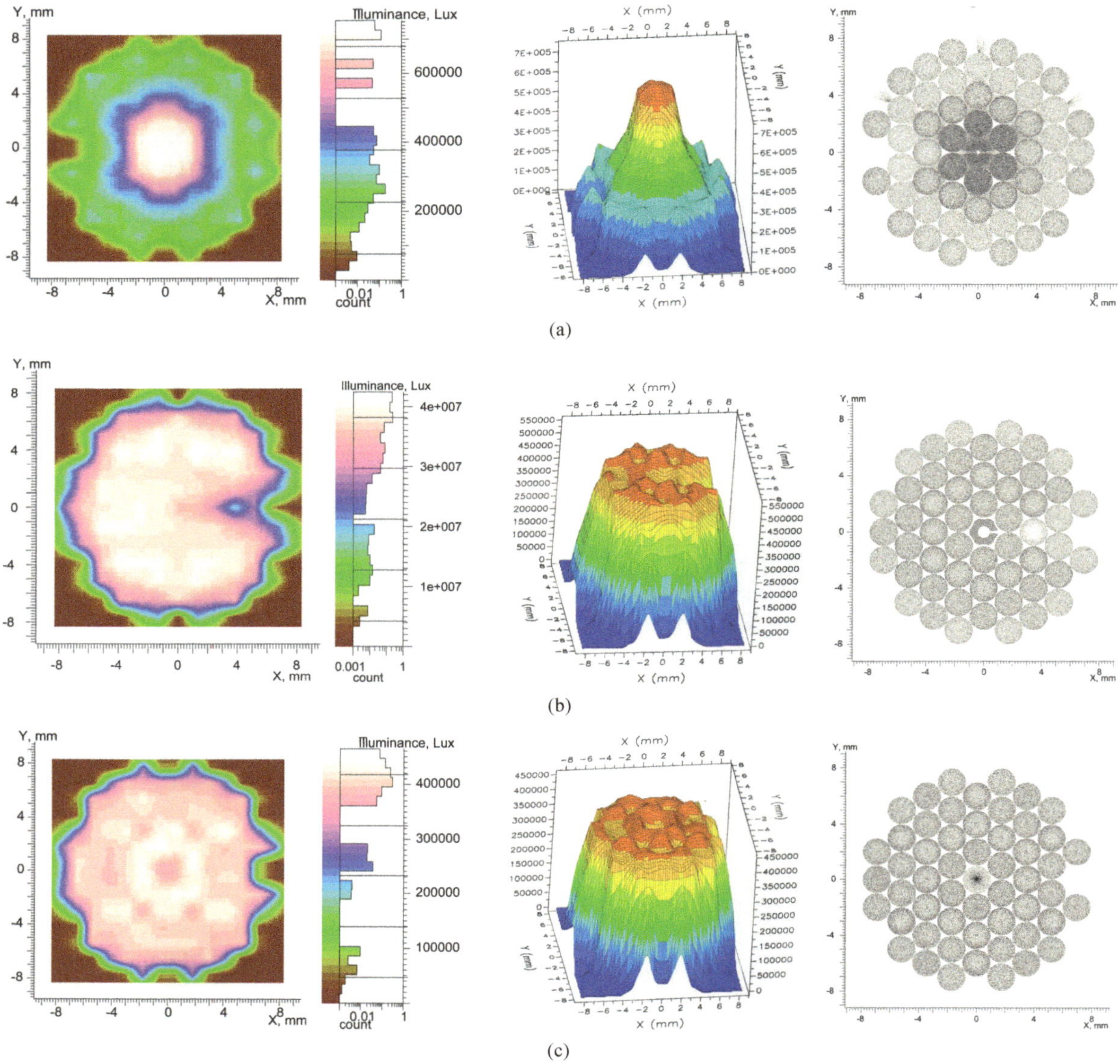

Figure 3. 2D illuminance, 3D illuminance, and rays scattered charts using (a) secondary circular reflector (b) secondary parabolic reflector, and (c) lenses.

5. Experimental Results

The system contains sun tracking, light collecting, light transmission, and light distributing modules. Experimental setup of the system using reflectors and lenses is shown in **Figures 4(a)** and **(b)**. Three presented designs were established separately. The illuminance is measured in sunny and cloudy weather conditions at different times of the day. Experiments have shown that the proposed daylighting system can illuminate the required region uniformly. Uniform illumination using parabolic reflectors and lenses is shown in **Figures 4(c)**, and **Figure 4(d)** shows the output illumination using secondary circular reflector that is not uniform. Therefore, the proposed system is acceptable for real implementation. Finally, it is concluded that both approaches using parabolic reflectors and lenses give high illuminance, because the light is distributed uniformly in both cases. Measurements of different parameters are illustrated in **Table 1**. Average input illuminance and output illuminance, which is measured at the end side of the fiber bundle, are mentioned in **Table 2**. Currently, the system is in working state. It has been installed at the university building to illuminate the room during the daytime.

6. Conclusions

As a result of the proposed system, a specific area of the building can be completely illuminated. Two efficient approaches have been presented. The study also demon-

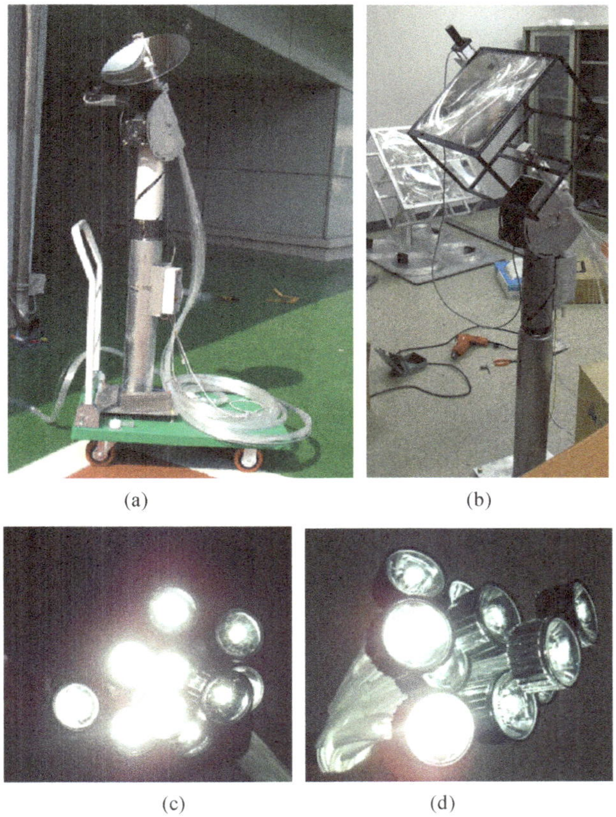

Figure 4. Hardware schematic using (a) reflectors and (b) lenses. Experimental results showing (c) uniformly illumination and (d) non-uniformly illumination.

Table 1. Measurements of different parameters.

Title	Value (mm)
Diameter of the primary parabolic reflector	320
Focal length of the primary parabolic reflector	128
Diameter of the secondary parabolic reflector	13.75
Focal length of the secondary parabolic reflector	5.5
Distance between parabolic reflectors	122.5
Size of the Fresnel lens	300 × 300
Radius of the Fresnel lens	163.02
Focal length of the Fresnel lens	330
Diameter of the plano-concave lens	16.1
Focal length of the plano-concave lens	21.5
Distance between lenses	319.4

Table 2. Illuminance from simulation.

Title	Value (klx)
Input illuminance	1.0
Output illuminance using secondary circular reflector	248.7
Output illuminance using secondary parabolic reflector	313.7
Output illuminance using lenses	279.4

strates that more sunlight can be guided into optical fibers using a large sunlight collecting system. Finally, it is concluded that illuminance of more than 500 lx can be delivered uniformly into the building. A practical daylighting system has been explained and verified by real experiments and simulations.

7. Acknowledgements

This research was supported by Basic Science Research Program through the National Research Foundation of Korea (NRF) funded by the Ministry of Education, Science and Technology (2010-0012926).

REFERENCES

[1] A. Dunne, "Some Effects of the Quality of Light on Health," *Journal of Orthomolecular Medicine*, Vol. 4, No. 4, 1989, pp. 229-232.

[2] I. Ullah and S. Shin, "Development of Optical Fiber-Based Daylighting System with Uniform Illumination," *Journal of the Optical Society of Korea*, Vol. 16, No. 3, 2012, pp. 247-255.

[3] I. Ullah and S. Shin, "Development of Optical Fiber-Based Daylighting System with Collimated Illumination," *Proceedings of the* 2012 17*th Opto-Electronics and Communications Conference* (*OECC*), Busan, 2-6 July 2012, pp. 596-597.

[4] V. E. Gilmore, "Sun Flower over Tokyo," Popular Science, Bonnier Corporation, New York, 1988.

[5] C. Tsuei, W. Sun and C. Kuo, "Hybrid Sunlight/LED Illumination and Renewable Solar Energy Saving Concepts for Indoor Lighting," *Optics Express*, Vol. 18, No. S4, 2010, pp. A640-A653.

[6] P. Nostell, A. Roos and B. Karlsson, "Ageing of Solar Booster Reflector Materials," *Solar Energy Materials and Solar Cells*, Vol. 54, No. 1-4, 1998, pp. 235-246.

[7] H. Abdul-Rahman and C. Wang, "Limitations in Current Day Lighting Related Solar Concentration Devices: A Critical Review," *International Journal of the Physical Sciences*, Vol. 5, No. 18, 2010, pp. 2730-2756.

Significance of Storage on Solar Photovoltaic System —A Residential Load Case Study in Australia

Mohammad T. Arif, Amanullah M. T. Oo, A. B. M. Shawkat Ali, G. M. Shafiullah

Central Queensland University, Power and Energy Centre, Institute for Resources Industries and Sustainability, Rockhampton, Queensland 4702, Australia.

ABSTRACT

Energy storage is an essential part in effective utilization of Renewable Energy (RE). Most RE sources cannot provide constant energy supply and introduce a potential unbalance in generation and demand, especially in off-peak periods when RE generates more energy and in peak period when load demand rises too high. Storage allows intermittent sources like solar Photovoltaic (PV) to address timely load demand and adds flexibility in load management. This paper analyses the significance of storage for residential load considering solar PV as RE generator. The significance of storage was evaluated in off-grid or stand alone and grid connected configurations. Moreover it outlined the significance of storage in terms of environment and economics by comparing the Renewable Fraction (RF), Greenhouse Gas (GHG) emission, Cost of Energy (COE) and Net Present Cost (NPC). Investigation showed that storage has positive influences on both (off-grid and grid connected) configurations by improving PV utilization. It was found that in grid connected configuration storage reduced 46.47% of GHG emission, reduced COE, NPC and improved RF compared to the system without storage.

Keywords: Storage; Photovoltaic; Residential Load; GHG; NPC; COE

1. Introduction

Storage system stores excess energy and releases stored energy when load demand goes high; it also minimizes the intermittent nature of solar energy. Storage offers substantial added values to the energy sector. Storage can improve electric grid system reliability, efficiency and flexibility by using it for frequency regulation [1]. Storage also stabilizes the cost of electricity and helps to reduce Greenhouse Gas (GHG) emission [2]. Fast response storage such as batteries, flywheels, and compressed air is as effective as conventional generation for the supply of regulation services [3]. However the use of storage depends on the load demand and generated electricity from RE sources. Natural factors limit solar PV to generate electricity according to the load demand. This study considered typical residential load in Australia to investigate the significance of storage. Residential solar PV system is very popular both in metropolitan area and regional areas in Australia and worldwide. Moreover Australian Government added carbon tax and provided rebates for residential solar PV projects to make RE popular to general people. However this home based solar PV systems are unable to meet the resident's desire as presently installed systems are mostly without storage that are unable to provide electricity during night and cloudy days. Therefore for effective solar PV systems, storage should be integrated with it.

Load profile of a residential house varies according to the residents work time pattern. In the Capricornia region of Rockhampton in Australia, the working nature of the residents of Kawana suburb is such that most of the residents start work between 7:00 AM to 8:00 AM and return home between 5:00 PM to 6:00 PM during weekdays. Overall load demand on distribution network is very high in the evening and also in the morning however the residential solar PV generates electricity mostly during the working period which is of minimum use for the residents. Considering this scenario, this paper investigates how storage can improve this situation and help residential solar PV to support most of the time. Significance of storage is demonstrated in this paper by evaluating the benefit of storage in terms of environmental and economic perspective.

2. Background

Every kWh of electricity from conventional sources can be replaced by RE which eventually can reduce overall GHG emission. In Australia, Government's Mandatory Renewable Energy Target (MRET) is to ensure 20% electricity or additional 45,000 GWh from RE sources by 2020 [4-6]. Moreover national GHG emission target is 60% below 2000 level by 2050 [5]. According to the National GHG inventory (Kyoto protocol accounting framework), energy sector is the major contributor to the GHG emission [7]. It is found that Energy sector alone contributes 62.71% of GHG emission in Queensland and overall 73.93% in Australia [5].

Many studies explained the effectiveness of RE in replacing fossil-fuel to produce electricity. Compared to most other countries, Australia's solar resource is equal to world's best. Annual average solar exposure is greater than 2200 kWh/m^2/year [8]. Recent PV collectors can track the sun to allow collection of a greater amount of energy. However the main drawbacks of solar PV are diurnal solar cycle, irregular solar radiation, seasonal variation and geographic locations. Moving clouds can produce fast and short irradiance, which causes voltage and power fluctuations at the Point of Common Coupling (PCC) and can introduce feeder losses [9,10]. The fluctuation level of solar energy is such that it can drop energy in very short time as shown in **Figure 1** [11].

PV systems are a small part of present electricity infrastructure and have little effect on the overall power quality or reliability of grid power. However, when PV penetration reaches sufficiently high level, the intermittent/transient nature of PV generation begins to have noticeable negative effects on the entire grid [12]. Grid energy storage is mostly important for matching supply and demand over 24 hours period of time. Storage systems and storage integrated PV applications are described in the next section.

2.1. Storage Systems

Storage improves the capacity of RE systems. The ability to store large amounts of energy would allow electrical utilities to have greater flexibility in their operation, because with this option the supply and demand do not have to be matched instantaneously [13]. From different form of energy storage systems, Pumped hydro, Compressed Air Energy Storage (CAES), Thermal Energy Storage (TES), Superconducting Magnetic Energy Storage (SMES), Flywheel, Hydrogen, Capacitors and Batteries are used in different renewable energy conservation process. Advantage and limitations of various storage systems are listed in **Table 1** [14-19]. Although large scale storage is still expensive but significant research is underway for inexpensive & efficient large scale storage systems [20] suitable for large scale RE applications.

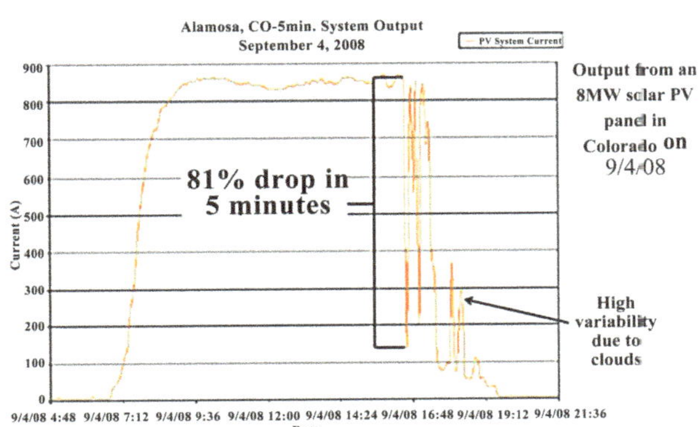

Figure 1. Variability nature of solar energy [11].

2.2. Storage Integrated PV Applications

PV arrays can be considered for different kinds of Application *i.e.* from small stand-alone to large scale gridconnected solar power plants, also from ground to roof installation or building integrated photovoltaic application. Stand-alone solar power system is suitable for applications in remote areas, which is located far away from utility grid. Grid-connected solar PV application interacts with utility grid by transforming DC to AC by an inverter. AC electricity is consumed by AC home appliances and excess energy stored in storage for home-use, also can be sold back to the utility.

A recent study on high penetration of PV on present grid mentioned that energy storage is the ultimate solution for allowing intermittent sources to address utility base load needs [12]. Storage integrated PV systems should be beneficial in operational, financial and environmental perspective. Therefore a model has been developed to assess economical & financial benefit of storage with solar PV and explained in the next section.

3. Model Evaluation

Simulation was performed to identify the optimized configuration of PV with storage, grid or diesel generator for the residential load. Model was developed using HOMER optimization tool as shown in **Figure 2**. The model was evaluated considering the project life time of 20 years.

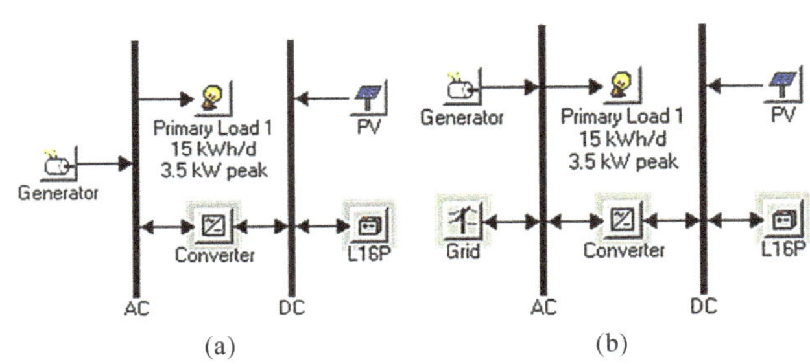

Figure 2. Simulation model configuration: (a) Off-grid configuration; (b) Grid-connected configuration.

Table 1. Energy storage systems [14-19].

Type of storage	Energy efficiency (%)	Energy density (Wh/kg)	Power density (W/kg)	life (cycles or years)	Discharge at rated capacity (hours)	Response time (s)	Self discharge
Pumped hydro	70 - 80	0.3	-	20 - 60 years	1 - 24+	10	Negligible
CAES	40 - 50	10 - 30	-	20 - 40 years	1 - 24+	360	Low
TES	75	-	-	30 years	-	>10 s of minutes	-
SMES	90	10 - 75	-	>100,000	2.7×10^{-7} - 0.0022	0.01	10% - 15%
Flywheel (steel)	85 - 95	5 - 30	1000	>20,000	2.7×10^{-7} - 0.25	0.1	Very high
Super capacitor	80 - 95	2 - 5	800 - 2000	10 years	2.7×10^{-7} - 1	0.01	5% - 20%
Lead-acid	65 - 80	20 - 35	25	200 - 2000	0.0027 - 2+	<1/4 cycle	Low
Ni-Cd	60 - 90	40 - 60	140 - 180	500 - 2500	0.0027 - 2+	<1/4 cycle	0.2% - 0.3%
Li-MH	50 - 80	60 - 80	220	<3000			High
Li-ion	70 - 85	100 - 200	360	500 – 10,000	0.017 - 2+	<1/4 cycle	1% - 5%
NaS	70 - 89	120	120	2000 - 3000	0.0027 - 2+	<1/4 cycle	-
Hydrogen	50	100 - 150	-	-	-	360	Low
Fuel cell	-	-	-	>1000	0.0027 - 24+	<1/4 cycle	-

The performance matrices of the optimized model, required data, component costs and considerations are explained in following subsections.

3.1. Performance Matrices

The performance matrices considered are Net Present Cost (NPC), Cost of Energy (COE) as economical factor, Renewable Fraction (RF) and Greenhouse Gas (GHG) emission as environmental factor which compared in off-grid and grid connected configurations.

Net Present Cost (NPC): NPC is the present value of all costs over its lifetime minus the present value of all the revenue that it earns over its lifetime. Costs includes capital cost, replacement cost, operation and maintenance (O&M) costs, fuel cost, emission penalties and cost of buying power from the grid. Revenue includes salvage value and grid sales revenue. NPC is the main economic output of the model and ranked all optimized systems according to total NPC as represented [21,22] by Equation (1):

$$\text{NPC}(\$) = \frac{\text{TAC}}{\text{CRF}} \tag{1}$$

where TAC is the total annualized cost (which is the sum of the annualized costs of each system component). The capital recovery factor (CRF) is given by Equation (2):

$$\text{CRF} = \frac{i(1+i)^N}{(1+i)^N - 1} \tag{2}$$

where N denotes number of years and i means annual real interest rate (%). Model considered nominal interest rate. The overall annual interest rate considered as 6%.

Cost of Energy (COE): It is the average cost per kWh of useful electrical energy produced by the system. It can be calculated by dividing the annualized cost of producing electricity by the total useful electric energy production and represented in [21] by Equation (3):

$$\text{COE} = \frac{C_{ann,tot}}{E_{prim,AC} + E_{prim,DC} + E_{def} + E_{grid,sales}} \tag{3}$$

where $C_{ann,tot}$ is total annualized cost of the system ($/yr), $E_{prim,AC}$ is AC primary load served (kWh/yr), $E_{prim,DC}$ is DC primary load served (kWh/yr), E_{def} is deferrable load served (kWh/yr) and $E_{grid,sales}$ is total grid sales (kWh/yr).

Emission: Emission is widely accepted and understood environmental index. Greenhouse gases (CO_2, CH_4, N_2O, HFCs, PFCs, SF6) are considered one of the main causes for global warming. In addition SO_2 is another pollutant gas released by coal fired energy system. Emission is measured as yearly emission of the emitted gases in kg/year and emissions per capita in kg/kWh. It is considered as input in the model when calculating the other O&M cost. The other O&M cost is the sum of system fixed O&M cost, penalty for capacity shortage and penalty for emissions of pollutants. It was represented in [21] as shown in Equation (4):

$$C_{om,other} = C_{om,fixed} + C_{cs} + C_{emission} \tag{4}$$

where $C_{om,fixed}$ is system fixed O&M cost ($/yr), C_{cs} is the penalty for capacity shortage ($/yr) and $C_{emission}$ is the penalty for emission ($/yr).

Renewable Fraction (RF): RF is the total annual renewable energy production divided by the total energy production. A greater value of this fraction means the greater renewable energy generated. RF can be calculated by Equation (5) [23]:

$$RF = \frac{E_{PV}}{E_{TOT}} \quad (5)$$

where E_{PV} and E_{TOT} are the energy generated by photovoltaic and total energy generated respectively.

3.2. Model Data

Residential load data of Rockhampton (a regional subtropical town based in Queensland, Australia) was used in this analysis. This regional site has good solar radiation history and most the residents working nature is monotonous and many houses have roof top PV facility. Solar radiation data was collected from Bureau of Meteorology [24] for the site location. All required system components are discussed in the following sub-sections.

3.2.1. Data Collection

A three bed room house daily load in Capricornia region of Rockhampton was estimated as 15.0 kWh/day. This was done by multiplying the power rating of all the home appliances by the number of expected operating hours on an average day to obtain daily load in watt-hour (Wh). **Table 2** shows the list of electrical appliances used in a 3 bed room house and calculated daily average electricity consumption.

This 3 bed room residential load was considered for the analysis; therefore yearly electricity consumption becomes 5475 kWh/yr. For grid connected household appliances daily average load can also be obtained from monthly utility bills. According to Ergon Energy (utility operator in Queensland, Australia) electricity billing information, the daily average residential electricity consumption is 15.7 kWh/day [25]. Providing estimated hourly load data in the model, the daily load profile with seasonal variation is as shown in **Figure 3**. Daily electricity consumption pattern shows that 08:00 - 09:00 hrs and 18:00 - 22:00 hrs have the peak demand however from 18:00 - 19:00 hrs has the super peak load demand.

The load profile in **Figure 3** shows the daily consumption pattern and the maximum consumption period is 6:00 PM to 10:00 PM and in the morning 8:00 AM to 9:00 AM.

3.2.2. Solar Radiation and Power from PV Array

Solar radiation varies with time and season. In Australia yearly average sunlight hours varies from 5 to 10 hours/day and maximum area is over 8 hours/day [24] and in Rockhampton solar radiation varies from 8 to 9 hrs/day. Therefore to capture maximum RE and to provide electricity at night or cloudy day, a storage system should be incorporated.

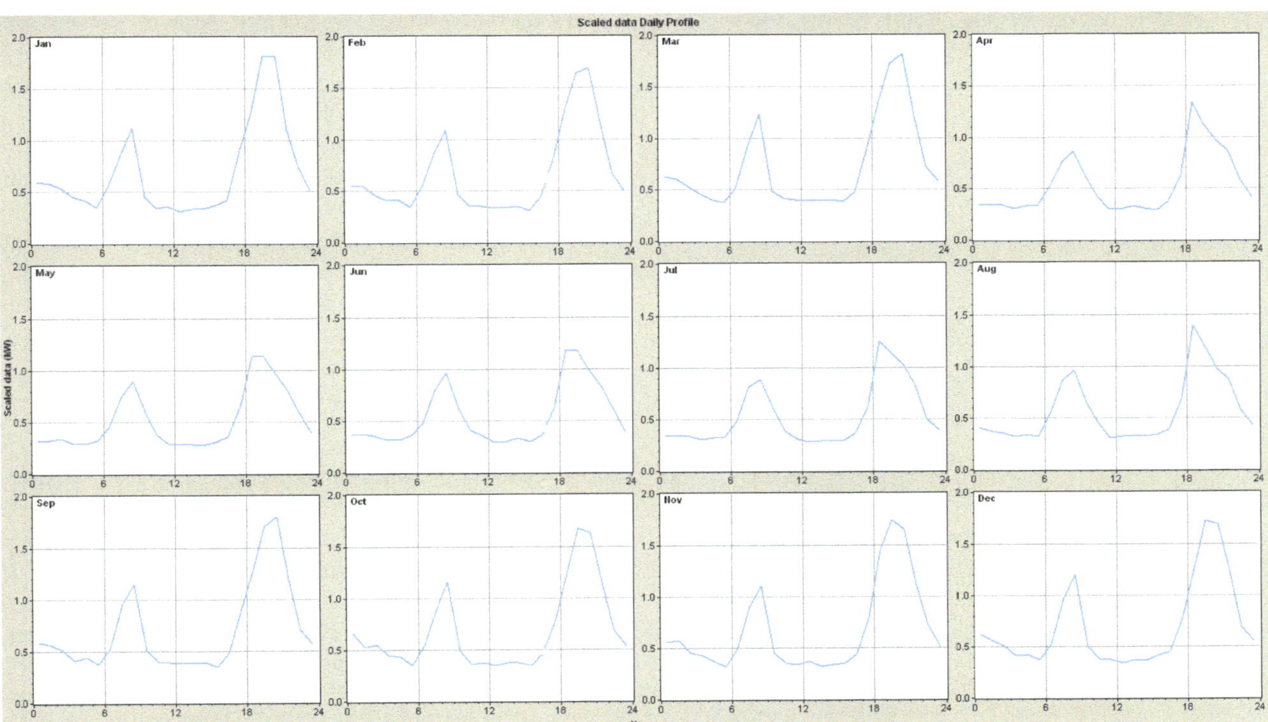

Figure 3. Daily electric load demand in each month.

Hourly solar radiation data was collected from Bureau of Meteorology [24] for the year 2009 and 2010 of station No. 039083 for the location of Rockhampton Aero Weather Station (−23.3753°N, 150.4775°E) which is at 10.4 m above the mean sea level. Daily average solar radiation in Capricornia region of Rockhampton city is as shown in **Figure 4**. It is found that annual average solar radiation is 5.48 kWh/m²/day and the lowest monthly average solar radiation 4 kWh/m²/day in May.

Table 2. Electrical appliances of a three bedroom house & daily average load.

Appliances	Rating	Qty	Average daily use (Wh/day)
Refrigerator (active/standby)	602 kWh/year	1	1650
Electric stove	2100 W	1	2100
Microwave oven	1000 W	1	500
Rice cooker	400 W	1	200
Toaster	800 W	1	80
Ceiling fan	65 W	5	1300
Fluorescent light	16 W	20	450
Washing machine (vertical axis)	500 W	1	71
Vacuum cleaner	1400 W	1	200
Air conditioner (window type)	1200 W	3	1200
TV 32" LCD (active/standby)	150/3.5 W	1	670
DVD player (active/standby)	17/5.9 W	1	50
Cordless phone	4 W	1	96
Computer (laptop)	20 W	1	80
Clothe iron	1400 W	1	350
Heater (portable)	1200 W	1	600
Hot water system	1800 W	1	5400
Total			14,997 Wh/day

Data source: product catalogue and [26].

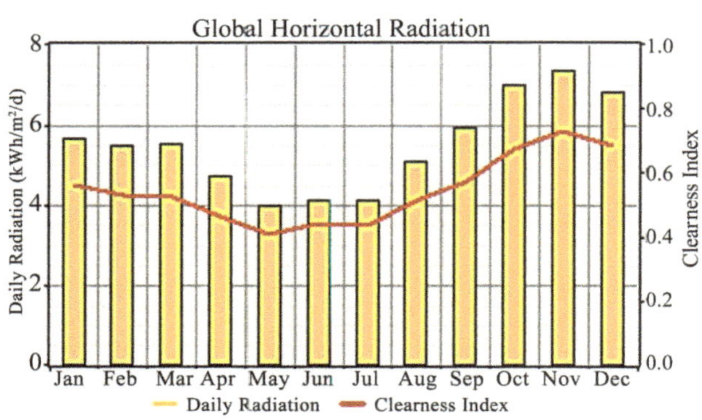

Figure 4. Daily average solar radiation in Rockhampton.

PV array can be modeled as a device that produces DC electricity in direct proportion to the global solar radiation. Therefore, the power output of the PV array can be calculated using Equation (6) [27,28].

$$P_{PV} = Y_{PV} f_{PV} \left(\frac{G_T}{G_{T,STC}} \right) \left[1 + \alpha_p \left(T_C - T_{C,STC} \right) \right] \quad (6)$$

where Y_{PV}—rated capacity of PV array which means power output under standard test conditions [kW]. f_{PV}—PV de-rating factor [%]. G_T—solar radiation incident on PV array in current time step [kW/m²]. $G_{T,STC}$—incident radiation under standard test conditions [1 kW/m²]. α_P—temperature coefficient of power [%/°C]. T_C—PV cell temperature in current time step [°C]. $T_{C,STC}$—PV cell temperature under standard test conditions [25°C]. Performance of PV array depends on de-rating factors like temperature, dirt and mismatched modules.

3.2.3. Storage

Battery is one of the fast response storage systems. For this analysis Trojan L16P Battery (6 V, 360 Ah) at 24 V DC system voltage was used. The efficiency of this battery is 85%, minimum state of charge 30%. Model considered PV array includes the battery charge controller with efficiency 95% and PV de-rating factor of 90% therefore overall charging efficiency considered (95 × 90)% = 85.5%.

Battery dispatch: Two battery dispatch strategy explained [29], named "load following" and "cycle charging". In load following strategy, a generator produces only enough power to serve the load, battery charging and supporting deferrable load depends on renewable power sources. In cycle charging strategy, whenever a generator operates it runs at its maximum rated capacity, charging the battery bank with any excess electricity until the battery reaches the specified state of charge. Load following strategy was considered for this analysis for the better utilization of solar energy.

3.3. System Components and Costs

Table 3 lists the required system components with related costs in Australian currency. PV array, battery charger, inverter, deep cycle battery, generator and grid electricity cost are included for the analysis. It was found that 1.52 kW PV array with inverter price is $3599 [30], also it was found that 1.56 kW PV array with inverter price is $4991 [31]. However model considered PV array includes the battery charger therefore inverter costs considered separately. SMA Sunny Boy Grid Tie Inverter (7000 Watt, SB7000US) price is $2823 [32], however Sunny Boy 1700 W inverter price is $699 [33] and in

Table 3. Technical data and study assumptions.

Description	Value/Information
PV array	
Capital cost	$3100.00/kW
Replacement cost	$3000.00/kW
Life time	20 years
Operation & maintenance cost	$50.00/year
Grid electricity	
Super peak (6 PM - 7 PM)	$0.65/kWh
Peak time (8 PM - 10 PM, 8 AM - 9 AM)	$.38/kWh
Off Peak (all other time)	$0.30/kWh
Emission factor	
CO_2	632.0 g/kWh
CO	0.7 g/kWh
Unburned hydrocarbons	0.08 g/kWh
Particulate matter	0.052 g/kWh
SO_2	2.74 g/kWh
NO_x	1.34 g/kWh
Inverter	
Capital cost	$400.00/kW
Replacement cost	$325.00/kW
Life time	15 years
Operation & maintenance cost	$25.00/year
Storage (Battery)	
Capital cost	$170.00/6V 360 Ah
Replacement cost	$130.00/6V 360 Ah
System voltage	24 volts
Generator	
Capital cost	$2200.00/kW
Replacement cost	$2000.00/kW
Operation & maintenance cost	$0.05/hr
Life time	15,000 hrs
Fuel cost	$1.54/ltr

verter efficiency was considered 94%.

Grid electricity cost in Rockhampton was found from Ergon Energy's electricity bill and for Tariff-11 it was $0.285/kWh (including GST & service). However carbon tax at the rate of $23 for every ton of emission increases the electricity bill as well as the cost of conventional energy sources, therefore off-peak electricity cost is considered as $0.30/kWh for this analysis. Trojan T-105 6 V, 225 AH (20 HR) flooded Lead-acid battery price is $124.79 [34]. Fuel cost for generator is considered at the available price in Rockhampton. **Table 3** shows the unit cost of each component.

The significance of storage was analyzed from the optimization result and evaluated environmental and economical advantages of storage in off-grid and grid-connected configurations in six different cases.

Configuration-1: Off-grid Configuration [as shown in Figure 2(a)]

- Case-1: Diesel Generator only;
- Case-2: PV with Diesel Generator;
- Case-3: PV with Storage and Diesel Generator.

Configuration-2: Grid-connected Configuration [as shown in Figure 2(b)]

- Case-1: Grid only;
- Case-2: PV with Grid and Diesel Generator;
- Case-3: PV with Storage, Grid and Diesel Generator.

4. Results and Discussion

Model was analyzed in six different cases and the output of the optimized model is explained below:

Configuration—1

Case 1. Diesel Generator in off-grid configuration:

In this configuration, total load of 5475 kWh/yr were supported only by 10 kW diesel generator which consumed enough fuel (8371 L/yr) and emitted significant amount of pollutant gas to the air. Generator requires frequent maintenance and fuel cost is also high, therefore NPC was high. COE also found very high, which is $5.238/kWh. This case configuration is the costliest and environmentally most vulnerable.

Case 2. PV with Diesel Generator in off-grid configuration:

In this configuration, solar PV used to support residential load and used diesel generator as backup power. To meet 5475 kWh/yr load demand, 12 kW PV system was used with 5 kW inverter and 10 kW diesel generator. It was found that, although PV generates electricity more than the total load demand but could not meet the load demand during night.

Total 12,489 kWh/yr of electricity was generated by PV and diesel generator, where 3581 kWh/yr or 29% was from diesel generator and 8908 kWh/yr or 71% was from PV array but most of the energy from PV array was wasted. Diesel generator directly supports 65.40% of load and PV array supports only 1894 kWh/yr or 34.60% of load through inverter. Therefore a significant amount of electricity from PV array was wasted and the excess electricity was 6893.10 kWh/yr or 55.19% of total electricity production but compared to the total PV array production the wasted electricity was 77.38%. Storage could reduce this great amount of electricity loss also could reduce the use of diesel generator.

Figure 5(a) shows the electricity generated from PV array and the diesel generator. Compared to the load demand, PV generates more electricity but PV was unable to meet timely load demand as shown in **Figures 5(b)-(d)**. **Figure 5(b)** shows the daily electricity generated by PV array and shows that PV was unable to generate electricity in the morning and night. During morning and night, load demand was supported solely by diesel gen-

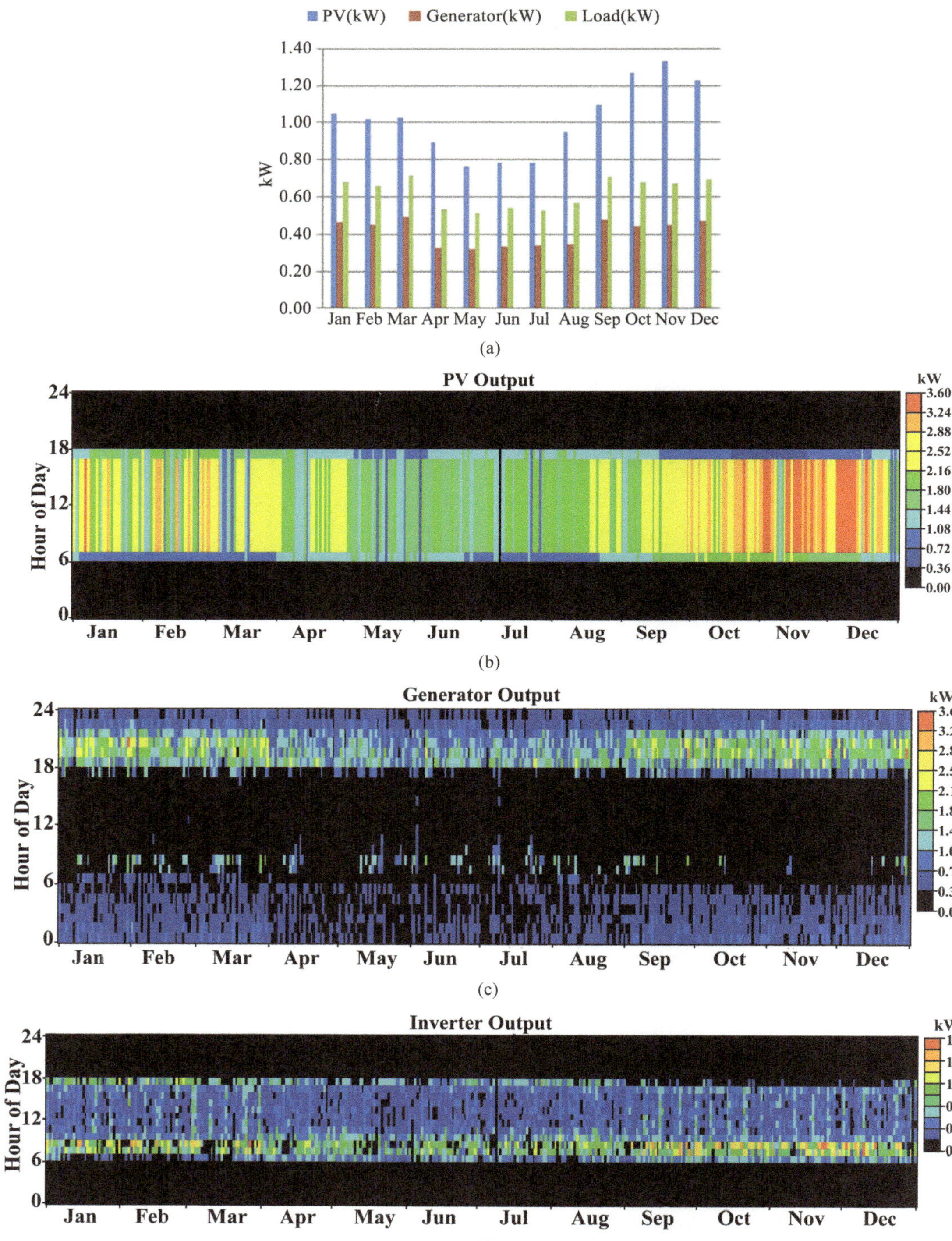

Figure 5. Configuration-1 Case-2 output: (a) PV and diesel generator output compared to load demand; (b) Electricity generated from PV array; (c) Electricity generated from diesel generator; (d) Inverter output.

erator as shown in **Figure 5(c)**. Inverter converts only a fraction of PV generated DC electricity to AC according to the load demand as shown in **Figure 5(d)**, therefore great amount of electricity simply wasted.

Case 3. PV with Storage and Diesel Generator in off-grid configuration:

In this case, model adds battery as storage. This optimized model used 10 kW PV, 32 number of Trojan L16P battery (@ 6 V, 360 Ah) at 24 V DC system voltage, 1 kW diesel generator and 5 kW inverter to support the load of 5475 kWh/yr. This model configuration nearly shaded out the use of diesel generator and only 3 kWh/yr load was supported by diesel generator. It was found that, PV generates electricity more than the load demand and battery stored enough energy to support the load during morning and night.

Total 7427 kWh/yr of electricity was generated from PV array and diesel generator where diesel generator generates only 3.37 kWh/yr. PV array generates 7423 kWh/yr or 99.95% of total production. PV array with storage supports 5472 kWh/yr or 99.95% of load demand and diesel generator supports only 0.05% of load demand. AC load supported directly by PV array through inverter during day time and by stored energy in battery during morning & night time. Inverter converts 5821 kWh/yr of DC to 5472 kWh/yr of AC. Battery stored 4040 kWh/yr of energy and supplied 3477 kWh/yr to support the load. Battery stored 54.42% of PV generated energy and supported 63.50% of load while PV directly supports 36.44% of load. However still 1039 kWh/yr of excess energy generated by the PV array, *i.e.*, 13.99% of total generated energy simply wasted which could be sold to the grid.

In this application scenario, utilization of inverter was better than earlier case scenario, because inverter not only converts the PV output but also converts stored energy from battery as shown in **Figure 6(d)** compared to **Figure 5(d)**. **Figure 6(a)** shows the electricity generated from PV array, diesel generator compared to load demand. Diesel generator only used in March for a short time where load to PV output ratio is highest. **Figure 6(b)** shows the daily electricity generated by PV array and it is clear that PV array is unable to support load in the morning and night when battery supports the load as shown in **Figure 6(c)**. It was found that battery SOC was from 30% to 100% in this case. However a moderate amount of energy was wasted in this case.

This system supports nearly 100% load demand by PV and battery therefore there is less emission which makes it environment friendly off-grid configuration.

Configuration—2

Case 1. Grid only configuration:

This is the present connection configuration for household electricity. Grid supplies the total load demand of 5475 kWh/yr. Grid electricity tariff varies with season, time and application [35,36]. In this grid connected configuration model, for residential load grid electricity cost was considered in 3 different price levels shown in **Table 3**. However it is mentioned earlier that currently available grid electricity cost for residential use is $0.285/kWh. Simulation result showed that average COE becomes $0.393/kWh. Grid electricity mainly comes from conventional sources therefore a good amount of pollutant gas emitted to the air.

Case 2. Grid connected PV with Diesel Generator:

In this case diesel generator was shaded out as grid electricity was much cheaper. PV still contributed a reasonable portion of load demand. This optimized model used 3 kW PV with 5 kW inverter and grid supply. Total 6924 kWh/yr of electricity was generated where PV array generates 2227 kWh/yr or 32.16% of total production. Grid supplied 4697 kWh/yr *i.e.*, 67.84% of total production or 85.79% of total load demand. PV array supported 778 kWh/yr or 14.21% of load demand, also 529 kWh/yr of PV array generated electricity was sold back to the grid. However 837 kWh/yr or 37.58% of PV generated electricity was wasted in timely demand mismatch which could be stored and supplied to the load.

Figure 7(a) shows total electricity production from PV and grid support compared to the load demand. It was found that PV generates electricity during day time and peak demand was met by grid supply therefore a significant amount of PV generated electricity was wasted as shown in **Figure 7(b)**. In this application scenario, inverter was minimally utilized only during day time as shown in **Figure 7(c)**.

Case 3. Grid connected PV with Storage and Diesel Generator:

In this case diesel generator was shaded out as overall generated electricity was much cheaper. PV array contributed a good portion of load demand with storage support. This case is very interesting compared to the earlier case that by only adding sufficient amount of storage it improved the contribution of PV array also made room for additional PV array to support the same total load. This optimized model used 5 kW PV with 12 number of L16P battery (@ 6 V, 360 Ah) at 24 V DC system voltage, 5 kW inverter and grid supply.

Total 5943 kWh/yr of electricity was produced where grid supplied 2232 kWh/yr or 37.55% of total production or 40.76% of total load demand. PV array produced 3712 kWh/yr which was 62.46% of total production. Battery stored 1860 kWh/yr or 50.11% of PV generated electricity and supported 1599 kWh/yr or 29.20% of total load demand. However PV array directly supported 1644 kWh/yr or 30.03% of load demand through inverter.

Figure 8(a) shows the electricity production from PV array and support from grid compared to load demand. PV cannot support load during morning and night as

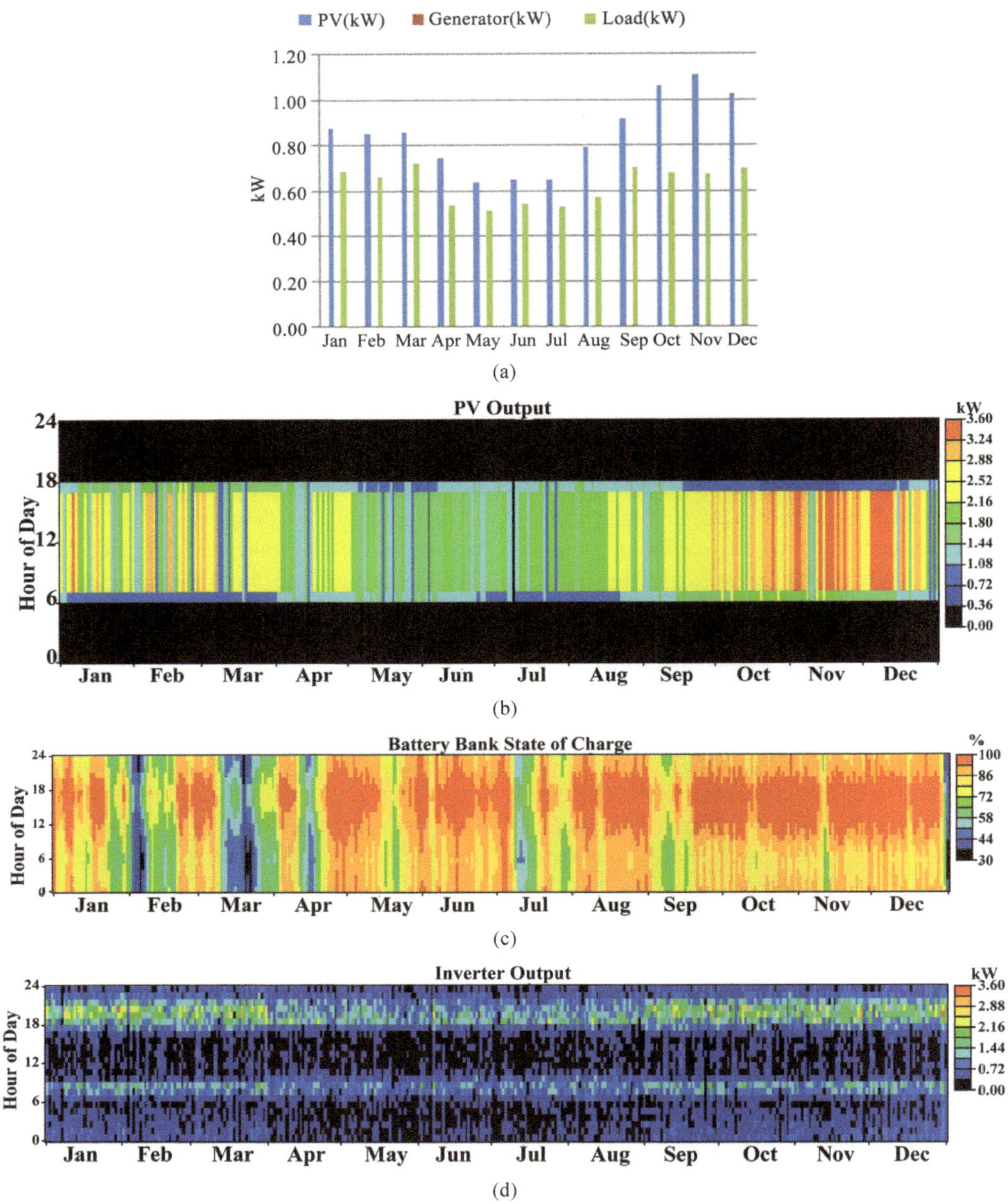

Figure 6. Configuration-1 Case-3 output: (a) PV & Diesel generator output compared to load demand; (b) PV array output; (c) Battery state of charge (SOC); (d) Inverter output.

shown in **Figure 8(b)**. **Figure 8(c)** shows the battery SOC and it shows that battery mostly utilized during peak demand time in the evening which is evident in **Figure 8(d)** as inverter was used to convert stored energy.

Therefore storage helped in supporting load in peak demand time when grid electricity is costly. **Figures 8(c)** and **(d)** showed that both these components (Battery and Inverter) were not completely utilized therefore more PV array could utilize these components to sell extra electricity to the grid and can reduce the overall cost of the system.

Summary:

From the discussed result above it was found that storage improved RE share in load support and reduced loss of energy as shown in **Table 4**. However the storage influences on the performance indexes, these are explained in Findings section.

Findings

The optimization was done in two configuration categories and 3 cases in each configuration and 4 different factors were compared in each case. These factors are GHG

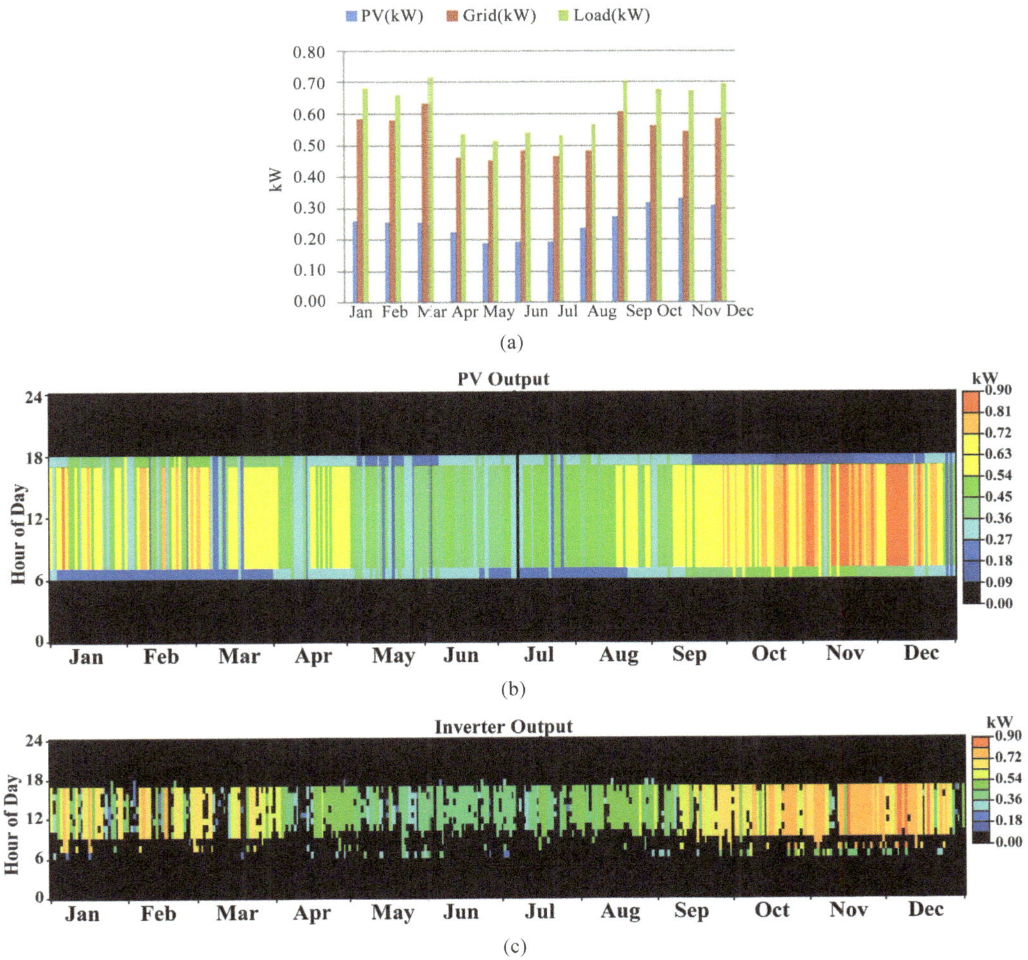

Figure 7. Configuration-2 Case-2 output: (a) PV & grid output compared to load demand; (b) PV array output; (c) Inverter output.

& Pollutant gas emission, RF, COE and NPC. The findings of these factors are explained below:

GHG & pollutant gas emission:

Optimized model results in different cases showed that in both configurations storage in case 3 helped in reducing GHG & other pollutant gas emission as shown in **Figure 9**.

Renewable Fraction (RF):

Results showed that in both configurations storage improved RE production. **Figure 10** shows that Storage helps in increasing RF.

Cost of energy:

Results showed that generator only configuration was most costly therefore COE was very high. However Storage sharply reduced the COE in off-grid configuration. Although overall grid only electricity in case-1 was little cheaper but storage still helped in reducing the COE by improving the PV penetration as shown in **Figure 11** where in case-2 COE was $0.579/kWh and in case-3 COE became $0.556/kWh.

Net Present Cost (NPC):

Results showed that in off-grid configuration, PV with diesel generator in case-2, a good amount of electricity was supplied from PV array which reduced NPC. Storage further accelerated the share of RE and reduced NPC as shown in case-3 in **Figure 12**.

In grid-connected configuration although NPC increased in case-2 while PV shared the load but in case-3 storage improved PV participation and reduced NPC as shown in **Figure 12**.

Payback period:

Payback is the number of years in which the cumulative cash flow switches from negative to positive. For this analysis it was found by comparing PV, storage, grid system with PV and grid base system. Payback period requires the collection of annualized cost of the system which are calculated by summing the capital cost, O&M cost, operating cost, fuel cost and replacement cost for each year. Annualized costs are then subtracted from each other for each consecutive year which provides savings

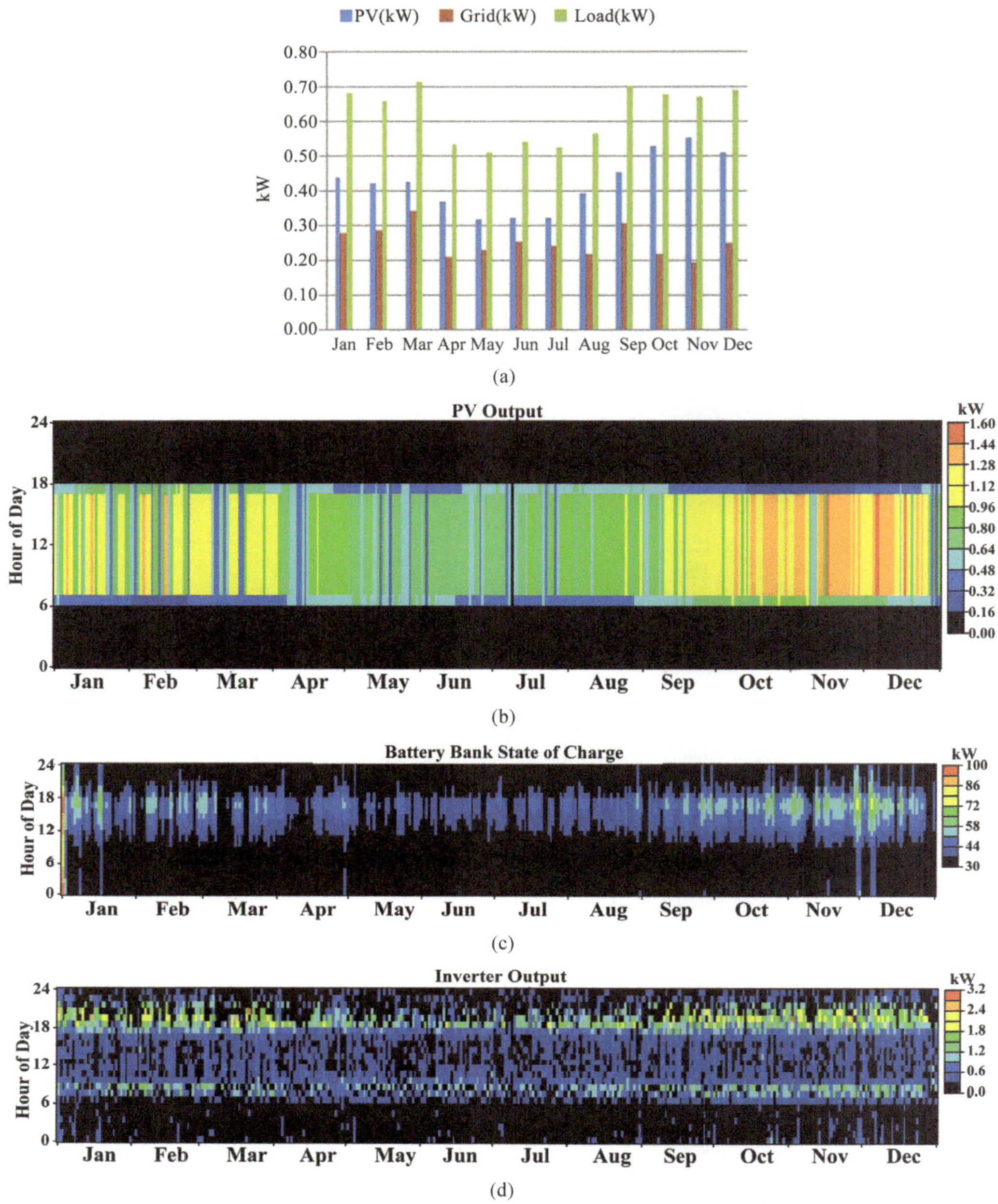

Figure 8. Configuration-2 Case-3 output: (a) PV & grid output compared to load demand; (b) PV array output; (c) Battery state of charge (SOC); (d) Inverter output.

Table 4. Model result and significance of storage.

	Simulation case	PV array	Inverter	Storage (@ 24 V system voltage)	RE generation & use		
					Load support	*Grid sales*	*RE loss*
Configuration-1	Case-2	12 kW	5 kW	-	34.60%	-	77.38%
	Case-3	10 kW	5 kW	69.12 kWh	99.95%	-	13.99%
Configuration-2	Case-2	3 kW	5 kW	-	14.21	9.66%	37.58%
	Case-3	5 kW	5 kW	25.92 kWh	59.23%	-	0.0021%

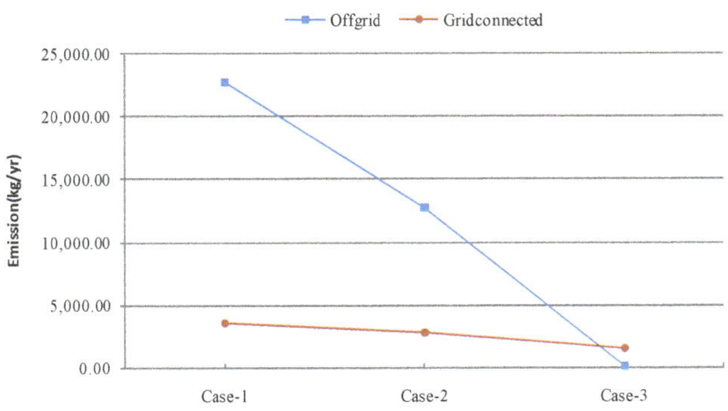

Figure 9. GHG emission in different case configurations.

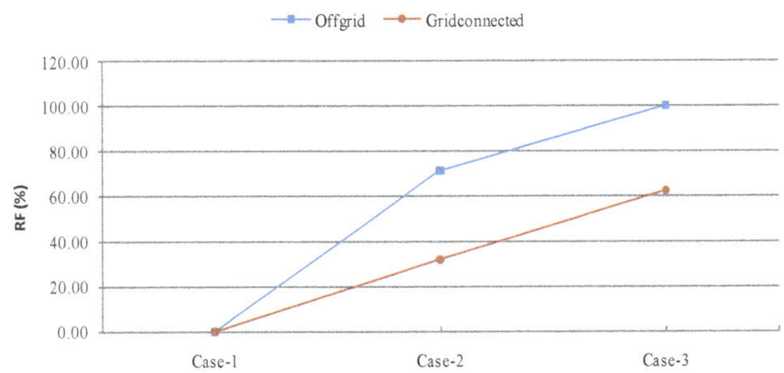

Figure 10. RF in different case configurations.

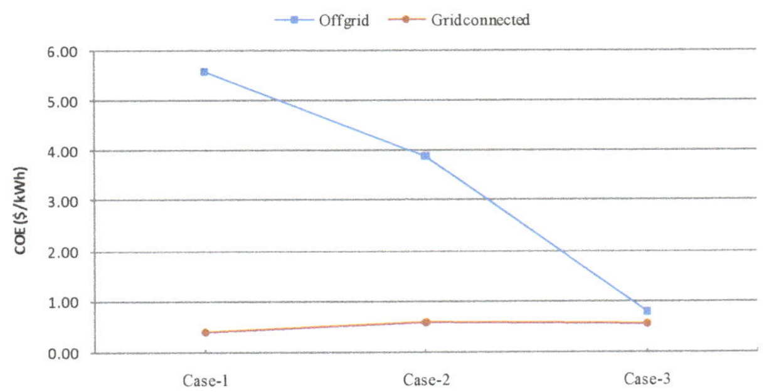

Figure 11. COE in different configurations.

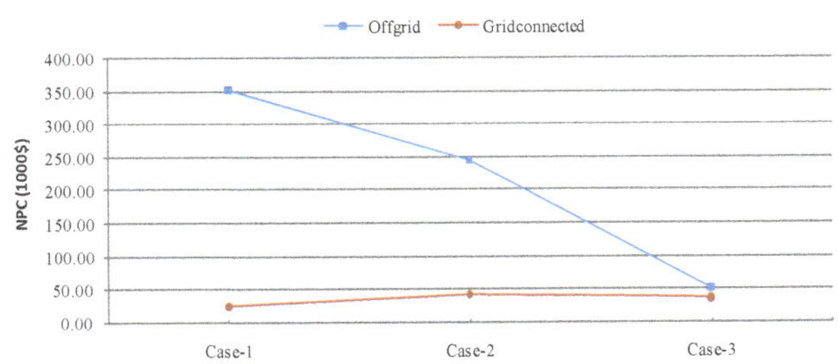

Figure 12. NPC in different configurations.

or loss for each year. Result showed that the cost of storage returned in 4.9 years considering the project life of 20 years as shown in **Figure 13**. In Australia solar bonus scheme awards the price of electricity fed into the grid from PV systems at a rate of $0.44/kWh [37,38] this ensures that the payback period is much shorter in Australia.

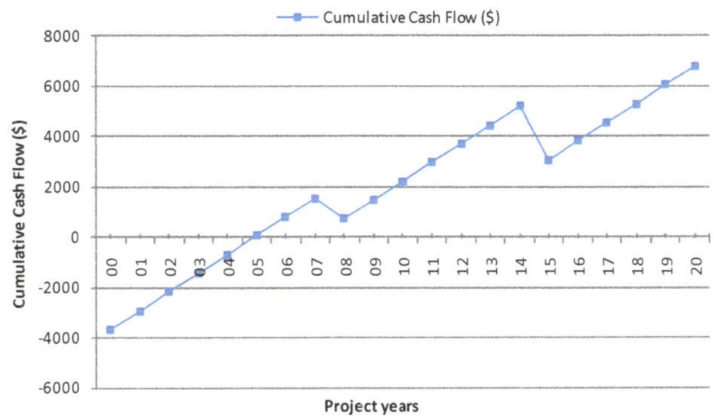

Figure 13. Storage pay back period.

5. Conclusions

Storage integrated PV system was analyzed for a residential load in Rockhampton, Queensland, Australia in off-grid and grid connected configurations. Analysis was conducted to observe the storage influences over the GHG emission, RF, COE and NPC indexes. It was found that storage greatly influenced the improvement of RE utilization.

It was clear from the analysis that storage helped significantly in reducing GHG & other pollutant gas emission, and reduced COE, improved RF also reduced the NPC. Comparing PV integrated system with and without storage, it was found that in the grid connected configuration, storage reduced 46.47% of GHG and pollutant gas emission and in stand-alone system it reduced 99.97%. In grid connected configuration, storage integrated PV system improved RF 93.78% and in standalone system 40.17% compared to without storage condition. Storage reduced COE by 3.97% in grid connected configuration and 79.45% in standalone configuration. Similarly storage reduced NPC by 13.76% in grid connected configuration and 79.46% in standalone configuration. Moreover storage integrated PV in grid connected configurations has reasonable payback period and the cost of storage returns in very short time. Therefore storage has a positive influence in environmental and economic indexes either in off-grid or grid-connected configurations.

Australia has comparatively good solar radiation rate which shows such impressive results in all four indexes. This result also identifies the scope of maximizing the utilization of RE by using storage to meet residential load demand.

REFERENCES

[1] R. Fioravanti, "The Importance of Energy Storage for System Regulation," 2012. http://www.elp.com/index/display/article-display/2450936349/articles/utility-automation-engineering-td/volume-14/issue-10/features/the-importance_of.html

[2] KEMA, "Large-Scale Electricity Storage," 2012.

http://www.kema.com/services/consulting/etd/es/large-scale-storage.aspx

[3] KEMA, "Benefits of Fast-Response Storage Devices for System Regulation in ISO Markets," 2012. http://www.kema.com/Images/FastResponseStorage11-25-08SCREEN.pdf

[4] DoRET, "Energy in Australia 2010," Department of Resources, Energy and Tourism, Australia, 2010.

[5] Queensland-Government, "The Queensland Renewable Energy Plan, a Clear Energy Future for Queensland," 2009. http://www.cleanenergy.qld.gov.au/queensland_renewable_energy_plan.cfm

[6] REN21, "Renewables 2010," *Global Status Report*, 2010. http://www.ren21.net/Portals/97/documents/GSR/REN21_GSR_2010_full_revised%20Sept2010.pdf

[7] DCC, "National Greenhouse Gas Inventory—Accounting for the KYOTO Target May 2009," Department of Climate Change, Australian Government, 2009. http://www.climatechange.gov.au/climate-change/~/media/publications/greenhouse-report/national-greenhouse-gas-inventory-pdf.ashx

[8] Energy Research Institute, "Australian Sustainable Energy, Zero Carbon Australia Stationary Energy Plan," 2010. http://www.energy.unimelb.edu.au/

[9] R. Albarracin and H. Amaris, "Power Quality in Distribution Power Networks with Photovoltaic Energy Sources," *IEEE Proceedings of the International Conference on Environment and Electrical Engineering*, Karpacz, 10-13 May 2009.

[10] R. K. Varma, M. Salama, R. Seethapathy and C. Champion, "Large-Scale Photovoltaic Solar Power Integration in Transmission and Distribution Networks," *IEEE Power & Energy Society General Meeting*, Calgary, 26-30 July 2009, pp. 1-4.

[11] S. Rahman, "The Smart Grid, Its Opportunities and Challenges," *Asia-Pacific Power & Energy Engineering Conference*, Chengdu, 29 March 2010.

[12] C. J. Hanley, D. T. Ton, G. H. Peek and J. D. Boyes, "Solar Energy Grid Integration Systems—Energy Storage (SEGIS-ES)," *Sandia Report*, SAND2008-4247, 2008.

[13] L. L. Grigsbay, "The Electric Power Engineering Handbook," CRC Press, New York, 2001.

[14] S. Vazquez, S. M. Lukic, E. Galvan, L. G. Franquelo and J. M. Carrasco, "Energy Storage Systems for Transport and Grid Applications," *IEEE Transactions on Industrial Electronics*, Vol. 57, No. 12, 2010, pp. 3881-3895.

[15] M. T. Arif, A. M. T. Oo and A. B. M. Shawkat Ali, "Estimation of Energy Storage and Its Feasibility Analysis, Energy Storage—Technologies and Applications," A. Zobaa, Ed., 2013. http://www.intechopen.com/books/energy-storage-technologies-and-applications/estimation-of-energy-storage-and-its-feasibility-analysis

[16] F. Rahman, S. Rehman and M. A. Abdul-Majeed, "Overview of Energy Storage Systems for Storing Electricity from Renewable Energy Sources in Saudi Arabia," *Renewable and Sustainable Energy Reviews*, Vol. 16, No. 1, 2012, pp. 274-283.

[17] M. Beaudin, H. Zareipour, A. Schellenberglabe and W. Rosehart, "Energy Storage for Mitigating the Variability of Renewable Electric Sources: An Updated Review," *Energy for Sustainable Development*, Vol. 14, No. 4, 2010, pp. 302-314.

[18] F. Diaz-Gonzalez, A. Sumper and O. Gomis-Bellmunt, "A Review of Energy Storage Technologies for Wind Power Applications," *Renewable and Sustainable Energy Reviews*, Vol. 16, No. 1, 2012, pp. 2154-2171.

[19] I. Hadjipaschalis, A. Poullikkas and V. Efthimiou, "Overview of Current and Future Energy Storage Technologies for Electric Power Applications," *Renewable and Sustainable Energy Reviews*, Vol. 13, No. 6-7, 2009, pp. 1513-1522.

[20] C. Wessells, "Nanoparticle Electrode for Batteries Could Make Large-Scale Power Storage on the Energy Grid Feasible," *Stanford University News*, 2011. at:http://news.stanford.edu/news/2011/november/longlife-power-storage-112311.html

[21] HOMER, "Analysis of Micro Powersystem Options." https://analysis.nrel.gov/homer/

[22] G. J. Dalton, D. A. Lockington and T. E. Baldock, "Feasibility Analysis of Stand-Alone Renewable Energy Supply Options for a Large Hotel," *Renewable Energy*, Vol. 33, No. 7, 2008, pp. 1475-1490.

[23] A. N. Celik, "Techno-Economic Analysis of Autonomous PV-Wind Hybrid Energy Systems Using Different Sizing Methods," *Energy Conversion and Management*, Vol. 44, No. 12, 2003, pp. 1951-1968.

[24] BoM, Bureau of Meteorology, Australian Government. http://reg.bom.gov.au/

[25] ErgonEnergy, "Understanding Your Electricity Bill, Information about Average Consumption," 2012. http://www.ergon.com.au/your-business/accounts--and--billing/understanding-your-bill

[26] G. M. Masters, "Renewable and Efficient Electric Power Systems," John Wiley & Sons, Inc., Hoboken, 2004.

[27] T. Lambert, "How HOMER Calculates the PV Array Power Output," *Homer Help File*, 2007. http://homerenergy.com/

[28] P. N. John, A. M. A. Haidar and M. Shawal, "Optimal Configuration Assessment of Renewable Energy in Malaysia," *Renewable Energy*, Vol. 36, No. 2, 2011, pp. 881-888.

[29] D. C. Barley and B. C. Winn, "Optimal Dispatch Strategy in Remote Hybrid Power Systems," *Solar Energy*, Vol. 58, No. 1, 1996, pp. 65-79.

[30] PV-Price, "Sun Solar System," 2012. www.sunsolarsystem.com.au

[31] PV-Price, "Goodhew Electrical and Solar," 2012. http://www.goodhewsolar.com.au/customPages/goodhew-electrical-%26-solar-offers-homeowners-the-most-afford

able-quality-solar-systems-on-the-market.?subSiteId=1

[32] Inverter-Cost, "SMA Sunny Boy Grid tie Inverter 7000 W SB7000US Price," 2012. http://www.google.com/products/catalog?hl=en&q=sunny+boy+grid+tie+inverter+price&gs_sm=3&gs_upl=23781784011185341111101011110122312055l0.3.711110&bav=on.2,or.r_gc.r_pw.,cf.osb&biw=1680&bih=831&um=1&ie=UTF-8&tbm=shop&cid=10871923140935237408&sa=X&ei=z0g6T-T0LayziQfnkZGQCg&ved=0CGkQ8wIwAQ

[33] Inverter-Cost, "SMA Sunny Boy 1700 Price," 2012. http://www.solarmatrix.com.au/special-offers/sunny-boy-1700?ver=gg&gclid=CMO0gIOzna4CFYVMpgod7T1OHg

[34] ALTE-Store, "Battery Price, Trojan T-105 6 V, 225 AH (20HR) Flooded Lead Acid Battery." http://www.altestore.com/store/Deep-Cycle-Batteries/Batteries-Flooded-Lead-Acid/Trojan-T-105-6V-225AH-20HR-Flooded-Lead-Acid-Battery/p1771/

[35] ErgonEnergy, "Electricity Tariffs and Prices." http://www.ergon.com.au/your-business/accounts--and--billing/electricity-prices

[36] RedEnergy, "Pricing Definition for Electricity Customers," NSW, 2011. http://www.redenergy.com.au/docs/NSW-Pricing-DEFINITIONS-0311.pdf

[37] Future_Sustainability, "Rebates for Solar Power." http://futuresustainability.rtrk.com.au/?scid=80507&kw=4858156&pub_cr_id=17164703877

[38] Economic Development & Innovation Department of Employment, Queensland, "Solar Bonus Scheme," Queensland Government, Office of Clean Energy. http://www.cleanenergy.qld.gov.au/demand-side/solar-bonus-scheme.htm?utm_source=WWW2BUSINESS&utm_medium=301&utm_campaign=redirection

Very Short-Term Generating Power Forecasting for Wind Power Generators Based on Time Series Analysis

Atsushi Yona[1], Tomonobu Senjyu[1], Funabashi Toshihisa[2], Chul-Hwan Kim[3]

[1]University of the Ryukyus, Okinawa, Japan; [2]Meidensha Corporation, Tokyo, Japan; [3]Sungkyunkwan University, Suwon City, South Korea.

ABSTRACT

In recent years, there has been introduction of alternative energy sources such as wind energy. However, wind speed is not constant and wind power output is proportional to the cube of the wind speed. In order to control the power output for wind power generators as accurately as possible, a method of wind speed estimation is required. In this paper, a technique considers that wind speed in the order of 1 - 30 seconds is investigated in confirming the validity of the Auto Regressive model (AR), Kalman Filter (KF) and Neural Network (NN) to forecast wind speed. This paper compares the simulation results of the forecast wind speed for the power output forecast of wind power generator by using AR, KF and NN.

Keywords: Very Short-Term Ahead Forecasting; Wind Power Generation; Wind Speed Forecasting; Time Series Analysis

1. Introduction

Wind generators are rapidly gaining acceptance as some of the best alternative energy sources, and an alternative to conventional power generation. Especially due to decreasing cost of the wind generated power, it has the potential to compete with traditional power plants. However, wind speed is not constant and the power output of wind generators is proportional to the cube of the wind speed. The intermittent nature of wind makes the power output fluctuation of wind generator. In electric companies, wind speed forecasting is an important tool for utilizing the hybrid power systems with storage battery, wind generators, solar cells, etc. For example, the amount of energy stored in the battery is easily decided by forecast data. These decisions are beneficial for effective operation of hybrid power systems and consequently, their profitability depends on the technique utilized in the forecasting powers. Although the technique to forecast power output of wind generators based on wind speed prediction is regarded as an effective method in practical applications, it requires large meteorological data. In addition, although meteorological agencies or weather services will provide free data for prediction, the implementation of the above-mentioned techniques provide the order of several minutes data or hourly data. Since the data is mostly gathered over a wide area by prediction, it becomes rather difficult to determine the exact values for the order of several seconds at the installation area of a wind generator. Auto Regressive Model (AR) [1], Kalman Filter (KF) [2-4] and Neural Network (NN) [5-16] have been known that techniques are convenient in forecasting applications. It is possible to forecast wind speed by using only meteorological data. There is a valid argument: AR and KF are well known models for 1 - 3 time steps ahead forecasting. However the forecast result of using NN is usually case-by-case with the characteristic (stochastic appearance) of wind speed.

This paper describes very short-term ahead power output forecasting of wind generators based on wind speed forecasting. For the purpose of comparing the forecasting abilities of AR, KF and NN in this paper, all input data are based on the same condition. These models are constructed by using historical data on wind speed and tested for the target time. In addition, these data are observed at only one specific site. The power output of the wind generator is determined by the forecast wind speed data. The abilities of these models are confirmed by comparing the results in 1 - 30 seconds ahead forecasting.

2. Methodology and Application Model

The concept of the wind speed forecast technique is shown in **Figure 1**. In this paper, it is assumed that 1 to 30 seconds are forecast period of the wind forecasting. AR, KF and NN are constructed by making use of the historical data of 1 - 30 seconds at before 1-second. The idea is assumed that these models are constructed respectively in regard to the application of time series. It means the previous 30 seconds data reflect the trend of wind characteristic to forecast models with simple approach. After the wind speed forecasting, these three outputs are utilized for determination of power outputs of wind power generator. Explanation of AR, KF and NN are given below.

2.1. Auto Regressive Model

A common approach for modelling of time series is the AR model. The method of least-squares is applied in the estimation method of AR model. To apply the least squares method for time-varying AR model estimation, AR parameters are assumed to be time-invariant. If we want to obtain the forecast value y_{k+i}, regression coefficient a_k and residual error b_k should be determined by:

$$y_{k+i} = a_k x_k + b_k \tag{1}$$

2.2. Kalman Filter

The Kalman filter is a recursive estimator. It means that only the estimated state from the previous time step and the current measurement are needed for the current state. The first task during update of measurement is to compute the Kalman gain K_k. If estimated value $\bar{x}_k$ and covariance matrix P_k for estimate error are obtained,

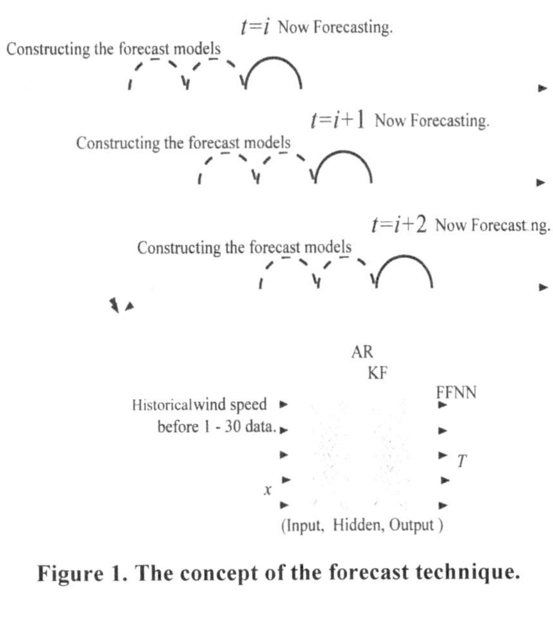

Figure 1. The concept of the forecast technique.

the next step is to measure the process to obtain observed value z_k, and update the estimated value $\hat{x}_k$ and covariance matrix P_k for posteriori estimation. This process is represented by:

$$\hat{x}_{k+1} = \bar{x}_k + K_k \left(z_k - H_k \bar{x}_k \right) \tag{3}$$

$$P_k = \left(1 - K_k H_k\right) \bar{P}_{k+1} \tag{4}$$

$$K_k = \bar{P} H_k^{\mathrm{T}} \left(H_k \bar{P}_k H_k^{\mathrm{T}} + R_k \right)^{-1} \tag{5}$$

where R is covariance for observation error, H is $m \times n$ matrix, which shows the relation between the state variable of a system and an observed value. 1 step-ahead estimation value $\bar{x}_{k+1}$ and covariance matrix $\bar{P}_{k+1}$ for estimation errors are calculated by:

$$\bar{x}_{k+1} = A_k \hat{x}_k + B_k u_k \tag{6}$$

$$\bar{P}_{k+1} = A_k P_k A_k^{\mathrm{T}} + Q_k \tag{7}$$

where A_k is $n \times n$ matrix, B_k is $n \times l$ matrix, Q_k is covariance of process error, $\bar{x}_{k+1}$ will be update of z_k after 2 steps ahead.

2.3. Neural Network

Figure 2 shows the Feed-forward Neural Network (FFNN), having l and m neurons in input layer and hidden layer respectively, and n neurons in output layer. These neurons are connected with a linear coupling, and x_1 - x_l are input data to NN. There are connection weights between each neuron. The output of the hidden neurons and output neurons are converted to nonlinear values by the sigmoid function. The sigmoid function is as follows:

$$f(x) = \frac{1}{1 + \exp(-x)} \tag{8}$$

where, x is the input data.

Back Propagation (BP) method is adopted for training the NN. Generally, BP is explained as follows. To begin with, output of the hidden neuron is transmitted to the output neuron. Then the output of the output neuron is

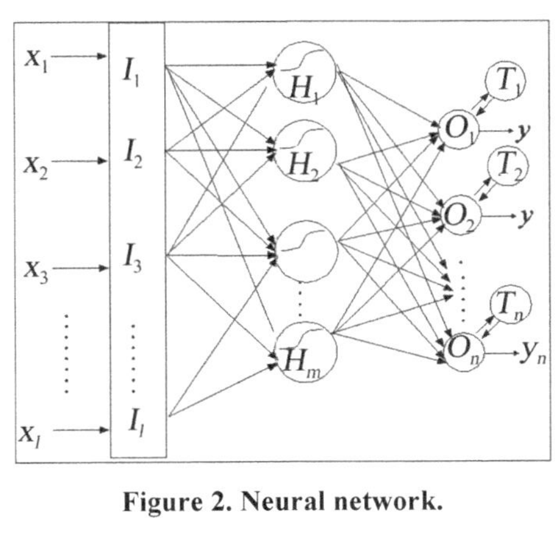

Figure 2. Neural network.

compared with target signal, T, as shown in **Figure 2**. Finally, to minimize the mean square error margin, each connection weight and the output value of each neuron are changed in the direction of a straight line from output layer to input layer. In this paper, Levenberg-Marquardt algorithm [17] was adopted for training each connection weight of neurons. The "momentum" and "training coefficient" terms are parameters of training; momentum promotes training speed and acts rapidly by changing each connection weight of neurons. The training coefficient parameter has to be large; however, if it is too large, the network becomes unstable. It is desired that the mean square error margin of NN model should not be unstable. The authors determined these parameters by a trial-and-error method.

3. Simulation Results

This section shows the simulation results of wind speed forecasting. Analysis results of forecasting error are also described below. The wind speed is important parameter to know the forecasted power output of wind turbine. Because of the accuracy of the forecasted power output depends on characteristic of fluctuating power output and also wind speed.

3.1. Wind Speed Forecasting

Figure 3 shows the results of 10 sec ahead wind speed forecasting at the time t = 121 - 900 for observation (1-second intervals), and calculates the forecast error (*MAE*, mean absolute error) by using AR, KF and NN as shown in **Figure 4**. In the case of **Figure 4**, mean value of wind speed is 14.3265 $m \cdot s^{-1}$, and the variance is 3.6736 $m \cdot s^{-1}$. *MAE* is given as:

$$MAE = \frac{1}{N}\sqrt{\sum_{K=1}^{N}\left(P_a^k - P_f^k\right)^2} \quad (9)$$

where N is number of data, P_f^k is forecast value, P_a^k is actual value, and k is forecast time. In the result of using AR, the forecast error (*MAE*) is 1.6169 $m \cdot s^{-1}$ in 10 sec ahead, while KF and NN cause 1.6120 $m \cdot s^{-1}$ and 1.6047 $m \cdot s^{-1}$, respectively. **Figure 4** shows the calculated *MAE* of 1 - 30 ahead wind speed forecast. In wind speed forecasting, as forecast time lengthens, forecast error increases. This phenomenon appears prominently in the result shown in **Figure 4**. There are no differences in calculated results of *MAE* with AR model, KF and NN. But the forecast results of these models vary on a case-by-case basis with the characteristic (stochastic appearance) of wind speed. In the case of AR model, **Figure 3(a)** shows that the forecast result of 10 sec ahead wind speed is influenced by gradient between forecast time and before 10 sec wind speed. In the case of NN, **Figure 3(b)** shows that the result of 10 sec ahead wind speed is influenced by before 1 - 30 seconds wind speed characteristic. In the case of KF, **Figure 3(c)** shows that the 10 sec ahead forecast result is similar to the result of using AR model, but the gradient between forecast time and before 10 sec wind speed have an insignificant effect

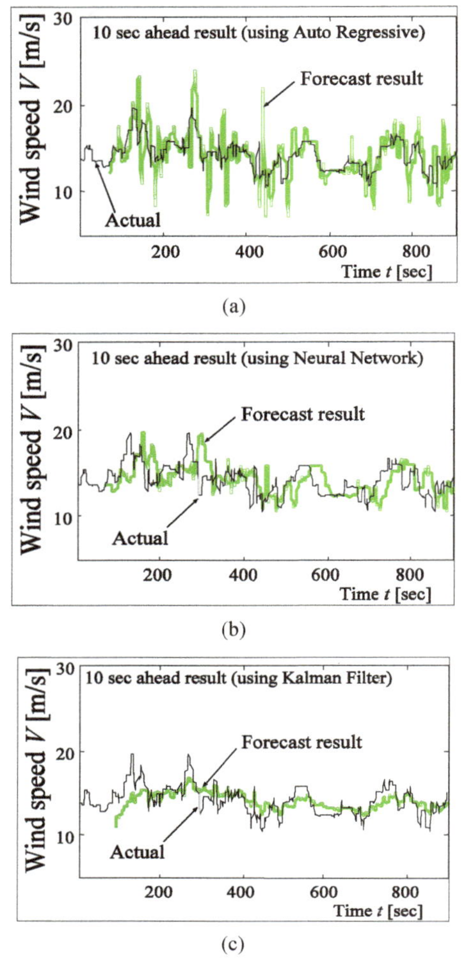

Figure 3. Simulation result for 10 sec ahead wind speed forecasting. (a) The result for using auto regressive model; (b) The result for using neural network; (c) The result for using Kalman filter.

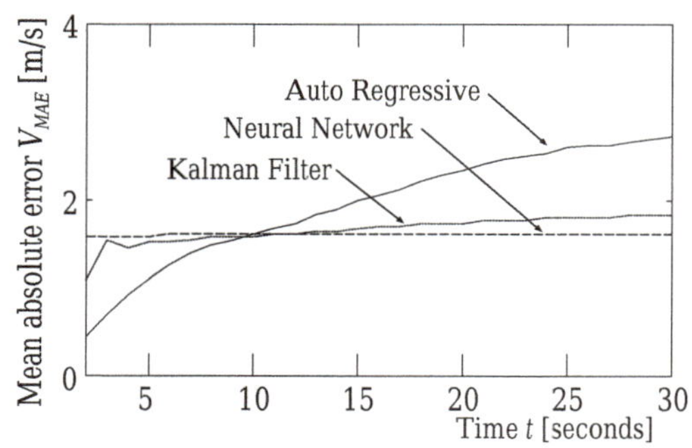

Figure 4. Mean absolute error V_{MAE} (wind speed forecasting).

on forecast result. In the simulation results of the forecasted wind speed, each of these models has its own merits and demerits as shown in **Figure 3**. But one of the criterions to decide the forecast performance of these models is described along with forecast time as shown in **Figure 4**.

3.2. Power Spectral of Wind Speed

Figure 5 shows the power spectral of wind speed at the time t = 121 - 900 s for observation in **Figure 3**. The power spectral (Van der Hoven model [18]) is as indicated below. The power spectral $S_{yy}(f_k)$ for wind speed is an exponential-type Fourier transform $Y(f_k)$ calculated by observed wind speed $y(t)$. Therefore, the power spectral $S_{yy}(f_k)$ is represented by:

$$S_{yy}(f_k) = f_k \times \left|\frac{Y(f_k)}{N}\right|^2 \times T \quad (10)$$

where $y(t)$ [m·s^{-1}] is observed wind speed, $S_{yy}(f_k)$ [m^2·s^{-2}] is the power spectral for wind speed, f_k [Hz] is frequency, T [sec] is observation time, and N is the number of data. Fourier transform $Y(f_k)$ is obtained by:

$$Y(f_k) = \sum_{t=1}^{N} y(t)\exp\left(-\frac{i2\pi kt}{N}\right) = \sum_{t=1}^{N} y(t)\exp(i2\pi kt\Delta t); \quad (11)$$

$$f_k = \frac{k}{N\Delta t}, \qquad k = 0,\cdots,N$$

The power spectral density is usually represented by:

$$S_{yy}(f_k) = \frac{1}{N}\times|Y(f_k)|^2, \quad 0 \le f_k \le f_N \quad (12)$$

Equation (10) supposes that wind speed has white noise properties, *i.e.*, (distribution = 1, power spectral density = 1). **Figure 5** shows that the power spectral of wind speed is low density at around 7 - 10 sec. Therefore, wind speed forecast errors of 7 - 10 sec ahead in **Figures 3** and **4** are influenced by characteristic of wind speed for the power spectral with low density.

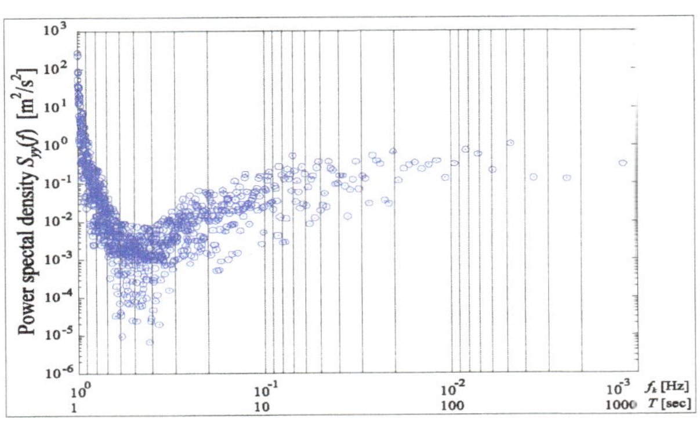

Figure 5. Power spectral density of wind speed (Van der Hoven model).

4. Power Output of Wind Power Generator

After the wind speed forecasting, the authors determined the power output of wind power generator. In this paper, the point under discussion was put on wind speed forecasting. To confirm the validity of the proposed method, the authors determined the power output. It means that power output for wind generator can be forecast by using only wind speed data. The energy of the wind is kinetic energy, and kinetic energy of mass m and speed V is defined by 0.5 $m\cdot V^2$. Thus the wind power energy received by the swept area of the blades has the following equation:

$$P = \frac{1}{2}mV^2\eta = \frac{1}{2}(\rho AV)V^2\eta = \frac{1}{2}\rho AV^3\eta \quad (13)$$

where m [kg] is mass, V [m·s^{-1}] is wind speed, η [%] is efficiency coefficient, ρ is air density 1.225 kg·m^3, and A [m^2] is swept area of the blades [3]. Power output of a typical wind turbine is generated starting at "cut-in wind speed", and not generated at over the "cut-out wind speed". Rated wind speed is different in each type of wind turbine. Usually, cut in wind speed is 3 - 5 m·s^{-1}, rated wind speed is 8 - 16 m·s^{-1}, and cut out wind speed is 24 - 25 m·s^{-1} [19-21]. In this paper, power output of wind power generator is determined by parameters shown in **Table 1** and **Figure 6**. As mentioned above, the

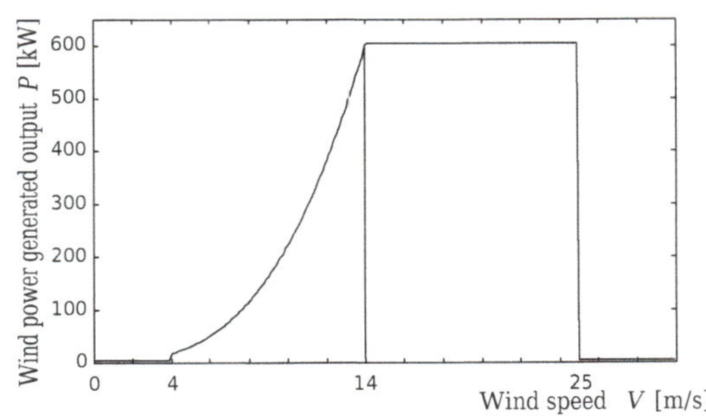

Figure 6. Power output curve of wind power generation system.

Table 1. Simulation parameters of tested wind power generator.

Blade radius	21 m
Swept area of the blades	1428 m^2
Efficiency coefficient	25%
Rated wind speed	14 m·s^{-1}
Rated output power	600 kW
Cut in wind speed	4 m·s^{-1}
Cut out wind speed	25 m·s^{-1}

power output of wind power generator is determined by using only wind speed data.

Figure 7 shows the forecasting results of power output of wind power generator. The results of **Figure 8** are calculated by wind speed forecasting shown in **Figure 3**. The calculated results of *MAE* with AR model, KF and NN are similar. But for the forecast result of the power output by using AR in **Figure 7**, the forecast error is 87.0152 kW in 10 sec ahead, while the results of KF and NN are 71.4639 kW and 68.7320 kW respectively. In fact, forecast error of power output increases by increasing forecast error of wind speed. Although our data sample is relatively small (representing only 900 wind speed data), adding data would further improve the model performance.

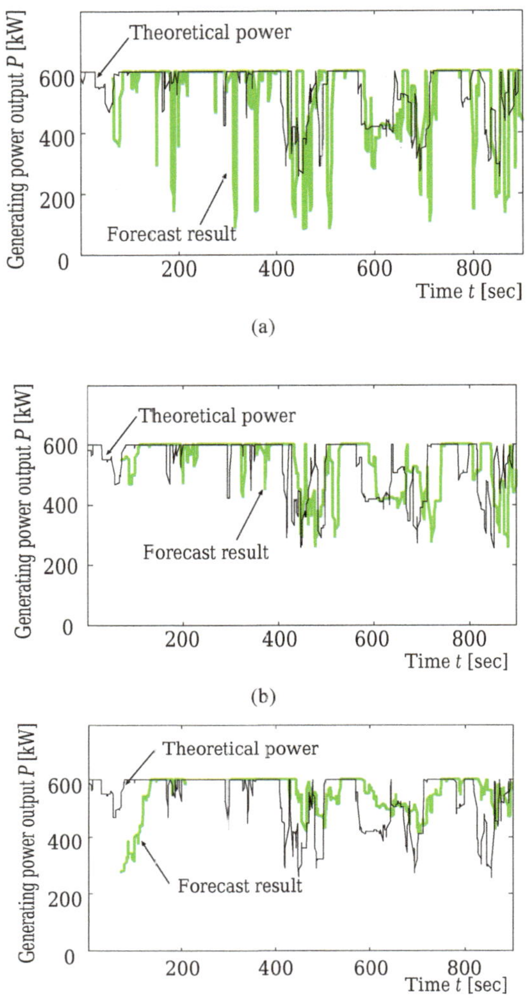

Figure 7. Simulation result of 10 sec ahead forecasting power output for wind generator. (a) The result for using auto regressive model; (b) The result for using neural network; (c) The result for using Kalman filter.

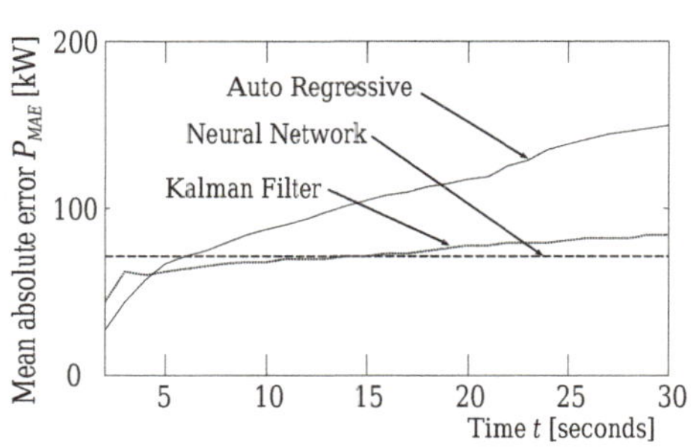

Figure 8. Mean absolute error P_{MAE} (power output for wind generator).

5. Conclusion

This paper shows the generating power forecasting of wind generators based on wind speed time series analysis by using AR, KF and NN. The constructed methods do not require complicated calculations due to wind speed data. At the time of wind speed forecasting, it is possible to shorten the calculation time by using historical wind speed data. The validity is confirmed by comparing forecasting results with AR, KF and NN. It is found that the proposed method shows good performance to forecast power output of wind generator. Simulation results lead to the conclusion that it is possible to build up the following hypotheses: if wind speed data are always available, the advantage of the proposed method in this paper is the ability to forecast and estimate sequences of seconds-scale power output for wind power generations from the wind speed only.

REFERENCES

[1] A. C. N. Salles, A. C. G. Melo and L. F. L. Legey, "Risk Analysis Methodologies for Financial Evaluation of Wind Energy Power Generation Projects in the Brazilian System," 2004 *International Conference on Probabilistic Methods Applied to Power Systems*, 16 September 2004, pp. 457-462.

[2] D. E. Gustafson and W. H. Ledsham, "Applications of Estimation Theory to Inverse Problems in Meteorology," 1978 *IEEE Conference on Decision and Control Including the 17th Symposium on Adaptive Processes*, Vol. 17, No. 1, 1978, pp. 405-412.

[3] L. Hui, K. L. Shi and P. G. McLaren, "Neural-Network-Based Sensorless Maximum Wind Energy Capture with Compensated Power Coefficient," *IEEE Transactions on Industry Applications*, Vol. 41, No. 6, 2005, pp. 1548-1556.

[4] G. Galanis, P. Louka, P. Katsafados, I. Pytharoulis and G. Kallos, "Applications of Kalman Filters Based on Non-Linear Functions to Numerical Weather Predictions," *Annales Geophysicae*, No. 24, 2006, pp. 1-10.

[5] M. A. Mohandes, S. Rehman and T. O. Halawani, "A Neural Networks Approach for Wind Speed Prediction," *Renewable Energy*, Vol. 13, No. 3, 1998, pp. 345-354.

[6] T. Toumiya, E. Sakaguchi, T. Tanaka and T. Suzuki, "Modeling of Wind Power Generating System with a Propeller Type Windmill by Neural Networks," *The Institute of Electrical Engineers of Japan*, Vol. 118-E, No. 7-8, 1998, pp. 794-801. (in Japanese)

[7] Y. Kemmoku, H. Ishii, H. Takikawa, T. Kawamoto and T. Sakakibara, "Method of Forecasting Wind Velocity of Next Day Using Weather Data over Wide Area," *Journal of Japan Solar Energy Society*, Vol. 27, No. 1, 2001, pp. 85-91. (in Japanese)

[8] P. Flores, A. Tapia and G. Tapia, "Application of a Control Algorithm for Wind Speed Prediction and Power Generation," *Renewable Energy*, Vol. 30, No. 5, 2004, pp. 523-536.

[9] S. Kelouwani and K. Agbossou, "Nonlinear Model Identification of Wind Turbine with a Neural Network," *IEEE Transactions on Energy Conversion*, Vol. 19, No. 3, 2004, pp. 607-612.

[10] A. Oztopal, "Artificial Neural Network Approach to Spatial Estimation of Wind Velocity Data," *Energy Conversion and Management*, Vol. 47, No. 4, 2006, pp. 395-406.

[11] J. L. Elman, "Finding Structure in Time," *Cognitive Science*, Vol. 14, No. 2, 1990, pp. 179-211.

[12] J.-H. Li, A. N. Michel and W. Porod, "Analysis and Synthesis of a Class of Neural Networks: Linear Systems Operating on a Closed Hypercube," *IEEE Transactions on Circuits and Systems*, Vol. 36, No. 11, 1989, pp. 1405-1422.

[13] G. N. Kariniotakis, G. S. Stavrakakis and E. F. Nogaret, "Wind Power Forecasting Using Advanced Neural Networks Models," *IEEE Transactions on Energy Conversion*, Vol. 11, No. 4, 1996, pp. 762-767.

[14] T. G. J. Barbounis, B. Thecharis, M. C. Alexiadis and P. S. Dokopoulos, "Long-Term Wind Speed and Power Forecasting Using Local Recurrent Neural Network Models," *IEEE Transactions on Energy Conversion*, Vol. 21, No. 1, 2006, pp. 273-284.

[15] Y. Kitamura and A. Matsuda, "Study on Raising Efficiency of Heat Accumulating Air Conditioning System Using Knowledge Processing Techniques," *Journal of Mitsubishi Research Institute*, No. 36, 2000, pp. 31-51. (in Japanese)

[16] B. Kermanshahi, "Recurrent Neural Network for Forecasting Next 10 Years Loads of Nine Japanese Utilities," *Neurocomputing*, Vol. 23, No. 1-3, 1998, pp. 125-133.

[17] M. T. Hagan and M. B. Menhaj, "Training Feed-Forward Networks with the Marquardt Algorithm," *IEEE Transactions on Neural Networks*, Vol. 5, No. 6, 1994, pp. 989-993.

[18] V. D. Hoven, "Power Spectrum of Horizontal Wind Speed in the Frequency Range from 0.007 to 900 Cycles," *Journal of Meteorology*, Vol. 14, No. 2, 1957, p. 160.

[19] M. A. H. El-Sayed, "Substitution Potential of Wind Energy in Egypt," *Energy Policy*, Vol. 30, No. 8, 2002, pp. 681-687.

[20] M. A. Elhadidy and S. M. Shaahid, "Decentralized/Stand-Alone Hybrid Wind-Diesel Power Systems to Meet Residential Loads of Hot Coastal Regions," *Energy Conversion and Management*, Vol. 46, No. 15-16, 2005, pp. 2501-2503.

[21] S. M. Shaahid, L. M. Al-Hadhrami and M. K. Rahman, "Economic Feasibility of Development of Wind Power Plants in Coastal Locations of Saudi Arabia—A Review," *Renewable and Sustainable Energy Reviews*, Vol. 19, No. C, 2013, pp. 589-587.

17

Hybrid Power Systems Energy Controller Based on Neural Network and Fuzzy Logic

Emad M. Natsheh, Alhussein Albarbar

Advanced Industrial Diagnostics Centre, School of Engineering, Manchester Metropolitan University (MMU), Manchester, UK.

ABSTRACT

This paper presents a novel adaptive scheme for energy management in stand-alone hybrid power systems. The proposed management system is designed to manage the power flow between the hybrid power system and energy storage elements in order to satisfy the load requirements based on artificial neural network (ANN) and fuzzy logic controllers. The neural network controller is employed to achieve the maximum power point (MPP) for different types of photovoltaic (PV) panels. The advance fuzzy logic controller is developed to distribute the power among the hybrid system and to manage the charge and discharge current flow for performance optimization. The developed management system performance was assessed using a hybrid system comprised PV panels, wind turbine (WT), battery storage, and proton exchange membrane fuel cell (PEMFC). To improve the generating performance of the PEMFC and prolong its life, stack temperature is controlled by a fuzzy logic controller. The dynamic behavior of the proposed model is examined under different operating conditions. Real-time measured parameters are used as inputs for the developed system. The proposed model and its control strategy offer a proper tool for optimizing hybrid power system performance, such as that used in smart-house applications.

Keywords: Artificial Neural Network; Energy Management; Fuzzy Control; Hybrid Power Systems; Maximum Power Point Tracker; Modeling

1. Introduction

Renewable energy sources have been found to be promising energy sources toward building a sustainable and environment friendly energy economy in the next decade. Among these renewable energy sources, solar and wind energy which are two of the most promising renewable power generation technologies. The growth of PV and wind power generation systems has exceeded the most optimistic estimation. However, geographic and seasonal climatic conditions affect the solar-wind energy output. Therefore, a backup power system is needed to improve the energy supply reliability. Thus, the PEMFC with a reversible energy storage system ideally fulfills the need for any start up power.

However, the price of PEMFC is considerably high and its membrane lifetime is rather short [1]. In addition, the dynamic interaction between the load demand and the renewable energy source can lead to critical problems of stability and power quality that are not very common in conventional power systems. Therefore, managing the flow of energy throughout the hybrid system is essential to increase the operating life of the membrane and to ensure the continuous energy flow.

In the literature, there are a few studies related to energy management of hybrid power system [2]. Among them, Wang and Nehrir [3], proposed a power management strategy for an ac-linked hybrid wind/PV/FC energy system. Ahmed *et al.* [4], presented a power management strategy which studied power fluctuations in a hybrid PV/wind turbine/FC power system. Onar *et al.* [5], proposed a power management strategy algorithm which dealt with a hybrid PV/wind/FC power system containing an ultra capacitor bank. However, all of the early mentioned methods have used conventional approaches for controlling hybrid power systems, such as linear PI controller which has afterwards been proven its instability in handling various changes in weather conditions [6,7]. This resulted in developing other approaches led to more robust algorithms with ability to handle various changes dynamically without any major problems; by establishing new management criteria depending upon informational data and the environmental changes.

Moreover, previous studies showed [8,9] that using the artificial intelligence in the hybrid vehicle can control the FC system within a specified high-efficiency region. Hence, this paper presents an adaptive management strategy for power flow in stand-alone hybrid power systems based on fuzzy logic and neural network. The proposed method introduces an online energy management by a hierarchical controller between four energy sources comprises PV panels, wind turbine, battery storage, and PEMFC. The overall aim is to optimize the active power flow between system power sources for different modes of operation, and to maintain the battery state-of-charge (SOC) at a reasonable level. The block diagram of the developed hybrid power system is shown in **Figure 1**.

The paper is organized as follows: Section 2 presents the dynamic model of the hybrid system. The proposed control structure is addressed in Section 3. Here PEMFC temperature controller based on fuzzy logic is implemented. To demonstrate the effectiveness of the proposed strategy, simulation results are presented in Section 4; and finally, in Section 5, conclusions are given.

2. Dynamic Modeling of Hybrid System

In this section, the dynamic simulation model is described for a hybrid power system consists of a PV panels, wind turbine, PEM fuel cell, lithium-ion (Li-Ion) battery bank, dc-dc and dc-ac converters. The mathematical models describing the dynamic behaviour of each of these components are given below.

2.1. The Photovoltaic Module

The general mathematical model for the solar cell has been studied over the past three decades [10]. The model of PV power plant used in this study is based on the dynamic PV model developed and validated in our previous work [11]. The following equation model the current of the PV cell [11]:

$$I_{pv} = I_{gc} - I_o \left[\exp^{\left(\frac{qv_d}{K_B F T_c}\right)} - 1 \right] - \frac{v_d}{R_p} \quad (1)$$

where I_{gc} is the light generated current, I_o is the dark saturation current dependant on the cell temperature, q is the electric charge (1.6×10^{-19} C), K_B is the Boltzmann's constant, F is the cell idealizing factor, T_c is the cell's absolute temperature, v_d is the diode voltage, and R_p is the parallel resistance.

2.2. The Wind Turbine

Several studies have been reported regarding to WT and

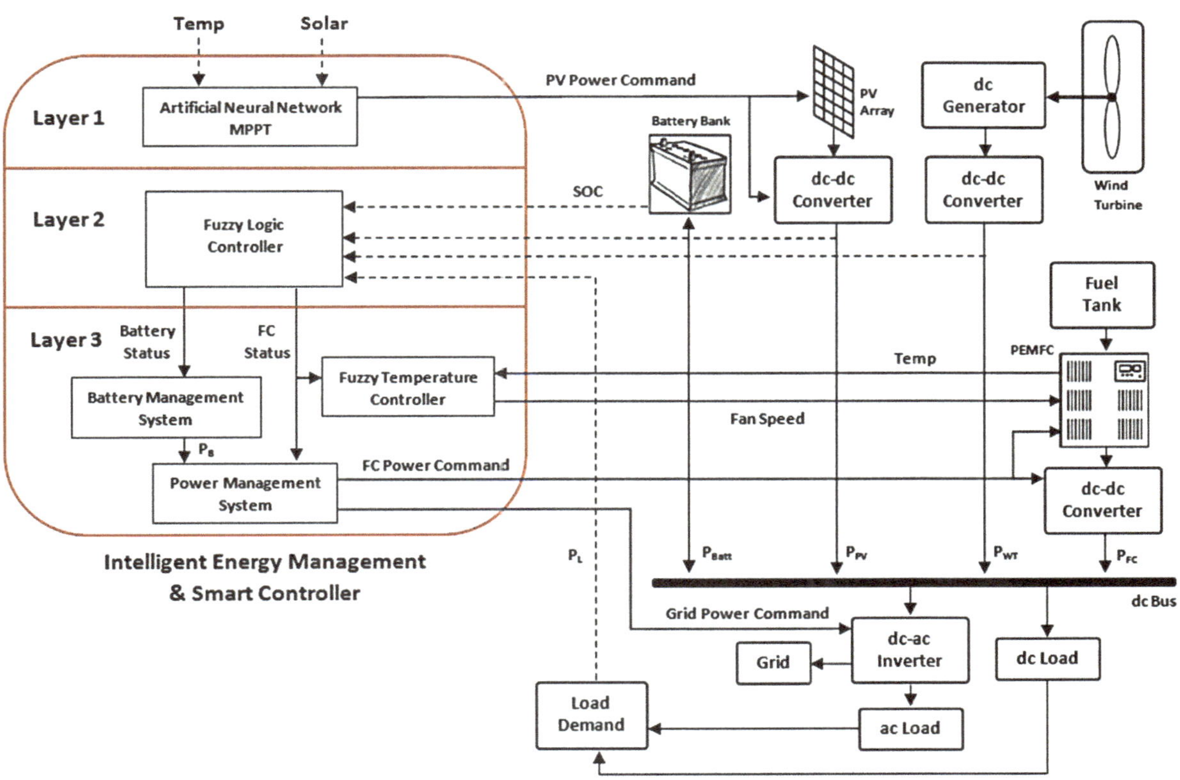

Figure 1. Block diagram of the proposed system.

wind generators [12]. In this study, the proposed WT model is based on the wind speed versus WT output power characteristics. The output power of the wind turbine is given by [13]:

$$P_m = c_p(\lambda, \beta)\frac{\rho A_S}{2}v_{\text{wind}}^3 \quad (2)$$

where P_m is the mechanical output power of the turbine, c_p is the performance coefficient of the turbine, β is the blade pitch angle, ρ is the air density, A_S is the swept area, and v_{wind} is the wind speed. The performance coefficient of the turbine (c_p) can be presented as in Equation (3). This equation based on the modeling turbine characteristics of [13].

$$c_p(\lambda, \beta) = c_1\left(\frac{c_2}{\lambda_i} - c_3\beta - c_4\right)e^{(c_5/\lambda_i)} + c_6\lambda \quad (3)$$

where

$$\frac{1}{\lambda_i} = \frac{1}{\lambda + 0.08\beta} - \frac{0.035}{\beta^3 + 1} \quad (4)$$

In this model, whereas the inputs are the wind speed and generator speed, the output is the torque applied to the generator shaft. The permanent magnet dc generator model is designed using the built-in SimPowerSystem library.

2.3. The Proton Exchange Membrane Fuel Cell

PEMFC is an electrochemical device that converts chemical energy into electrical energy. They show great performance in residential applications due to the low working temperature and fast start-up. The modified PEMFC model used in this study is based on the dynamic PEMFC stack model developed and validated in [14]. The following equation model the voltage of the fuel cell stack [14]:

$$V_{fc} = \left(E_{oc} - NA_T \ln\left(\frac{I_{fc}}{I_e}\right)\frac{1}{s\frac{T_d}{3}+1}\right) - RI_{fc} \quad (5)$$

where E_{oc} is the open circuit voltage, N is the number of cells, A_T is the Tafel slope, I_e is the exchange current, I_{fc} is the fuel cell current, T_d is the response time, and R is the internal resistance.

2.4. The Li-Ion Battery

In this section, the mathematical modeling of the Li-Ion battery bank used in the simulation program is introduced. The open voltage source is calculated with a nonlinear equation based on the actual SOC of the battery as follows [15]:

$$V_{batt} = E_{batt} - Ri \quad (6)$$

- During discharge:

$$E_{batt} = E_o - K\frac{Q}{Q-it}it - K\frac{Q}{Q-it}i^* + A\exp(-Bit) \quad (7)$$

- During charge:

$$E_{batt} = E_o - K\frac{Q}{Q-it}it - K\frac{Q}{it+0.1Q}i^* + A\exp(-Bit) \quad (8)$$

where E_{batt} is the no-load voltage, E_o is the battery constant voltage, K is the polarization constant, Q is the battery capacity, it is the actual battery charge, i^* is the low frequency current dynamics, A is the exponential zone amplitude, B is the exponential zone time constant inverse $(\text{Ah})^{-1}$, V_{batt} is the battery voltage, and i is the battery current. The main feature of this battery model is that the parameters can easily be deduced from a manufacturer's discharge curve [15]. The SIMULINK model for the Li-Ion battery has been validated in previous works [15,16].

2.5. The Power Conditioning Units

To connect a photovoltaic, wind turbine or PEMFC to an external system, it is necessary to boost their voltage or to increase their number. Therefore, a boost converter is used. A boost converter is a class of switching-mode power supply containing at least two semiconductor switches and at least one energy storage element. In addition, a capacitor is often added to the converter output to reduce the ripple of its output voltage. Equation (9) describes the relation between the input and the output voltage of a boost converter as a function of the duty cycle D:

$$\frac{V_{out}}{V_{in}} = \frac{1}{1-D} \quad (9)$$

On the other hand, an ac averaged switched model inverter has been implemented as in [13].

3. Energy Management and Control System

3.1. PV MPPT Based on Neural Network

Tracking the MPP of a photovoltaic array is an essential part of a PV system. As such, many maximum power point tracking (MPPT) techniques have been developed and implemented. Among these techniques, hill-climbing MPPT such as perturb and observe (P&O), which is a simple algorithm that does not require previous knowledge of the PV generator characteristics and is easy to implement with analogue and digital circuits [11]. However, the main problem of this technique can be seen when solar radiation rapidly change. In addition, P&O technique may cause many oscillations around the MPP, and this slows down the response of the system. In this model, intelligent control technique using artificial neural network is associated to an MPPT controller in order to increase the tracking response and consequently increase the tracking efficiency. **Figure 2** shows the structure of the proposed PV control system.

The neural network control (NNC) is used to estimate the PV array operating voltage (V_{ref}) which corresponds to P_{max} at any given solar radiation and cell temperature. It consists of three layers. The input layer is composed of three nodes in inputs which are: the solar radiation, the cell temperature and the cell's open circuit voltage at a 25°C and 1 kW/m^2. The hidden layer is composed of four nodes whose activation function is the hyperbolic tangent sigmoid transfer function. The output layer is composed of one node that is the optimum operating voltage whose function of activation is of the linear type. Once the neural network model architecture is defined, data are collected and fed to the model. The network is then trained to recognize the relationships between the input and output parameters. The proposed NNC uses the Levenberg Marquardt (LM) training algorithm [17]. In this algorithm, the interlayer connection weight and the processing element's thresholds are first initialized at small random values. The network is then presented with a set of training patterns. Each set is composed of three inputs (solar irradiance, temperature, cells open circuit voltage), and one output. The solar irradiance and temperature data are gathered from a 28.8 kW solar power system located on a tower block on campus in MMU, central Manchester [16]. The output or rather the targeted output (ideal V_{ref}) is generated from an applied Matlab code which analyzes the output P-V characteristics of the validated PV model.

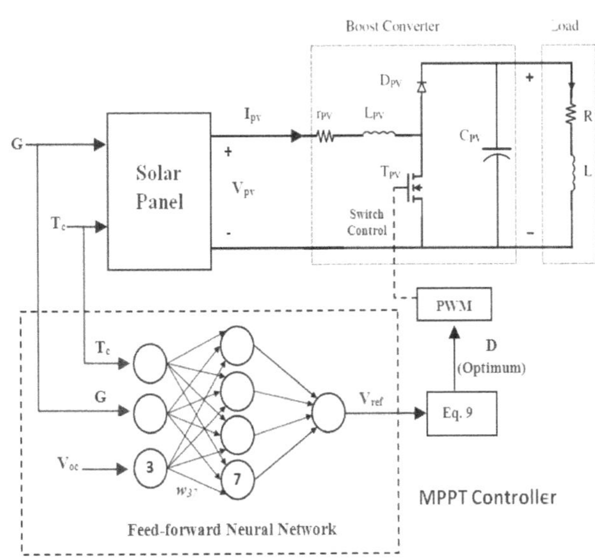

Figure 2. The proposed PV control system.

Prior to conducting the network training operation using the LM algorithm, a training set of 2000 cases were obtained from four different PV panels, namely: Lorentz, Sharp's-NUS0E3E, BP-485J, and Astronergy-CHSM-6610P. The cell's open circuit voltage (V_{oc}) is used as a reference variable to select from among the four PV panels. This training data set covers the different solar radiation and temperature conditions that could possibly take place. The training patterns were presented repeatedly to the neural network model and the adjustment was performed after each iteration whenever the network's output is different from the desired output. This process continues until the mean square error (MSE) converged and is measure at less than 0.01.

3.2. Advanced Fuzzy Energy Distribution Strategy

The energy management strategy (EMS) should determines the split power between the photovoltaic, wind turbine, fuel cell stack and battery while satisfying the load power requirement with respect to dynamic restrictions to the battery and fuel cell stack. The EMS developed in this work is shown in **Figure 3**.

Frequent power demand variations and unpredictable load profile are unavoidable. Adding to this, the nonlinear subsystems add to the complexity of the structure of hybrid system. Hence, an advance fuzzy expert system according to the weather variations, load demand and SOC is presented. **Figure 4** shows the proposed fuzzy expert system.

A fuzzy logic controller (FLC) is used to decide the optimum operation of the PEMFC/battery system. As shown in **Figure 3**, there are five possible operating modes. When in hybrid mode, PEMFC power changes smoothly so that the FC operating point deviation is minimized.

The battery management system maintains the SOC at a reasonable level (40% - 80%). Also, it protects against voltage collapse by controlling the power level required from the battery (P_B) (see **Figure 3**). An S-R flip-flop type has been used for storing battery status (BS). The power management system controls the reference power of the PEMFC stack (FC_P_{ref}) by splitting the power demand (P_L) as a function of the available power of the battery and the PV/WT system (P_H) (see **Figure 3**). The FLC relates the outputs to the inputs using a list of if-then statements called rules. The if-part of the rules describes the fuzzy sets (regions) of the input variables. In this work, the fuzzy variables P_H, P_L, and SOC are described by fuzzy singleton, *i.e.* the measured values of these variables are used in the interface process without being fuzzified. Specifically, the fuzzy rules are in the form:

Rule i: IF P_H is A_i and P_L is B_i and SOC is C_i, **THEN** FC_E is D_i and BS is E_i.

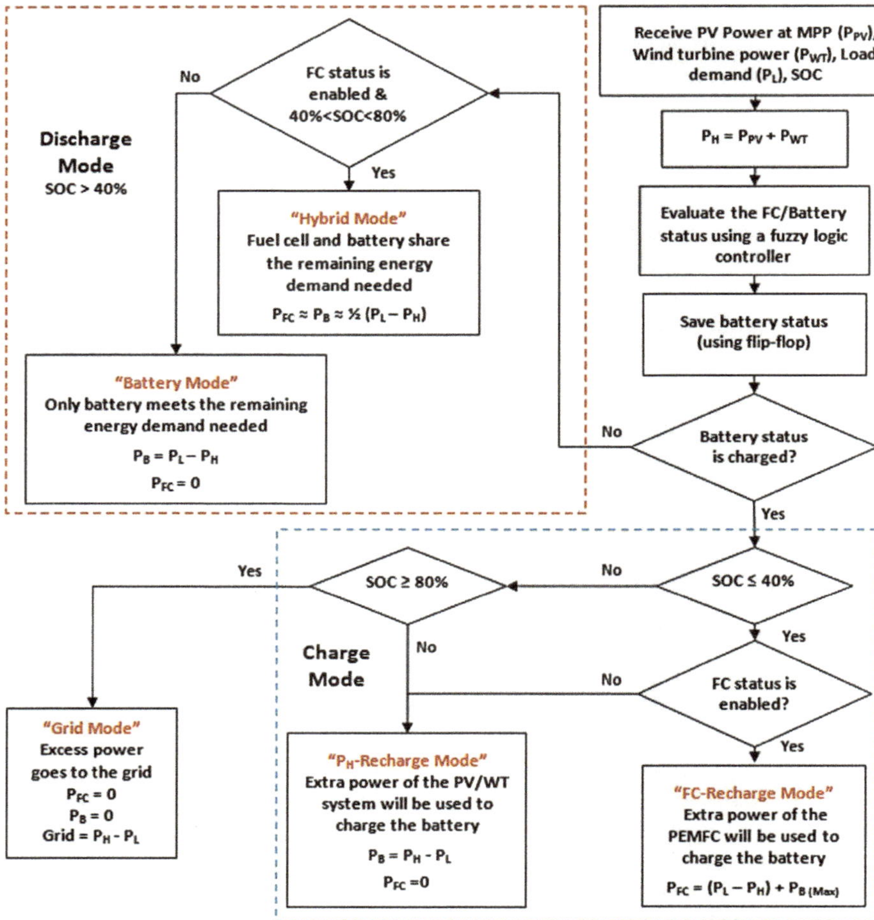

Figure 3. Proposed energy management system algorithm.

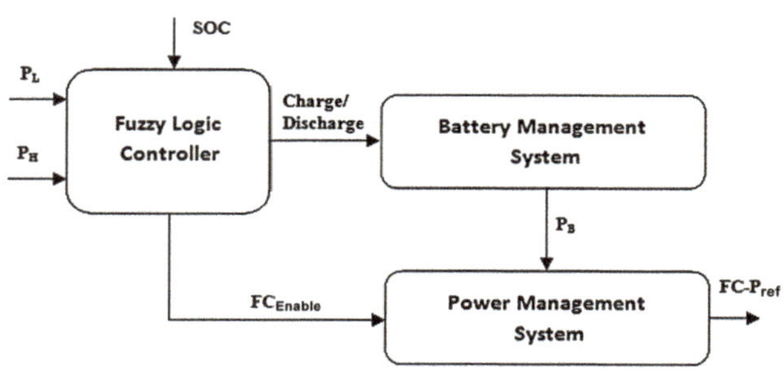

Figure 4. Proposed fuzzy expert system for EMS.

Where A_i, B_i, and C_i are fuzzy subsets in their universes of discourse, and D_i, E_i are fuzzy singletons. Each universe of discourse is divided into three fuzzy subsets: L (Low), M (Medium), and H (High). The fuzzy subsets and the shape of membership function are shown in **Figure 5**.

These degrees of membership are evaluated to obtain the output controller, and the then-parts of all rules are averaged and weighted by these degrees of membership. The core of the rule set of the fuzzy controller is shown

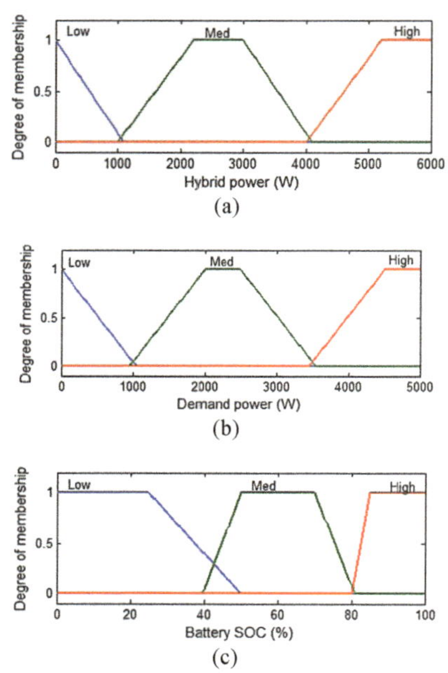

Figure 5. Membership functions of the FLC. (a) PV/WT power; (b) Load demand; (c) Battery SOC.

in **Table 1**.

3.3. PEMFC: Temperature Controller

The operating temperature of the stack is an important controlled variable, which impacts the performance of the PEMFC. Hence, in order to improve the generating performance of the PEMFC and prolong its life, it must be controlled efficiently. Consequently, in this work, the PEMFC stack model described by Oliver in [14] is modified to include a fuzzy logic temperature controller.

The main function of this control system is to maintain the temperature stable and equal to the stack operating temperature. The FLC is used to control the airflow from the heater and cooler fans, based on the stack temperature. In case of cold start-up, the heater fan (F1) needs to be on; until the stack temperature is equal to the room temperature. On the contrary, if the stack temperature is hot, the cooler fans (F2) speed needs to be at the high level. Triangular membership functions and centroid defuzzification method are adopted. The membership functions are shown in **Figure 6**.

Table 2 shows the rule base of the FLC.

Over time, the control system calculates the cooling/heating effect inside the stack, taking into account the thermal isolation and cooling/heating activities. And based on these equations the output temperature will be calculated as follow [18]:

$$\Delta T_{FC} = (T_{FC} - T_A) \times K_T + (T_{FC} \times K_{Fan}) \quad (10)$$

$$T_{FC}(t) = \int \Delta T_{FC} \mathrm{d}t \quad (11)$$

where T_A is the ambient temperature, T_{FC} is the FC stack temperature, K_T is the thermal isolation multiplier, and K_{Fan} is the cooling/heating factor.

4. Simulation and Discussion

4.1. The Evaluation of the Proposed MPPT

Using the 400 cases allocated for the testing set, the modelinput parameters were entered consecutively for each case and a prediction for the optimal V_{ref} was obtained. The prediction results were then compared with the actual results of the 400 cases. The statistical analysis of these results indicates that the R^2 value for the testing set was 0.99. This result demonstrates that the MPPT-NNC model developed in this work can predict the optimal V_{ref} for any trained PV panel with high accuracy. The graphical results of the first 31 cases in the testing set are shown in **Figure 7**.

Figure 8 compares the optimum power (P_{mpp}) obtained from the classical P&O algorithm and the proposed ANN.

From **Figures 8(a)** and **(b)**, it is shown that by using the ANN the optimum power will be steadier and close to the theoretical power when compared to the classical P&O algorithm.

4.2. The Proposed Hybrid System

In the simulation process, the aim is to observe the proposed system behavior under different operating condi-

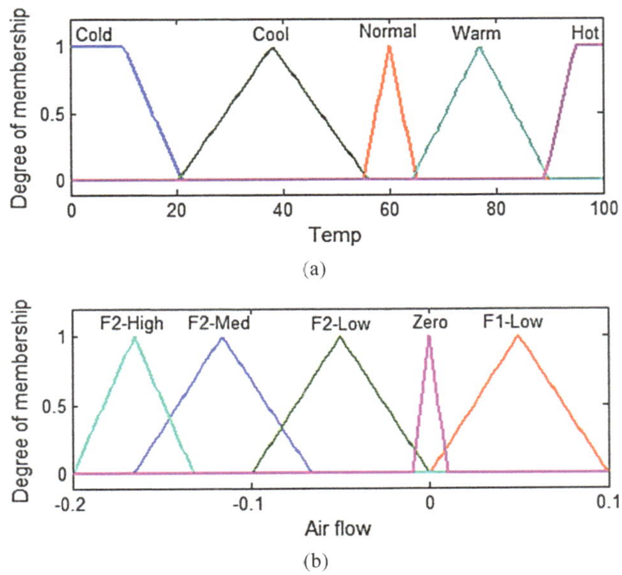

Figure 6. Membership functions of the FLC temperature. (a) Stack temperature; (b) Cooling/heating air flow.

Table 1. Fuzzy logic rules.

1-3	IF P_H is L and P_L is L\M\H and SOC is L, THEN FC_E is ON and BS is CO	4-6	IF P_H is M and P_L is L\M\H and SOC is L, THEN FC_E is OFF\ON\ON and BS is CO
7-9	IF P_H is H and P_L is L\M\H and SOC is L, THEN FC_E is OFF\OFF\ON and BS is CO	10-12	IF P_H is L and P_L is L\M\H and SOC is M, THEN FC_E is OFF\ON\ON and BS is CD
13-15	IF P_H is M and P_L is L\M\H and SOC is M, THEN FC_E is OFF\OFF\ON and BS is CD	16-18	IF P_H is H and P_L is L\M\H and SOC is M, THEN FC_E is OFF and BS is CD
19-21	IF P_H is L and P_L is L\M\H and SOC is H, THEN FC_E is OFF and BS is DO	22-24	IF P_H is M and P_L is L\M\H and SOC is H, THEN FC_E is OFF and BS is DO
25-27	IF P_H is H and P_L is L\M\H and SOC is H, THEN FC_E is OFF and BS is DO		

Note: CO means charge only, DO means discharge only, and CD means charge or discharge.

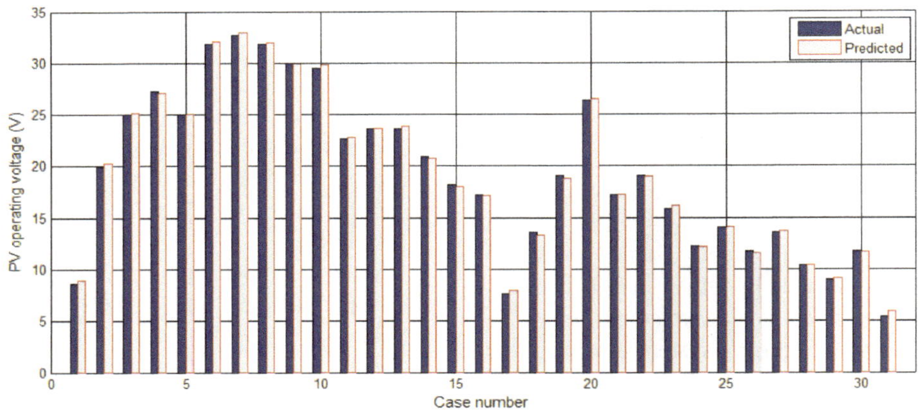

Figure 7. Actual and predicted PV operating voltage for the first 31 cases in the testing set.

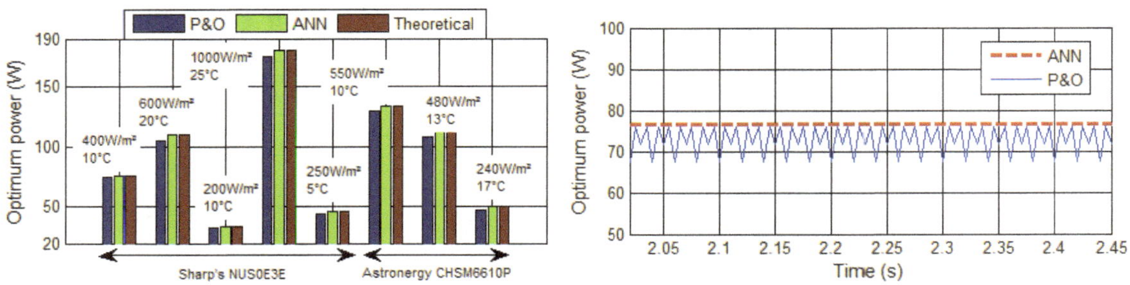

(a) (b)

Figure 8. Comparing the P_{mpp} obtained from the classical P&O algorithm and the proposed ANN. (a) P_{mpp} at different conditions; (b) P_{mpp} at 400 W/m^2, 10˚C (Sharp's NUS0E3E).

Table 2. FLC rules for the temperature controller.

Stack Temperature	Fan Airflow
Cold	F1-Low
Cool	Zero
Normal	F2-Low
Warm	F2-Medium
Hot	F2-High

tion. The solar radiation, panel temperature, wind speed and user load profiles are all used to test the performance of the proposed hybrid system model, as shown in **Figures 9(a)-(c)**.

Simulation results are obtained by developing a detailed MATLAB/SIMULINK software package.

The specifications of the renewable power sources used are given in **Table 3**. The specifications of the Li-Ion battery and PEMFC stack system are given in **Tables 4** and **5**. Finally, **Table 6** shows the power conditioning units' parameters include dc-dc and dc-ac converters.

Simulation results are obtained for the time interval between 0 and 600 sec. The power demand profile has a significant effect on determining the EMS. A practical load profile for two family members in resident appli-

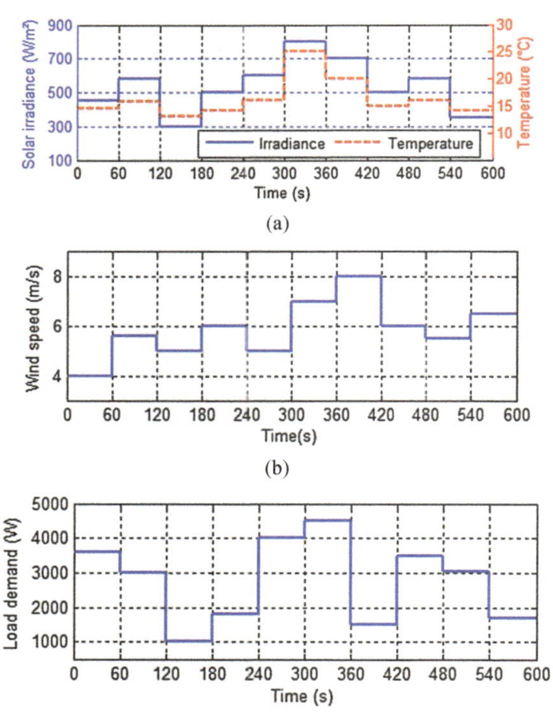

Figure 9. (a) Solar radiation and panel temperature profiles; (b) Wind speed profile; (c) Load demand profile.

cation [2] is established, as shown in **Figure 9(c)**. From this load profile it is evident that the average power demand is less than 2.8 kW. However, due to weather condition, the output power of the PV/WT system varies from 1.1 to 3.6 kW, as shown in **Figure 10**. Therefore, a PEM fuel cell with a reversible ESS was added to the renewable power system. The battery bank system is capable of sustaining the extra load of 1.3 kW for 1 h during peak load demand periods. While, the 6 kW PEMFC can meets the remaining extra power and protect the battery from overcharging.

Moreover, according to the proposed EMS algorithm, battery SOC level is determines the operation of the PEMFC, during peak load demand periods. Therefore in the following subsections two cases have been tested during the simulation. The solar radiation, wind speed and user load profiles are the same for the two cases (see **Figure 9**).

4.2.1. First Case

The initial SOC of the battery storage is 33.5%. Hence, when the SOC is low (0 - 120 sec), the peak load demand is satisfied by the PEMFC stack, and the extra power will be used to charge the battery (SOC goes high), as shown in **Figure 11(a)**. However, during that time, the PEMFC stack power is very high. Consequently, in order to maintain the stack temperature stable and equal to the stack operating temperature (65˚C), a fuzzy logic temperature controller has been added, as shown in Section 3.3. **Figures 11(b)** and **(c)** show the cooling activity and stack temperature during that period. On the other hand, when the SOC is above 42.5%, then the remaining energy is satisfied by either the battery bank or by the hybrid PEMFC/battery system, as shown in **Figures 11(a)** and **(d)**.

The power met by the battery introduces a current change at the battery bank terminal as depicted in **Figure 11(e)**. When the battery current is positive (time intervals 240 to 360 sec and from 420 to 540 sec), energy is transferred to the load by the battery bank. When the battery current is negative, the battery bank is recharged. **Figure 11(f)** shows the SOC of the battery during the simulation.

4.2.2. Second Case

The initial SOC of the battery storage is 87%. The system response to changes in solar irradiance and load profile is presented in the following figures. The peak load demand is satisfied by the battery bank as shown in **Figure 12(a)**. During this case the PEMFC is off; since the battery SOC is high. **Figure 12(b)**, shows the variation of

Table 3. PV/WT system specifications.

PV sharp's (NU-180) at (1 kW/m^2, 25˚C)		Wind turbine kestrel e400i	
Open circuit voltage	30 (V)	Peak output power	3200 (W)
Voltage at P_m	23.7 (V)	Base wind speed	11 (m/s)
Short circuit current	8.37 (A)	Base rotational speed	520 (rad/s)
Current at P_m	7.6 (A)		

Table 4. Li-Ion battery parameters.

Nominal [voltage, capacity]	200 (V), 5.8783 (Ah)
Fully charged voltage	232.8 (V)
Maximum capacity	6.5 (Ah)
Exponential zone [voltage, capacity]	216.07 (V), 0.319 (Ah)
Internal resistance	0.307 (Ω)

Table 5. PEMFC stack parameters.

Voltage at 0 A and 1 A	65 (V), 63 (V)
Nominal operating point	133.3 (A), 45 (V)
Maximum operating point	225 (A), 37 (V)
Number of cells	65
Nominal air flow rate	300 (l/m)
Nominal supply pressure [Fuel, Air]	1.5 (bar), 1 (bar)
Nominal composition [H_2, O_2, H_2O (Air)]	99.95, 21, 1 (%)

Table 6. Power conditioning units' parameters.

dc-dc Converter		dc-ac Inverter	
Rated voltage	211 (V)	AC line RMS voltage	120 (V)
Resistance	0.02 (Ω)	AC line frequency	60 (Hz)
Capacitance	10 (μF)	Output port series resistance	0.2 (Ω)
Inductance	200 (μH)	Switching loss current	0.03 (A)

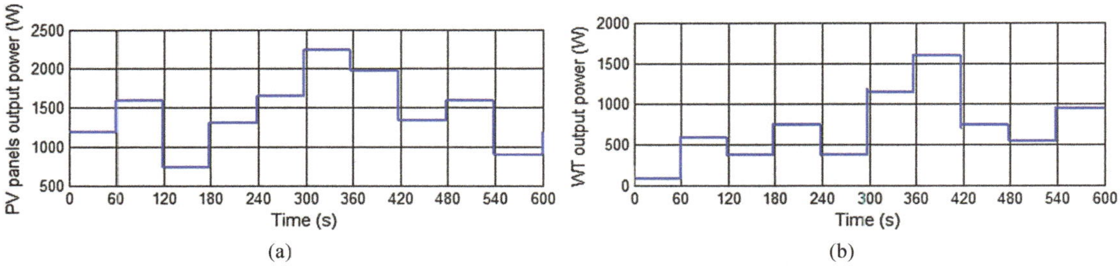

Figure 10. (a) Total power of the solar power plant; (b) Total power of the wind turbine.

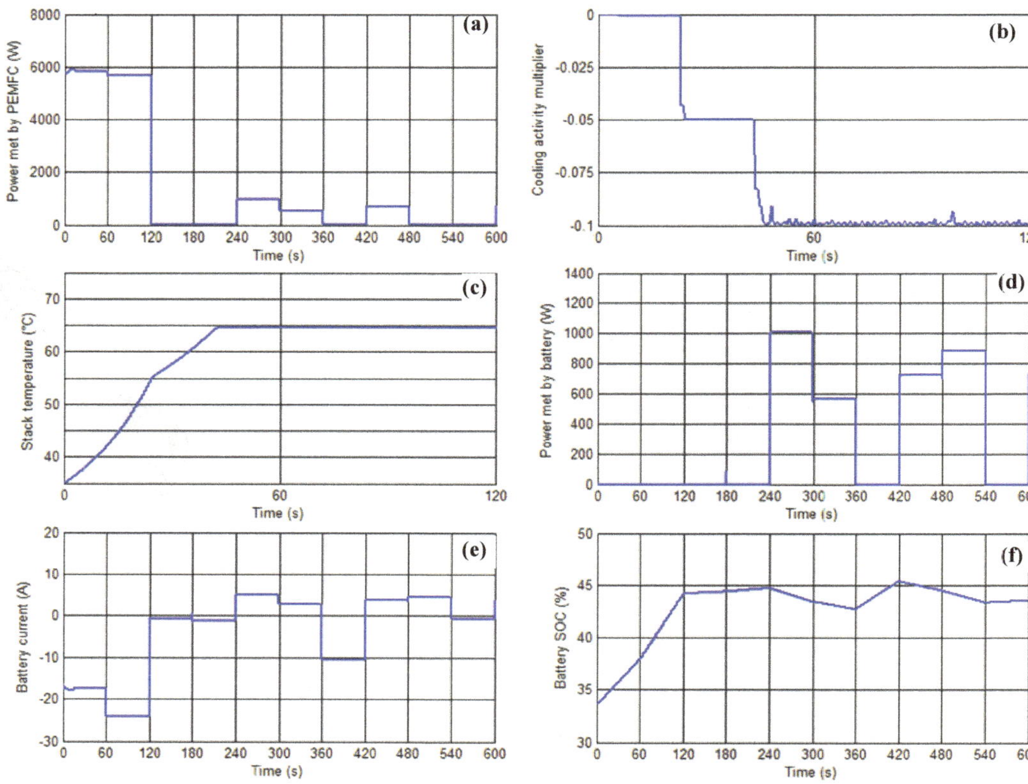

Figure 11. System response during the first case. (a) Power satisfied by PEMFC; (b) Cooling air flow; (c) PEMFC stack operating temperature; (d) Power satisfied by battery; (e) Battery current change with respect to load profile; (f) Battery SOC.

the battery bank current between positive (discharging) and negative (charging) according to the required load demand. **Figure 12(c)** shows the SOC of the battery during the simulation.

Consequently, it's clear that during peak load demand the PV/WT system supplies the available power and the battery bank supplies the remaining extra power (0 - 120 sec, 240 - 360 sec, and from 420 - 540 sec). On the other hand, when the produced power by the PV/WT system is more than the required power of load, and the SOC of the battery is below 80%, the extra power of the PV/WT will be used to charge the battery and the SOC goes high (360 - 380 sec). Otherwise, if the SOC is above 80%, based on the EMS it is not in the safe charge mode. In this case the excess power goes to the grid as shown in **Figure 12(d)**.

The simulated grid side inverter voltage and current waveforms are shown in **Figure 12(e)**. It can be seen that the line current is pure sinusoidal and in phase with the grid voltage.

5. Conclusions

An optimized online energy management based on a hierarchical controller has been implemented for standalone hybrid power systems. The proposed method includes an NNC in the first layer, to achieve the MPP for different types of PV panels. The statistical analysis of the results indicates that the R^2 value for the testing set was 0.99. This indicates that the ANN-based model developed in this work can predict the MPP for a PV panel with high accuracy. Moreover, results of the proposed

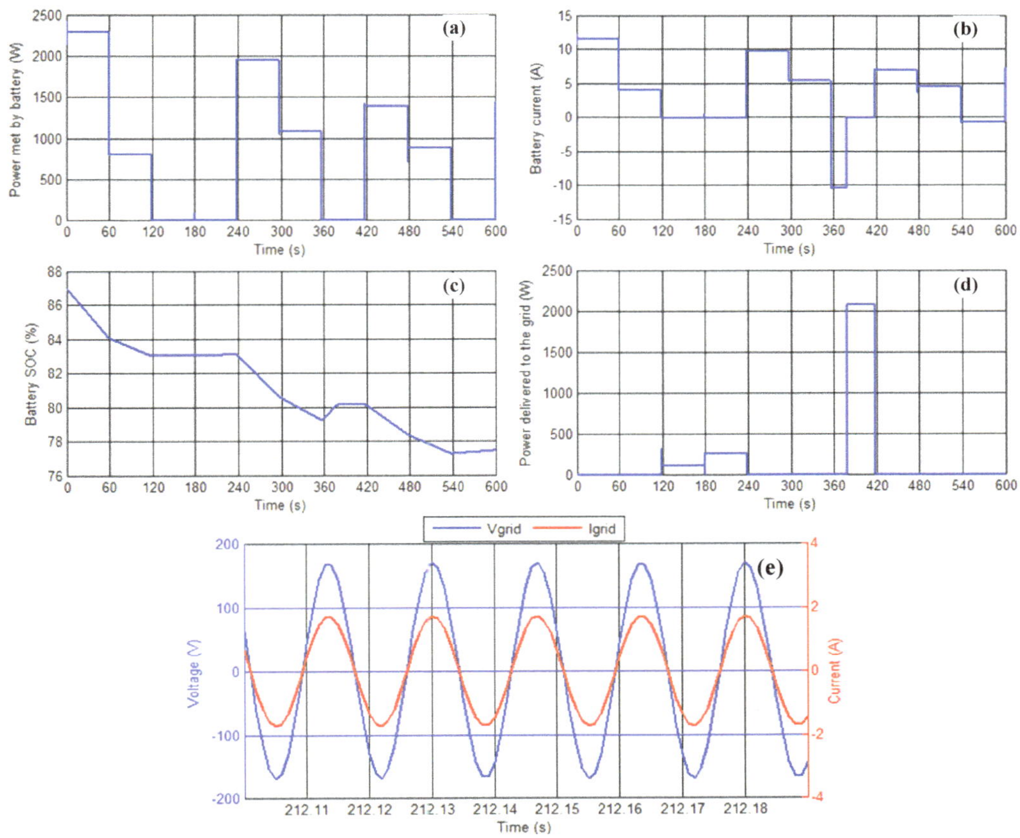

Figure 12. System response during the second case. (a) Power satisfied by battery; (b) Battery current change with respect to load profile; (c) Battery SOC; (d) Power delivered to the grid; (e) Grid voltage/current waveforms.

NNC shows better transient performance, with low overshoot, compared to the results obtained from literature [2-6].

Simulation results were obtained by developing a detailed dynamic hybrid system model. Real-time measured parameters and practical load profile are used as inputs for the developed management system. It was found that the hybrid topology exhibits excellent performance under various operating conditions, and maintain the SOC between 40% - 80%.

REFERENCES

[1] Y. Wang, K. S. Chen, J. Mishler, S. C. Cho and X. C. Adroher, "A Review of Polymer Electrolyte Membrane Fuel Cells: Technology, Applications, and Needs on Fundamental Research," *Applied Energy*, Vol. 88, No. 4, 2011, pp. 981-1007.

[2] E. Dursun and O. Kilic, "Comparative Evaluation of Different Power Management Strategies of a Stand-Alone PV/Wind/PEMFC Hybrid Power System," *Electrical Power and Energy Systems*, Vol. 34, No. 1, 2012, pp. 81-89.

[3] C. Wang and M. H. Nehrir, "Power Management of a Stand-Alone Wind/PV/Fuel Cell Energy System," *IEEE Energy Conversion*, Vol. 23, No. 3, 2008, pp. 957-967.

[4] N. A. Ahmed, M. Miyatake and A. K. Al-Othman, "Power Fluctuations Suppression of Stand-Alone Hybrid Generation Combining Solar Photovoltaic/Wind Turbine and Fuel Cell Systems," *Energy Conversion and Management*, Vol. 49, No. 10, 2008, pp. 2711-2719.

[5] O. C. Onar, M. Uzunoglu and M. S. Alam, "Modeling, Control and Simulation of an Autonomous WT/PV/FC/Ultra-Capacitor Hybrid Power System," *Journal of Power Sources*, Vol. 185, No. 2, 2008, pp. 1273-1283.

[6] A. Tofighi and M. Kalantar, "Power Management of PV/Battery Hybrid Power Source via Passivity-Based Control," *Renewable Energy*, Vol. 36, No. 9, 2011, pp. 2440-2450.

[7] W. Zhou, C. Lou, Z. Li, L. Lu and H. Yang, "Current Status of Research on Optimum Sizing of Stand-Alone Hybrid Solar Wind Power Generation Systems," *Applied Energy*, Vol. 87, No. 2, 2010, pp. 380-389.

[8] X. Li, L. Xu, J. Hua, X. Lin, J. Li and M. Ouyang, "Power Management Strategy for Vehicular Applied Hybrid Fuel Cell/Battery Power System," *Journal of Power Sources*, Vol. 191, No. 2, 2009, pp. 542-549.

[9] C.-Y. Li and G.-P. Liu, "Optimal Fuzzy Power Control and Management of Fuel Cell/Battery Hybrid Vehicles," *Journal of Power Sources*, Vol. 192, No. 2, 2009, pp. 525-533.

[10] M. G. Villalva, J. R. Gazoli and E. R. Filho, "Comprehensive Approach to Modeling and Simulation of PV Arrays," *IEEE Power Electronics*, Vol. 24, No. 5, 2009, pp. 1198-1208.

[11] E. M. Natsheh and A. Albarbar, "Photovoltaic Model with MPP Tracker for Stand-Alone/Grid Connected Applications," *IET Conference on Renewable Power Generation*, Edinburgh, 6-8 September 2011, pp. 1-6.

[12] H. D. Battista, R. J. Mantz and F. Garelli, "Power Conditioning for a Wind Hydrogen Energy System," *Journal of Power Sources*, Vol. 155, No. 2, 2006, pp. 478-486.

[13] E. M. Natsheh, A. Albarbar and J. Yazdani, "Modeling and Control for Smart Grid Integration of Solar/Wind Energy Conversion System," *IEEE PES International Conference and Exhibition on Innovative Smart Grid Technologies*, Manchester, 5-7 December 2011, pp. 1-8.

[14] N. M. Souleman, O. Tremblay and L.-A. Dessaint, "A Generic Fuel Cell Model for the Simulation of Fuel Cell Power Systems," *IEEE Power & Energy Society General Meeting*, Quebec, Montreal, 26-30 July 2009, pp. 1-8.

[15] O. Tremblay, L.-A. Dessaint and A.-I. Dekkiche, "A Generic Battery Model for the Dynamic Simulation of Hybrid Electric Vehicles," *IEEE Vehicle Power and Propulsion Conference*, Arlington, 9-12 September 2007, pp. 284-289.

[16] E. M. Natsheh and A. Albarbar, "Solar Power Plant Performance Evaluation: Simulation and Experimental Validation," *Journal of Physics*: *Conference Series*, Vol. 364, No. 1, 2012, pp. 1-13.

[17] M. Negnevitsky, "Artificial Intelligence: A Guide to Intelligent Systems," Addison Wesley, Boston, 2004.

[18] MathWorks, "Documentation Center," 2012. http://www.mathworks.co.uk/help/stateflow/gs/a-look-at-the-physical-plant.html

Design of a Monitoring System of Micro-Grid

Xiangyang Zhao, Shiyang Liu

School of Automation Science and Electrical Engineering, Beihang University, Beijing, China.

ABSTRACT

This paper describes a micro-grid system and its monitoring system. This micro-grid system consists of generation systems, consumer electrical equipments, auxiliary equipments and the monitoring system. All the equipments have 485 communication interfaces. In order to monitor and manage this micro-grid system, we built a monitoring system, which contains modular instrument system and industrial personal computer. In order to keep real time, we adopt some measures in software and hardware. We adopt LABVIEW and its program modules in software and adopt modular instrument system in hardware. Supporting by the software and hardware, the micro-grid system can be safe and stable.

Keywords: Micro-Grid; Real Time; The Monitoring System; Communication

1. Introduction

With the rapid development of the economy and the growth of power demand, in the past few decades the China power department has paid most of the attention to the large concentrated power, coal-fired power, hydroelectric power, nuclear power, the construction of ultra high voltage and distance transmission power grids The electric power industry has made great progress and development. With the constant expansion of power grid, the shortcomings of large scale power system are increasingly prominent. Especially in recent years, large area outages have occurred several times all around the world, and the vulnerability of electric grid has exposed.

In order to solve the problems mentioned above, the distributed power generation is put on the agenda. Distributed power generation has less pollution, higher reliability, higher efficiency, flexible installation and many other advantages.

Although the distributed power generation has a lot of advantages, there are also many problems [1]. For example, it is expensive when the distributed power single accesses the grid. Then the distributed power has some difficulties on control. The distributed power generation is an uncontrolled power supply relative to the large power grid. In order to coordinate the contradiction between the large power grid and the distributed power supply, considering the value and benefits between the distributed energy power grid and the user, scholars have put forward the concept of the micro-grid at the beginning of this century.

As an important component of the distributed power generation, the micro-grid is a necessary development trend for the future [2-5]. It can not only improve the power supply reliability, but also ease up the pressure on energy conservation and environmental protection [6-8]. This paper describes a micro-grid system and its monitoring system which manages this micro-grid system.

2. The Micro-Grid System

2.1. The Composition of the Micro-Grid System

Figure 1 shows the micro-grid system which contains

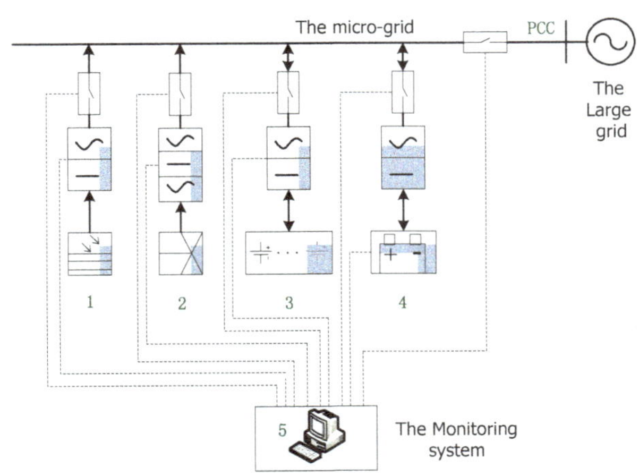

Figure 1. The composition of the micro-grid system.

five parts. This micro-grid system consists of photovoltaic (pv) power generation system, wind power generation system, super capacitor energy storage system, lithium battery energy storage system, modular instrument system based on PXI and industrial personal computer.

This micro-grid contains two kinds of generation systems. The first is photovoltaic (pv) power generation system. It consists of solar-cell panel, grid inverter, and intelligent switch. The second is wind power generation system, which contains wind generators, converter, and intelligent switch. The demand for more power, combined with interest in clean technologies has driven researchers to develop distributed power generation systems using renewable energy sources, which the two systems both belong to.

In order to achieve the technical characteristics of micro-grid better, electric energy storage technology is regarded as one of the core technologies in micro-grid, which plays an important role in stabilizing electric power system and utilizing renewable energy with high efficiency [9]. There are two energy storage systems in this micro-grid system. One is lithium battery energy storage system which has big storage capacity and can be charged and discharged many times. It can keep the micro-grid operate even if the two generation systems can not generate for some time because of the bad environment or other reasons. The other is super capacitor energy storage system, which can charge and discharge rapidly, so it can help solve some transient problems, which happen when load operates suddenly, or the micro-grid operates from grid mode to island mode.

The dual mode inverter is indispensable to the micro-grid, which is in the lithium battery energy storage system [10-12]. When the grid operates normally, it works in grid-connected mode, and the inverter operates in current-control mode as a current-source, controlling the active power and reactive power, so we also call it PQ mode. In this mode, the large grid controls frequency and voltage.

When the large grid failures, the micro-grid will open the grid switch, and work in the island mode. It keeps the voltage and the frequency stability as a voltage source. We call it VF mode. As the grid return to normal, the dual mode inverter closes the grid switch, and works from VF mode to PQ mode.

In order to keep the micro-grid safe and stability, we need to organically combine all the devices in the micro-grid. So we built a monitoring system. The monitoring system masters real-time state of all equipments and controls any equipment in any time. For example, when the micro-grid works in island mode, the two energy storage systems are both full of electricity, and the load has low electricity consumption, we need to control the two generation systems to reduce the output to keep the load work normally.

2.2. The Monitoring System

The monitoring system monitors real-time running status of all the equipments, and controls all the equipments according to the safe and stable operation requirements. It is very important to ensure real-time for the monitor system. In this micro-grid system, we ensure the system real time through the software and hardware [13].

In the hardware, the whole monitoring system contains PXI communication and control interface equipment and industrial personal computer. The monitoring system gathers equipment information through the 485 communication. In order to guarantee system real time, we adopt point-to-point 485 communication structure instead of the traditional 485 bus structure. It depends on PXI communication and control interface equipment to improve baud rate to 38,400 bps under the data transmission in stable condition. In addition, we adopt the special communication module to ensure the real time and stable. The PXI communication and control interface equipment contains embedded controller, 18-slot backplane, and Industrial RS485/422 serial interface.

The embedded controller is a high-performance Intel Core 2 Quad Q9100-based embedded controller. With its 2.26 GHz quad-core processor, it is ideal for high-performance modular instrumentation and data acquisition applications. The industrial RS485/422 serial interface offers all the features of high-performance NI PXI-843x modules plus high-voltage port-to-port isolation. The indus trial RS485/422 serial interface features high-speed communication with RS485 and RS422 devices at rates of up to 3 Mbit/s and offers configurable baud rates for data transmissions between 57 baud and 3,000,000 baud to within 1 percent accuracy for nonstandard baud rates and 0.01 percent accuracy for standard baud rates.

In software, we adopt LabVIEW as programming software which is based on a data flow graphical development platform, and it provides intuitive graphical user interface and programming way. Compared with the traditional text programming, LabVIEW is a natural kind of parallel structure of the programming language. In order to insure the real time of the system, especially for real-time of orders, we use multithreading and the parallel structure. When the monitoring system communicates with different equipments, the advantage of parallel structure is more obvious. In order for a system to be a real-time system, all parts of it need to be real-time. For example, even though a program runs in a real-time operating system, it does not mean that the program behaves with real-time characteristics. The program may rely on something that does not behave in real-time, which then causes the program not to behave in real-time.

The status of the grid operation displays in the interface of the industrial personal computer, which contains warning information, failure message, and so on. The workers can fully grasp the operation condition of the micro-grid and control all the equipments through the monitoring interface.

3. Software of the Monitoring System

The program has four functions, collecting data, uploading data, receiving the command, and controlling the devices. **Figure 2** shows the basic flowchart of PXI. First, the program initialization contains setting initial value to the corresponding variables. Then, it does not connect to the network, until the network connection is normal. Finally, it begins to work, gathering data, uploading data, receiving the command, and controlling the deceives.

This system gathers data by 485 communications interface, and the modbus protocol. The 485 communication interface is a kind of interface standards, which determines the interface electrical standards and just is the physical layer of a standard. The modbus protocol is used for electronic controller of the universal language. Through this protocol, controller between each other, controller and other equipment in the network can communicate. In order to send data to industrial personal computer, which will store the data for inquiring, the basic program adopts TCP/IP communication module, sending and receiving data.

4. The Operation of the Monitoring System

So far, the monitoring system has operated for about one month, and has finished part of the performance tests. The monitoring system can start and shutdown the micro-grid system automatically according to a button in the interface. Now, the system is working normally, and will accept the other aspects of the test on the next step.

5. Conclusion

In this paper, it describes a micro-grid, which contains the hardware construction system and the monitoring system. The real time control strategy and the management method of micro-grid were successfully employed in the micro-grid. The monitoring system has used 485 interfaces to communicate with all the facilities. We have taken the corresponding measures to solve the real-time problems from the software and hardware. However, it is need to continue to watch the power grid of the operation, perfect the automatic operation strategy, improve the system reliability and so on.

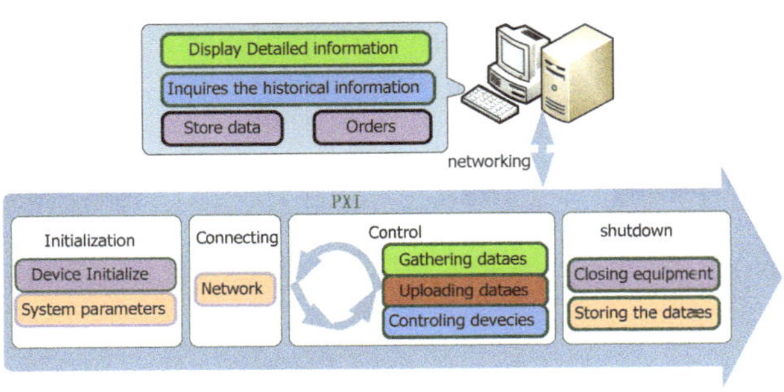

Figure 2. The structure of the monitoring system.

REFERENCES

[1] J. Driesen and R. Belmans, "Distributed Generation: Challenges and Possible Solutions," 2006 *IEEE Power Engineering Society General Meeting*, Montreal, October 2006, p. 8.

[2] C. X. Wu, F. S. Wen and Y. L. Lou, "The Existed Problems and Possible Solutions of Micro-Grid Based on Distributed Generation," *Electric Utility Deregulation and Restructuring and Power Technologies*, Nanjing, 6-9 April 2008, pp. 2763-2768.

[3] R. H. Lasseter, "Certs Microgrid," *IEEE International Conference on System of Systems Engineering*, Vol. 1, 2007, pp. 1-5.

[4] X.-S. Zhou, L.-Q. Cui and Y.-J. Ma, "Research on Control of Micro Grid," *Measuring Technology and Mechatronics Automation* (*ICMTMA*), Shanghai, 6-7 January 2011, pp. 1129-1132.

[5] T. Rihit, M. Ned and H. Chris, "Seamless Transfer of Grid-Connected PWM Inverters between Utility-Interactive and Stand-Alone Modes," *Seventeenth Annual IEEE Applied Power Electronics Conference and Exposition*, Vol. 2, 2002, pp. 1081-1086.

[6] Z. X. Lu, C. X. Wang and Y. Min, "Overview on Microgrid Research," *Automation of Electric Power Systems*, Vol. 31, No. 19, 2007, pp. 100-107.

[7] J. Driesen and F. Katiraei, "Design for Distributed Energy Resources," *IEEE Power Energy Magazine*, Vol. 3, No. 6, 2008, pp. 30-40.

[8] K. Sheng, L. Kong and Z. P. Qi, "A Survey on Research of Microgrid—A New Power System," *Relay*, Vol. 35, No. 12, 2007, pp. 75-81.

[9] T. Shimakage, J. Sumita, N. Uchiyama, T. Kato and Y. Suzuoki, "Supply and Demand Control of Distributed Generators in a Microgrid," 2008 *IEEE Telecommunications Energy Conference*, San Diego, 14-18 September 2008, pp. 1-5.

[10] Z. L. Yang, H. Liao, C. S. Wu and H. H. Xu, "Analysis and Selection of Switch for Double Modes Inverter in Micro-Grid System," *International Conference on Electrical Machines and Systems*, Wuhan, 17-20 October 2008, pp. 1778-1781.

[11] G. P. Wang, L. Z. Yi, S. F. He, B. Zhang, Z. Z. Yao and B. Wang, "Double-Mode Modulation Based on Limited Trajectories for ANN-SVPWM in PV Grid-Connected Inverter," *Control and Decision Conference* (*CCDC*), Xuzhou, 26-28 May 2010, pp. 4272-4277.

[12] H. R. Gu, Z. L. Yang, D. Y. Wang and W. Y. Wu, "Research on Control Method of Double-Mode Inverter with Grid-Connection and Stand-Alone," *Power Electronics*

and Motion Control Conference, 14-16 August 2006, pp. 1-5.

[13] T.-Y. Lee, K.-H. Ha, H.-J. Yoo, J.-W. Seo and M.-C. Shin, "Research for Data Acquisition Equipment with Micro-Grid System," *Electrical and Computer Engineering*, Dhaka, 20-22 December 2008, pp. 712-715.

Holonic Architecture of the Smart Grid

Ebisa Negeri[1], Nico Baken[1], Marjan Popov[2]

[1]Network Architectures and Services, Delft University of Technology, Delft, The Netherlands; [2]Electrical Sustainable Energy, Delft University of Technology, Delft, The Netherlands.

ABSTRACT

With the growing concerns about sustainable energy, energy efficiency and energy security, the electrical power system is undergoing major changes. Distributed energy sources are becoming widely available at the lower parts of the grid. As a result, more and more end consumers are transforming from passive consumers to active "prosumers" that can autonomously generate, store, import and/or export power. As prosumers increasingly dominate the power system, the system demands capability that allows enormous number of stakeholders with heterogeneous types to exchange power on the grid. Unfortunately, the classical power system cannot efficiently handle this scenario since it was designed for centralized power distribution. Thus, restructuring the rather old power system is indispensable. In this paper, we apply the holonic approach to structure the smart grid as a system that is bottom-up organized from autonomous prosumers that are recursively clustered at various aggregation layers. Based on this, we present a control architecture of the smart grid using holonic concepts. Our control architecture is characterized by autonomy of the prosumers, distributed control, recursive self-similar control structures at different aggregation levels. Further, we present a service oriented architecture (SOA) framework that models the control functions that make up the holonic control architecture. Our proposed control architecture is tested using a simulation set-up.

Keywords: Smart Grid; Architecture; Holon

1. Introduction

The electric power system, a vital system supporting the economy of every nation, is undergoing tremendous transformations driven by the growing concerns for increasing the share of green power supply, as well as enhancing the reliability, security and efficiency of the system. As opposed to the classical power system where power is generated at large-scale power plants and transported in a unidirectional way to the distributed consumers, distributed energy sources (DES), such as solar panels and wind turbines, are increasingly penetrating into the lower parts of the power grid. With the availability of the DES, the end customers of the power system are transforming from passive consumers to active "prosumers" that can generate power on site and feed their surplus production back to the grid. According to the European parliament, all new buildings to be built after 2019 will have to produce their own energy on site[1]. Moreover, the foreseen massive presence of electric vehicles (EV) will have a huge impact on the power system. The flexible appliances, energy storage systems, as well as the grid-to-vehicle (G2V) and the vehicle-to-grid (V2G) features of the EVs will offer the prosumers the flexibility to manage the shape of their loads. Given their generation capacity and their flexibility to shape their load profiles, the prosumers tend to autonomously manage their own resources and exchange/trade power with their surrounding network. As prosumers dominate the system, the power system faces an era of "prosumerization" whereby enormous number and types of stakeholders can autonomously generate, store, import and/or export power. While in the classical power system power is transported in a unidirectional way top-down from the centralized plants to the distributed consumers, in the prosumer-dominated power system power will be primarily exchanged locally between the prosumers in a bidirectional way.

Apparently, restructuring the rather old power system is crucial in order to accommodate the dynamics. Accordingly, an advanced future power grid paradigm is envisaged, which is commonly referred to as the smart grid. The anticipated benefits of the smart grids are listed

[1]European Parliament, "All New Buildings to be Zero Energy from 2019," Committee on Industry, Research and Energy, Brussels, 2009.

in the reports of the International Energy Agency (IEA) [1], the National Institute of Standards and Technology (NIST) [2] and the European Smart Grids Technology Platform [3]. Enhancing the reliability, security, efficiency, and resilience of the power system, accommodating all generation and storage options, enabling informed participation of customers, accommodating new products, services and markets are among the benefits. The challenge still remains on how to organize the smart grid in order to realize these benefits.

In this work, we focus on the future prosumer-dominated power system, and study how to organize the smart grids to deliver the expected benefits. We envisage to organize smart grids bottom-up from prosumers. While prosumers autonomously manage their own resources, they can join other prosumers whose profile "complement" their own to exchange power with them. Further, a network of such prosumers might constitute a larger prosumer cluster to gain the power of the collective and bargain with the rest of the grid as a unit. Yet, several clusters of prosumers can form a network at a next aggregation level to form a bigger prosumer cluster and benefit the power of the bigger collective. In this context, it seems that the future power system builds bottom-up through recursive clustering of prosumers at various aggregation levels forming networks of networks. We also propose a suitable control architecture for such a bottom-up organized smart grid. To this end, we embark on concepts from holons and holarchy, which will be described in Section 3.

The contributions of this paper are summarized as follows:

- We employ the holonic concepts to organize the smart grids enabling it to foster seamless accommodation of the foreseen dominance of prosumers.
- We propose a holonic control architecture for our envisaged smart grids.
- We present an SOA framework to support our proposed control architecture and validate it with a simulation set up.

The remainder of this paper is organized as follows. In Sections 2 and 3, we present the related work and the concepts used in our work, respectively. The envisaged smart grid holarchy and the holonic control architecture are described in Sections 4 and 5, respectively. After presenting our simulation scenario and results in Section 5, we provide discussions and concluding remarks in Section 6.

2. Related Work

Recently, various new concepts and developments have been proposed to manage the new trends in the power system. The concept of virtual power plants (VPP) was introduced as a cluster of distributed power sources which are collectively managed [4]. The Microgrid concept [5] was also suggested as a low voltage (LV) distribution system comprising of distributed sources, controlled loads and storage systems that are coordinated to achieve a controllable operation either as an island or connected to the medium voltage (MV) grid.

Provoost *et al.* [6] introduced the autonomous network (AN), which is a part of the system but its behavior is more or less independent from the rest. AN differs from the microgrid because AN is larger in size and complexity, as well as its aim is primarily optimizing its normal operation. Overbeeke and Roberts [7] proposed the *active networks* concept, where the network is subdivided into cells that are self managing, but not necessarily self supplying. This concept involves interconnection between the cells that provides more than one power flow paths, and allows rerouting to avoid congestion and to isolate faults. Further, system services are traded along the connections between the cells.

The smart grid concept represents the overall picture of the future power network that is supported by intelligent distributed devices and communication technologies. Tsoukalas and Gao [8] modeled the smart grid as an energy internet comprising of local area grids (LAGs), that are demand based autonomous entities consisting of a convenient mixture of different customers. Huang *et al.* [9] presented the Future Renewable Electric Energy Delivery and Management (FREEDM) as a power distribution system that interfaces with residential customers and industry customers having distributed renewable energy sources and distributed storage devices. The key technology features of the system are the plug-and-play interface, energy router, and pen-standard based operating system. The proposed system relies on a flat and distributed management architecture.

A market based control concept, named PowerMatcher, is suggested by Kok *et al.* [10] for supply and demand matching in electricity networks with large penetration of distributed sources. In the PowerMatcher, each device uses its agent to buy or sell power in the electronic market that is implemented in a distributed manner using a tree structure. On the other hand, service oriented architecture (SOA) is gaining attention in power system control. Vaccaro *et al.* [11] have proposed a web-service-based framework for integrated microgrid (MG) control, modeling, and monitoring. At the core of their framework is the MG engine which executes the MG management in a geographically distributed scenario.

Although their scale, complexity and intelligence varies, all the above concepts and technologies have their own contributions in managing the distributed sources. One concept could be more applicable in some settings than the other, and vice versa. Moreover, some of the concepts could be combined since they complement each other. The smart grid could be an umbrella to appropriately combine the concepts, thereby achieve a heterogeneous and intelligent power system that delivers its expectations. For instance, the FRIENDS, VPP, microgrid, AN, and active networks could form sub-networks of the smart grid, while FREEDM provides the power delivery infrastructure supporting the sub-networks and PowerMatcher delivers the transaction services.

In this work, we study how the smart grids can be organized to accommodate the foreseeable dominance of prosumers. We endeavor to propose an architecture of the smart grid that tends to converge the various proposed concepts. We model the heterogeneous types of components of the smart grid as prosumers. Based on this, we propose to organize the smart grid bottom-up from prosumers, by recursively clustering them at various aggregation layers. Moreover, we also propose a control architecture that fits our proposed organization of smart grids. Our prosumer model and control architecture are built based on the concept of holons and holarchy that is quite suitable to capture the properties of the future smart grids.

3. Related Concepts

3.1. Prosumer

In this work, *prosumer* is a general term that refers to a system that autonomously manages its resources and is capable of bidirectional exchange of power with its surrounding network. A prosumer could be as simple as a household that autonomously manages its resources (such as appliances, distributed sources, storage systems, electric vehicles, etc.). Other systems such as a virtual power plant, a microgrid, an energy community, an autonomous network, a power plant, etc. can also be regarded as prosumers. A prosumer can comprise sub-parts that are themselves prosumers, while at the same time can become a part of a larger prosumer. With the aid of the intelligent information and communication technology (ICT) capabilities of the smart grid, a prosumer can orchestrate its resources as well as coordinate itself with its surunding network to achieve desirable internal and system wide goals.

3.2. Holon and Holarchy

The word holon was coined by Koestler [12], and it refers to a logical entity that is both a whole and a part. For example, a cell in your body is a holon because it is a disnct living entity that has a distinct cell wall defining its interface with the rest of the world. However, a cell is composed of smaller holons such as nucleus and chromosome that are also separate entities. Yet, a group of cell holons together form a higher aggregation layer tissue holon, a group of tissue holons form an organ holon, and so on.

A holon is basically autonomous, cooperative and recursive. Holons exhibit self-similar recursive structure and are organized in various aggregation layers to form a *holarchy* as shown in **Figure 1**. Holarchies are recursive in the sense that a holon in a holarchy can in itself be an entire holarchy of other holons. In a holarchy, each holon can function autonomously, thereby enabling formation of a distributed system. While functioning autonomously, a holon can cooperate with other holons in the holarchy to achieve mutual goals. In a dynamic environment, holons can also be dynamically reorganized, thereby making the holarchy more robust to changes adapting itself to its environment. These properties of the holonic concept make it quite appealing to organize very complex systems. The holonic approach has been proposed to organize various systems, some of them are organizational modeling, manufacturing, sensor networks, traffic light management, etc. ([13-17]).

3.3. Service Oriented Architecture (SOA)

SOA is an architecture that models an application as a set of loosely coupled, flexible, reusable and adaptable functional units called services, whereby the services are interrelated through well-defined interfaces and contracts. The main advantages of SOA include platform-independence of the interfaces that aids interoperability, the flexibility and reusability of the services, and its easy adaptation to evolutionary changes in the application. While SOA can be implemented using different technologies, the web services are the most popular one because of wide acceptance of its standards. Web service technology exposes *web services*, which are sets of operations on the web; thereby allowing high degree of integration among them. These features of SOA and web services technologies are attractive to model the control functions of the holonic control, as will be discussed later.

4. The Holonic Smart Grid

While the smart grid has diverse types of components, such as households, microgrids, energy communities, etc.,

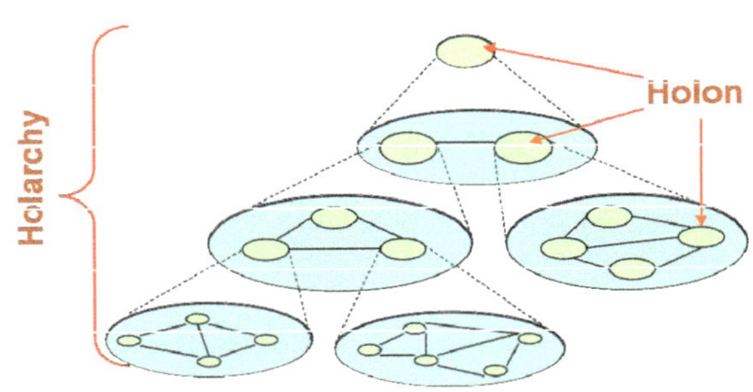

Figure 1. A holarchy—organization of holons.

each of the components can be seen as a prosumer. The prosumers tend to autonomously manage their resources. Moreover, they tend to aggregate together and form a bigger cluster to facilitate local power exchange as well as to gain a larger power of the collective to trade power with their surrounding. For instance, a group of prosumer households can cluster together to form a neighborhood energy community, which is a larger prosumer. Likewise, a group of neighborhood energy communities can form a district energy community (yet a larger prosumer) at the next aggregation level. This clustering could further be recursively repeated at various aggregation levels. Moreover, the prosumers can dynamically reorganize to efficiently adapt to the change in their environment. Apparently, the autonomy, aggregation into layers, recursive property, and dynamic adaptation of the prosumer based smart grids closely matches the properties of a holonic system. Accordingly, we model each prosumer as a holon and the entire smart grid as a holarchy. The smart grid holarchy has the following major features that contribute to the overall efficiency of the system.

4.1. Autonomy/Self Management

As prosumers own production capacity and acquire the flexibility to manage their own load profiles, they tend to autonomously manage their resources to optimize their interest. This is in line with the growing interest towards active participation of customers. A prosumer holon might also choose to be part of a bigger prosumer holon and exchange power with its surrounding, or operate as a self-supplying islanded unit. The autonomy of the prosumer holons aids a distributed control capability of the system that would be otherwise very difficult to control centrally. Autonomy might also increase the consciousness of the prosumer holon, thereby making it more cautious about its consumption. Autonomy of the prosumers might make coordination not easy to achieve, but this can be achieved using machine learning [18], game theory [19] and mechanism design [20] based approaches.

4.2. Recursive Aggregation

As mentioned earlier, aggregation is important for prosumers to facilitate local power exchange between themselves, and to gain a bigger bargaining power with the rest of the grid. Accordingly, the prosumer holons at a lower aggregation level may be self-organized to constitute a bigger prosumer holon. For the same reasons, the newly formed prosumer holon can still be connected to its peers in a network in the next higher aggregation layer to form yet a bigger prosumer holon. Further bottom-up grouping of such holons in networks-of-networks continues *recursively* in higher aggregation levels and eventually constitute the overall holarchy of the smart grid (**Figure 2**). Simon [21] has shown that this recursive clustering at different aggregation layers is inherent behavior of all complex systems. We refer to the smaller holons that form a bigger holon as the sub-holons of the bigger holon, whereas, the bigger holon on the next aggregation layer that contains a holon is referred to as its super-holon.

The bottom-up organization at different aggregation layers provides efficient structure that encapsulates and simplifies the coordination of the system, making it scalable. Moreover, the aggregation of prosumer holons into super-holons might increase the reliability of the collective as the profiles of the individual holons might complement each other. Further, the local power exchanges between the prosumer holons reduce both the transportation losses and the costly investment costs to upgrade the transmission capacity to meet the ever increasing demands.

4.3. Dynamic Reconfiguration

Over an interval of time, a prosumer holon may be in the state of either self sufficiency, surplus production, or excess demands. Thus, the prosumer holons can leave one super-holon to join another one over a period of time to look for a set of holons that complement their profiles, thereby optimizing their resource utilization. Erol-kantarci *et al.* [22] have shown that a set of microgrids could be dynamically reorganized during a day into clusters according to their profiles to optimize the utilization of their renewable resources. Moreover, prosumer holons can join a super-holon for some time intervals during a day and operate in islanded mode at other intervals, for example in response to potential faults. Such flexibilities lead to a dynamic reconfiguration of the smart grid holarchy over a period of time. The dynamic reconfiguration contributes to the self-healing, dynamic recovery, and increase efficiency of the prosumer holon.

4.4. Intelligence and Communication

A prosumer holon needs to coordinate its components, as well as coordinate itself with its surrounding to optimize utilization of resources and to ensure system stability.

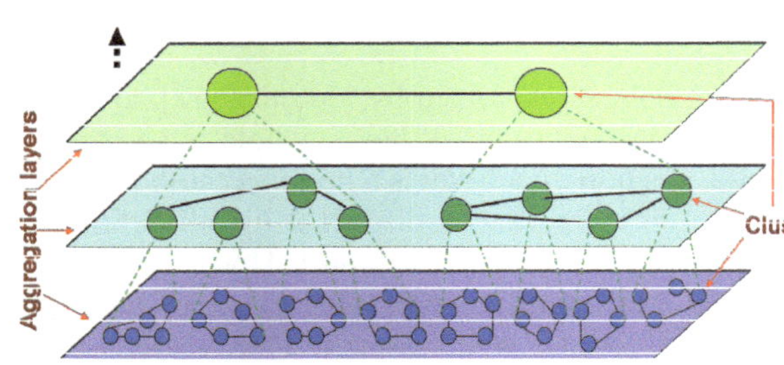

Figure 2. Bottom-up organization of the prosumers to form the smart grid holarchy.

This coordination involves large information flows and requires intelligent information processing units that constitute the "nerve system" of the "organic" power system. Accordingly, each prosumer holon requires communication capabilities and intelligence to make decisions based on the gathered information. Indeed, an appropriate communication infrastructure needs to be in place. The IEC-61850 standard [23] is developed for communication in substations. A review of the types of intelligence required as well as the candidate communication technologies for smart grids is presented in [24].

4.5. Security and Privacy

Cyber security is regarded as one of the biggest challenges in smart grids [2,3]. The enormous exchanges of information and control messages raise security concerns for the prosumers. Thus, the information and control message exchanges need to be protected against unauthorized access and guarded from malicious hackers. Therefore, the system requires a secure communication protocol. Moreover, privacy is becoming a major concern as the data exchanged might reveal personal information such as habits, activities, and preferences of the individual. For instance, the study in [25] has revealed that personal information can be estimated with high accuracy based on the meter reading of a household, even with relatively unsophisticated hardware and algorithms. Several solutions have been proposed to overcome security and privacy issues in smart grids, and their review is presented in [24]. The IEC61850 standard also supports secure communication in substations.

5. The Holonic Control Architecture

Given the nature of the envisioned smart grid, its control architecture needs to exhibit the following properties: 1) sufficient autonomy of each prosumer holon to manage its own resources; 2) recursive structure where a prosumer is composed of smaller prosumers at the lower aggregation layer, and at the same time be part of a larger prosumer holon at a higher aggregation layer; 3) layered structure to facilitate coordination between different aggregation layers; 4) dynamic reconfiguration to runtime topology change. The *holonic* architecture, as explained before, nicely fits the envisioned system because it possesses all of the properties required by the control architecture.

Our holonic control architecture maintains the proposed structure in the envisioned power system. In the control domain, a *control-holon* represents the control system of a prosumer holon. Thus, holonic control units will be recursively organized in a bottom-up structure to form a complete control holarchy of the smart grid. When the prosumer holon has prosumer sub-holons, its control-holon is also composed of the corresponding control-sub-holons (*i.e.* control systems of sub-holons).

In addition to controlling the internal dynamics of its prosumer holon, a control-holon needs to interact with the control-holons of its peer prosumer holons that are in the same super-holon as well as with that of its super-holon. Moreover, a control-holon needs to interact with the network operator operating at its aggregation level, regarding the power quality, infrastructure constraints, etc. Thus, the control functions of the control-holons are basically the same independent of their aggregation levels. This reveals the self-similarity (recursive property) of the control holarchy. **Figure 3** shows the functional architecture of a single control-holon. Corresponding to each of the control functions in the control architecture shown in **Figure 3**, we define a corresponding web service in the SOA model together with its possible operations. The control functions and their corresponding web services

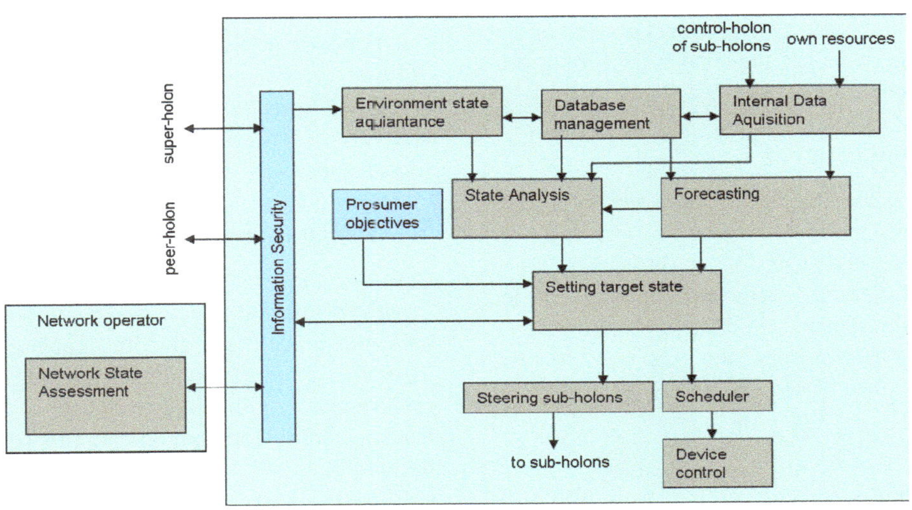

Figure 3. Architecture of a control holon.

are described subsequently.

5.1. Environmental State Acquaintance

A prosumer holon needs to be well acquainted with its environment in which it is operating to make optimal decisions. Acquiring information about the load profiles of its peer prosumers, market prices, incentives, etc. is helpful to plan optimal transactions with its environment. When a super-holon sets a target state (of, for instance, the aggregate load profile), it might generate control signals, incentives, etc. for its sub-holons, based on their mutual agreements, to achieve the desired goal. Therefore, this control function is responsible for acquiring such information and presents it to the relevant control functions in a useful shape. The web service corresponding to this control function is *EnvironmentSA-srv*. To provide its functionality, the web service involves several operations. Some of the operations that compose the *EnvironmentSA-srv* web service are *GetLoadProfile-opr*, to acquire the load profile in its environment, and *GetMarketState-opr*, to acquire the market price in its environment.

5.2. Internal State Acquisition

A prosumer holon needs to have up-to-date information about its own constituents in order to make optimal decisions. This function handles acquiring internal data about the parameters of the resources of the prosumer (such as distributed generations, storage systems, etc.), their current available capacities, the state (such as, the load profiles, surplus productions, etc.) of the prosumer sub-holons, etc. Corresponding to this control function, we define the *InternalDA-srv* web service. Below are some of the operations constituting the web service:

- *GetResourcePar-opr*: to get the parameters of a resource, such as the storage capacity of a battery.
- *GetResourceState-opr*: to acquire the current state of a resource, e.g. the current charge level of a battery.
- *GetSubProfile-opr*: to get the load profile of a prosumer sub-holon.

5.3. Database Management

This function is responsible for storing, retrieving and updating the relevant internal and external data. The web service that provides this function is *DatabaseMgt-srv*. Accordingly, the following operations can be identified as some of its components: *RdData-opr*, *RtData-opr*, *UpData-opr*, and *DlDalata-opr*, that are used to read, write, update, and delete data, respectively.

5.4. Forecasting

This control function takes care of forecasting both the internal profiles such as generation profile of its local power sources, its local power demands, the load profile of its sub-holons (if any), as well as the load profile of its environment. The equivalent web service is *Forecast-srv*, which could compose the operations *ForecastProduction-opr*, *ForecastDemand-opr* and *ForecastEnv-opr* that are responsible for forecasting the local production, local demand, and the load profile of its environment, respectively.

5.5. State Analysis

Based on the inputs from the *environment state acquaintance*, the *internal data acquisition*, and the *forecasting* control functions, this control function analyzes the resulting state of the system. This allows the control holon to assess the possible load profiles, the flexibility in the load profiles, the synergetic potential of the various components of the resources and the load profiles, etc. We define *StateAnalysis-srv* as the web service that takes care of this control function. The following are some of the operations that could compose this service:

- *AssFlexibility-opr*: to assess the flexibility of the load profiles.
- *AssSynergy-opr*: to assess the synergetic potential of the resources.
- *CompState-opr*: to compute possible states of the aggregate load profiles of the holon.

5.6. Setting Target State

After acquiring the state estimation information, this function sets the target state for the prosumer. The target state specifies the target load profiles, amount of power to be imported/exported, whether to operate connected or as an island (where applicable), etc. To set the target set, the control function uses as input, among others, the objectives of the prosumer, which could be cost minimization, increasing comfort, increase composition of green energy, etc. The web service that handles this function is *TargetState-srv*, and some of the operations that compose it are *SetTargetProduction-opr* and *SetTargetDemand-opr* that take care of setting the target production and demand profiles, respectively, of the prosumer holon.

5.7. Steering Sub-Holons

Based on the target state proposed, this control function generates relevant signals for the sub-holons that would help to achieve the goal. The signals generated could be incentives or control signals depending on the type of the terms and conditions in the contract of the prosumer with its sub-holons. Corresponding to this control function, we define a web service called *SteerSubHolon-srv*. The operations that constitute the service could be *GenControl-opr* and *GenIncentive-opr* to generate downstream control signals and incentive signals, respectively, for its sub-holons.

5.8. Scheduler

The scheduler determines the optimal load profiles of the schedulable resources of the prosumer to achieve the target state. Its equivalent web service is *Scheduler-srv*. An operation that could make up the web service is *GenSchedule-opr*, which is used to generate the optimal schedule of the resources.

5.9. Device Control

The device control function translates the schedule of the resources into the actual control signals of the devices that are scheduled. The *DeviceControl-srv* web service takes care of this, while the operation *SetDeviceControl-opr* could be employed to set the device controls accordingly.

5.10. Information Security

The control function maintains the security of the information and control message exchanges of the control-holon with its environment. Its corresponding web service is *Security-srv*, which could have the following operations.

- *EncryptInfo-opr:* to encrypt the information to be sent.
- *DecryptInfo-opr*: to decrypt the received information.
- *CheckAuthorization-opr*: to check the authorization of the sender of the information.

5.11. Network State Assessment

This control function takes care of assessing the state of the power network over which the prosumer holons are interconnected. This includes analyzing the voltage levels at various nodes in the network, the loading of the infrastructures, etc. It also deals with contingency analysis, alarms about faults and potential risks in the environment. This functionality is provided by the network operator through the web service *NtkStateAssessment-srv*, which could be composed of operations such as, *AssessNetk-opr*, to assess the state of the network, and *SdAlarms-opr, to* send alarms to the corresponding holons.

6. Simulation Setup

We considered a system where our proposed control architecture can be applied. Our experimental setting consists of prosumer holons that are aggregated into three layers. At the lowest level are a set of prosumer households (PH) that form a neighborhood energy community at the next aggregation level. The neighborhood energy community (NEC) is in turn part of a district energy community (DEC) at the next higher aggregation layer. Although their aggregation levels differ, the PHs, the NEC, and the DEC are all autonomous prosumer holons. The control-holon of the PH autonomously manages its resources to achieve a desired load profile of the PH. The control-holon of the NEC coordinates the control-holons of its PHs, by providing incentives, to achieve a desirable aggregate load profile that maximizes the benefit of the NEC in response to the incentive received from the control-holon of DEC.

The power flow in the power grid is managed by a network operator (NO). The NO interacts with the relevant control holons to make sure that the power flows in the network do not violate the infrastructure constraints. Further, it generates appropriate control signals for the relevant control-holon whenever a fault occurs in the network.

The sketch of the interconnection between the described holons is depicted in **Figure 4**. The figure shows both the physical grid connection (solid lines) and the corresponding network of the control-holons (doted lines). The control-holons (CH) in the figure could be implemented in intelligent devices that are capable of both making intelligent decisions and communicating across a communication infrastructure. For example, the CH of a prosumer household can be implemented in the home energy management system [26]. In our experimental setting, the control-holons are hosted by computers that are interconnected by IP network.

A Case Example

We present a case example that is implemented in the setting described above. Load management aiming at peak load reduction in a NEC that is composed of PHs is considered. Each of the PHs owns a solar panel, a micro-CHP unit, an electric vehicle and other electricity consuming appliances. The DEC wants to minimize its aggregate peak load and uses a control signal to steer each of its constituent NECs to reduce its peak load. The NEC

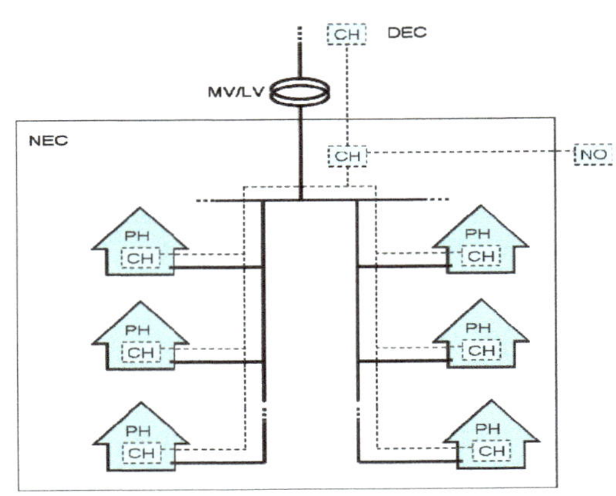

Figure 4. A sketch of the simulation setup (CH = control holon).

provides incentives to the PHs to yield peak reduction. For this purpose, we use the dynamic pricing based incentive mechanism that was proposed in [27]. The PHs tends to minimize their cost. Hence, each PH schedules the (dis)charging of its electric vehicles to minimize its cost in response to the pricing incentive, and then sends its resulting load profile back to the NEC. After receiving the response of all the PHs, the NEC modifies the pricing incentive based on the aggregate load profile and sends it to again to the PHs. This procedure is repeated for a fixed number of iterations. Eventually, the NEC chooses the price vector that yields the lowest peak load, and the corresponding load profiles are adopted. The control agent of the NO employs a low voltage power grid simulator tool called Gaia [28], which is commonly used by network operators in the Netherlands, such as Alliander, to assess the state of the network. For our simulations, we obtained data representing field measurements of the power demand of the appliances as well as the power production of the solar panels and the micro-CHPs from Alliander[2]. Each of these measurements is taken once every 15 minutes over a duration of one day (24 hours). Following this, a time step in our simulation refers to a 15 minutes time interval, and we have 96 time steps during 24 hours. While the patterns of the daily commute distance, the arrival at and departure from home of the EV drivers are derived from the 2009 mobility research of the Netherlands [29], the parameter specifications of the EVs are adopted from [27]. The plots of some of these data are depicted in **Figures 5** and **6**. Using the proposed control architecture, the following control steps could be involved to achieve the peak minimization in the aggregate load profile of the NEC.

1) The control-holon (CH) of the DEC, the superholon of the NEC, generates and sends a control signal to the CH of the NEC to reduce its peak load. The CH of the DEC uses the SteerSubHolon-srv web service to do so.

2) The CH of the NEC receives the signal from its super-holon using the EnvironmentSA-srv web service, and sets a target state that reduces peak load accordingly, using the TargetState-srv.

3) The CH of the NEC receives the load profile of its sub-holons (PHs) using the internalDA-srv web service.

4) The CH of the NEC uses the SteerSubHolon-srv web service to generate a price vector based on the aggregate load profile of its sub-holons, and sends it to the CHs of the PHs.

5) The CH of each PH receives the price vector from its super-holon using the EnvironmentSA-srv web service, and sets target state that minimizes its cost based on the received price vector, using the TargetState-srv web service.

[2]Alliander is a distribution network operator in The Netherlands owning approximately 40% of the distribution networks.

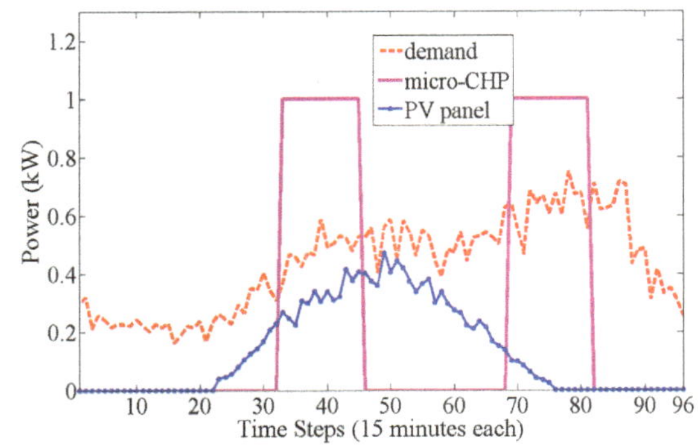

Figure 5. A sample prosumer household demand and production profiles.

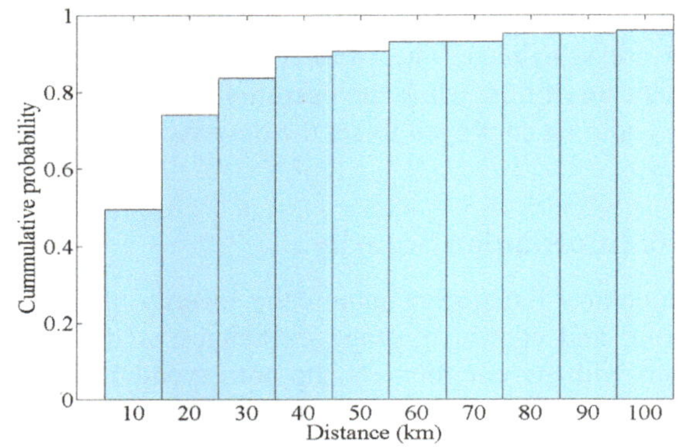

Figure 6. Distribution of the distance traveled per day per car driver.

6) The CH of each PH receives the data about the capacity and state of its resources using the InternalDA-srv web service. It then sets its target state, using the Target-Statesrv web service to minimize its cost based on the received price vector, and subsequently schedules, using the Scheduler-srv web service, the (dis)charging of its electric vehicle accordingly. Next, it sends the resulting net load profile of the PH back to the CH of the NEC.

7) Steps 3 - 6 represent the negotiation that takes place between the NEC and the PHs to arrive at a situation that yield mutual benefits for the parties. These steps are repeated for a fixed number of iterations. Eventually, the CH of the NEC selects the price vector that yielded the lowest aggregate peak load, together with the corresponding load profiles.

8) The CH of the NEC requests the NO to analyze the network state under the resulting load profile.

9) The NO employs the NtkStateAssessment-srv web service to analyze the resulting network state by using Gaia tool. Afterwards, it sends the result of the analysis back to the CH of the NEC.

10) If the NEC receives a "feasible state" network state report, then the resulting load profile is reported to the CH of the DEC and the selected price vector is sent

to the PHs as a final price vector. Subsequently, the CHs of the PHs use the Device Control-srv web services to control their electric vehicle (dis)charging according to the schedule corresponding to the final price vector.

11) If the received network state is "infeasible state", then the CH of the NEC will check the network state with the load profiles resulting from the other price vectors until it receives a "feasible state" reply. If none of the load profile yields a healthy network state, then it will generate and send a control signal to the PHs to reduce peak load.

After implementing the web services using the Java Web Services, the above simulation is executed with a NEC consisting of five PHs, and the resulting aggregate load profile is depicted in **Figure** 7. As can be observed form the figure, coordination of the NEC using our holonic control architecture has yielded a significant reduction (as large as 54%) in aggregate peak load compared to the uncoordinated scenario where the PHs charge their EVs as soon as they return home from their previous commute. While in our case the *Steer Sub Holon-srv* service uses the dynamic pricing model proposed in [27] to provide the incentives for the PHs in the NEC, other incentive mechanisms can also be used to achieve a desired aggregate load profile.

7. Discussions and Conclusions

The trends of transition in the electric power system are leading to an era of prosumerization whereby myriad of stakeholders autonomously generate and exchange power with each other on the power grid. While the classical power system cannot handle the dynamics since it was designed for centralized generation and unidirectional top-down power flows, the future smart grids are expected to efficiently manage the foreseeable era of prosumerization. To this end, we have proposed to organize the smart grids based on the concepts of holons and holarchy that effectively capture the features of the transitions.

The major features of our holonic smart grids include

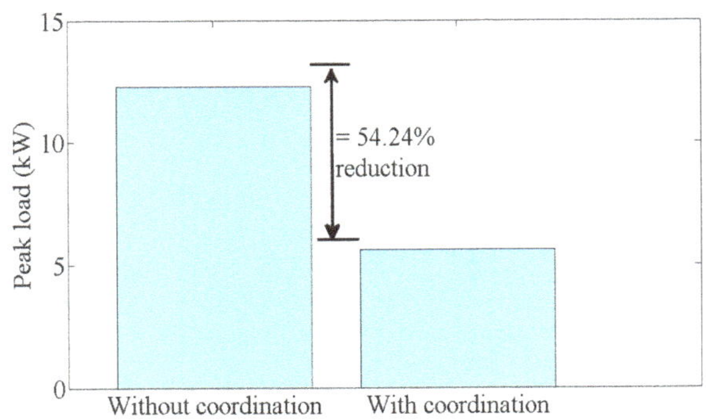

Figure 7. The peak load of a NEC under two scenarios: without coordination and with coordination employing our holonic control architecture.

sufficient autonomy of each prosumer to manage its own resources; the recursive structure where a prosumer is composed of smaller prosumers at the lower aggregation layer, and at the same time be part of a larger prosumer at a higher aggregation layer; dynamic reconfiguration to runtime topology change; intelligence and communication capability of prosumers, and maintaining security and privacy of prosumers. The prosumer holons are recursively aggregated into clusters to eventually constitute the smart grid holarchy. Corresponding to the envisaged holonic smart grid, we have proposed a holonic control architecture that conveniently fits the aforementioned features. The control architecture is composed of control-holons corresponding to the prosumer holons that are recursively organized to yield control-holarchy of the smart grid. After specifying relevant control functions of our control-holon, we have defined the corresponding web services based on Service Oriented Architecture (SOA) model. Eventually, the SOA model was implemented using Java web services and tested in a simulation setup yielding interesting results.

In order to implement our proposed smart grid holarchy, there are technical, marketing and regulatory requirements that need to be met. The requirements are in line with the trends in the smart grid developments. One of the requirements is the intelligence and communication capabilities of the prosumers. The current developments in smart meters promise availability of intelligent units that are capable of two way communication [30]. Various developments exist with regards to communication technologies that can be deployed in smart grids, comparisons of the candidate communication technologies are presented in [24]. Another requirement is supporting the bidirectional power flows in the grid. The Future Renewable Electric Energy Delivery and Management (FREEDM) [9] is among the promising developments that address this issue. The capability of dynamic reconfiguration is another requirement. The work presented in [22], which allows a dynamic reconfiguration of a set of microgrids is an interesting development in this regards. With the myriad of prosumers trading electrical power, the challenge lies in how to coordinate the market. The PowerMatcher technology proposed in [10] seems a suitable candidate to adopt for this purpose. Moreover, several solutions have been proposed that can be employed to overcome the security and privacy concerns of the prosumers [24]. Another important requirement is the presence of a suitable regulation that allows the autonomous prosumers to trade power on the grid. Given the trends in the deregulation of electricity markets over the recent past years and the ongoing efforts, although the extents of deregulation vary from one country to another, one can be optimistic to foresee a conducive scenario.

In general, our proposed smart grid holarchy is a suitable architecture to accommodate the foreseeable era of prosumerization. Moreover, the existing developments and the ongoing efforts in various aspects of the smart grids are in line with the requirements for its implementation.

REFERENCES

[1] International Energy Agency, "Technology Roadmaps Smart Grids," 2011. http://www.iea.org/publications/freepublications/publication/smartgrids_roadmap.pdf

[2] National Institute of Standards and Technology, "NIST Framework and Roadmap for Smart Grid Interoperability Standards, Release 1.0," 2010. http://www.nist.gov/public_affairs/releases/upload/smartgrid_interoperability_final.pdf

[3] European Smart Grids Technology Platform, "Vision and Strategy for Europe's Electricity Networks of the Future," 2006. ftp://ftp.cordis.europa.eu/pub/fp7/energy/docs/smartgrids_en.pdf

[4] K. Dielmann and A. van der Velden, "Virtual Power Plants (VPP)—A New Perspective for Energy Generation?" *Proceedings of the 9th International Scientific and Practical Conference on Modern Techniques and Technologies*, 2003, pp. 18-20.

[5] N. Hatziargyriou, N. Jenkins, G. Strbac, J. A. P. Lopes, J. Ruela and A. Engler, "MICROGRIDS—Large Scale Integration of Micro-Generation to Low Voltage Grids," CIGRE C6-309, 2006.

[6] F. Provoost, J. Myrzik and W. Kling, "Setting Up Autonomous Controlled Networks," 39*th* *International Universities Power Engineering Conference* (*UPEC*), Vol. 3, 2004, pp. 1190-1194.

[7] F. Overbeeke and V. Roberts, "Active Networks as Facilitators for Embedded Generation," *Cogeneration and On-Site Power production*, Vol. 3, No. 2, 2002, pp. 37-42.

[8] L. Tsoukalas and R. Gao, "From Smart Grids to an Energy Internet: Assumptions, Architectures and Requirements," *3rd International Conference on Electric Utility Deregulation and Restructuring and Power Technologies*, Nanjing, May 2008, pp. 94-98.

[9] A. Huang, M. Crow, G. Heydt, J. Zheng and S. Dale, "The Future Renewable Electric Energy Delivery and Management (FREEDM) System: The Energy Internet," *Proceedings of the IEEE*, Vol. 91, 2011, pp. 133-148.

[10] J. Kok, C. Warmer and I. Kamphuis, "Power Matcher: Multiagent Control in the Electricity Infrastructure," *Proceedings of the 4th International Joint Conference on Autonomous Agents and Multiagent Systems*, Utrecht, 25-29 July 2005, pp. 75-82.

[11] A. Vaccaro, M. Popov, D. Villacci and V. Terzija, "An Integrated Framework for Smart Microgrids Modeling, Monitoring, Control, Communication, and Verification," *Proceedings of the IEEE*, Vol. 99, 2011, pp. 119-132.

[12] A. Koestler, "The Ghost in the Machine," Penguin Group, New York, 1990.

[13] G. Zhang and W. Li, "Flexible Holonic Organization Modeling and Cultural Evolution," *4th IEEE International Conference on Wireless Communications, Networking and Mobile Computing*, Dalian, 12-14 October 2008, pp. 1-5.

[14] H. Van Brussel, J. Wyns, P. Valckenaers, L. Bongaerts and P. Peeters, "Reference Architecture for Holonic Manufacturing Systems: PROSA," *Computers in Industry*, Vol. 37, No. 3, 1998, pp. 255-274.

[15] M. Moghadam and N. Mozayani, "A Street Lighting Control System Based on Holonic Structures and Traffic System," *IEEE 3rd Inernational Conference on Computer Research and Development* (*ICCRD*), Shanghai, 11-13 March 2011, pp. 92-96.

[16] Y. Ye, V. Hilaire, A. Koukam and C. Wandong, "A Holonic Model in Wireless Sensor Networks," *IEEE International Conference on Intelligent Information Hiding and Multimedia Signal Processing*, Harbin, 15-17 August 2008, pp. 491-495.

[17] N. Baken, "Holons and Holarchies," TEDx Amsterdam 2009. http://www.tedxamsterdam.com/2009/video-nico-bakenon-holons-and-holarchies/

[18] D. O'Neill, M. Levorato, A. Goldsmith and U. Mitra, "Residential Demand Response Using Reinforcement Learning," *1st IEEE International Conference on Smart Grid Communications*, Gaithersburg, October 2010, pp. 409-414.

[19] C. Ibars, M. Navarro and L. Giupponi, "Distributed Demand Management in Smart Grid with a Congestion Game," *1st IEEE International Conference on Smart Grid Communications*, Gaithersburg, 4-6 October 2010, pp. 495-500.

[20] E. Gerding, V. Robu, S. Stein, D. Parkes, A. Rogers and N. Jennings, "Online Mechanism Design for Electric Vehicle Charging," *10th International Joint Conference on Autonomous Agents and Multi-Agent Systems*, Taipei, 2-8 May 2011, pp. 811-818.

[21] H. Simon, "Architecture of Complex Systems," *Proceedings of the American Philosophical Society*, Vol. 106, No. 6, 1962, pp. 467-482.

[22] M. Erol-kantarci, B. Kantarci and H. Mouftah, "Reliable Overlay Topology Design for the Smart Microgrid Network," *IEEE Transactions on Networks*, Vol. 25, No. 5, 2011, pp. 38-43.

[23] Alstom Grid, "Network Protection and Automation Guide," 2011.

[24] X. Fang, S. Misra, G. Xue and D. Yang, "Smart Grid—The New and Improved Power Grid: A Survey," *IEEE Communication Surveys and Tutorials*, Vol. 14, No. 4, 2011, pp. 944-980.

[25] M. A. Lisovich and S. B. Wicker, "Privacy Concerns in Upcoming Residential and Commercial Demand-Response Systems," *Proceedings of the Clemson University Power*

Systems Conference, 2008.

[26] D. M. Han and J. H. Lim, "Design and Implementation of Smart Home Energy Management Systems Based on Zigbee," *IEEE Transactions on Consumer Electronics*, Vol. 56, No. 3, 2010, pp. 1417-1425.

[27] E. Negeri and N. Baken, "Smart Integration of Electric Vehicles in an Energy Community," *Proceedings of the 1st International Conference on Smart Grids and Green IT Systems*, Porto, 19-20 April 2012, pp. 25-32.

[28] The Gaia Tool. http://www.phasetophase.nl/en_products/vision_lv_network_design.html

[29] MON, "Mobility Research of the Netherlands," 2009. http://rijkswaterstaat.nl/

[30] L. Wang, V. Devabhaktuni and N. Gudi, "Smart Meters for Power Grid—Challenges, Issues, Advantages and Status," *IEEE/PES Power Systems Conference and Exposition* (*PSCE*), Phoenix, 20-23 March 2011.

20

A Low-Cost, Smart AC Charging System for Electric Vehicle

Bo Wang

Shenzhen Power Supply Bureau Co., LTD, Shenzhen, China.

ABSTRACT

In industry development strategy of electric vehicles, apart from concerns on the development of electric vehicles, we also need to consider the issue of charging facilities construction. Firstly, through analysis, this paper discusses the importance of AC charging points for electric vehicle development. By studying existing AC charging points on the market, it presents a low-cost smart AC charging system to reduce the cost investigated by power companies and operational bodies when laying of a large number of AC charging points. Compared with the conventional one, the proposed system has prominent features of low cost, small footprint and low investment.

Keywords: Electrical Vehicle; Charging Facility; AC Charging Point

1. Introduction

Chinese automobile production, with an average annual growth rate of 27.6%, has surpassed the United States and Japan and has become the largest automobile producer and consumer in the world. However, the growth of Chinese traditional automobile is bound to increase our dependence on oil, and will further exacerbate the air pollution [1]. Chinese per capita possession of oil proven recoverable reserves is equal to only 7.7% of the world average. According to "BP World Energy Statistics 2009" [2], the world remaining proven reserves of crude oil in accordance with the 2008 annual extraction rate can also be mined for 42 years. Oil resources are depleting that constrained the development of the traditional vehicle. Chinese development and reform commission pointed out that Chinese current dependence on foreign oil had more than 52% [3]. The development of electric vehicles will significantly reduce oil consumption and reduce environmental pollution. The efforts to develop electric vehicles are the only way to ensure sustainable development of Chinese automobile industry. From the above analysis, it is clear that environmental protection and energy security issues will enable China to develop electric vehicles [4,5]. At present, the number of electric vehicles used is small, the corresponding lack of charging infrastructure is also more, and there is a low return on investment in electric vehicle. However, the development of electric vehicles will have a breakthrough in the future. So, apart from concerns on electric vehicle development, we also need to consider the issue of charging facilities construction in industry development strategy of electric vehicles [6].

To implement the national energy saving policy, promotion of electric vehicle application, guide and standardize the constructing electric vehicle charging facilities of the China Southern Power Grid, China Southern Power Grid Co. Ltd developed a series of standards for electric vehicle charging technology as follows [7]:

1) Q/CSG 11516.1-2010 General technical requirements for electric vehicle charging establishment;

2) Q/CSG 11516.2-2010 Code for design of electric vehicle charging station and charging point;

3) Q/CSG 11516.3-2010 Technical specification for electric vehicle off-board charger;

4) Q/CSG 11516.4-2010 Technical specification for electric vehicle A.C. charging point;

5) Q/CSG 11516.5-2010 Specification of charging interfaces for electric vehicle off-board charger;

6) Q/CSG 11516.6-2010 Communication protocols between off-board charger monitoring unit and battery management system for electric vehicle;

7) Q/CSG 11516.7-2010 Technical specification for supervisor system of electric vehicle charging station;

8) Q/CSG 11516.8-2010 Code for acceptance of electric vehicle charging station and charging point.

In 2010, China has adopted a recommended national standards (general technical requirements for electric vehicle charging stations), which lays the foundation for regulating and guiding the plan and design of electric vehicle stations in order to conduct energy supply infrastructure. These standards are of reference to electric vehicle charging stations of State Grid Corporation By 2009, State Grid Corporation had developed a series of standards for electric vehicle charging stations:

9) Q/GDW 233-2009 Electric vehicle off-board charger general requirements;

10) Q/GDW 234-2009 Electric vehicle off-board charger specification for electrical interfaces;

11) Q/GDW 235-2009 Electric vehicle off-board charger communication protocols;

12) Q/GDW 236-2009 Electric vehicle charging station general technical requirements;

13) Q/GDW 236-2009 Electric vehicle charging station guide of layout design;

14) Q/GDW 236-2009 Electric vehicle charging station power supply system criterion.

2. The Importance of AC Charging Piles for EV

The development of electric vehicles in China will gradually increase the demand for electric vehicle charging facilities. From Shenzhen Power Supply Bureau's operation analysis for the BYD E6 Taxi, in order to ensure the normal operation of the taxi companies, it generally needs to be charged with about 0.5 C charging current. The rated voltage of its battery pack using 96 3.3 V/200Ah batteries is about 316.8 V, the nominal battery capacity is about 200 Ah [8]. For the E6, 30 kW electric charging machine is probably needed to meet the needs for Shenzhen taxi operators.

The charging demand survey for electric taxi shows that while 90% users of electric vehicles would choose to charge their vehicles in two places: the parking lots and the home. For these users, they can use a slow charger. Generally, on board charger is used to charge the batteries of the electric vehicle. The charging time is usually 6 - 8 hours. The charging time is longer, but little damage to the batteries, compared with the rapid DC charge. So it can extend the life of batteries and can also reduce the impact on the power grid.

At present, the majority of our household electricity power is about 2 kW - 5 kW. However, the number of modern appliances and power consumption are increasing rapidly. The power consumption of many families has been close to the saturation power, even at night. The 2 - 3 kW charging power cannot be reserved for electric vehicles in most cases. And most of vehicles, especially in the city, are parked in a fixed area or commercial parking places, so their own existing AC power cannot be used for the electric vehicle charging. Moreover, there is also a great external cable security risk when charging Electric Vehicles.

While traditional power household appliances can be taken directly to a standard electrical plug, but electric vehicle charging is related to personal safety and protection, network communications, and many other issues, so if you want to meet the requirements of this part of consumers, the power companies and electric vehicles operating agency must provide AC charging piles to meet the national or/and enterprise standards. At present, in addition to national standards, the National Grid and Southern Power Grid have also introduced the corresponding technical specifications and standards for AC charging piles of electric vehicles.

For the convenient of electric vehicle users, slow charge piles can be constructed in the district, parking lots, shopping malls, and even in the roadside. Users can charge their electric vehicles using lots of slow charging piles during the day and the night. This method is less cost, but more effectiveness for our society.

3. The Structure of Chinese Current AC Charging Piles

As the demand for AC charging piles, many companies start to produce and sell slow AC charging piles in the market, and their general structure is shown in **Figure 1**. Through our analysis, the popular chargers include lots of modules, such as payment transactions, card readers, keyboards, display modules, printers and the corresponding drive modules, etc. [9].

4. A Low Cost Smart AC Charging System

4.1. System Structure and Composition

The block diagram of low-cost smart charging system is shown in **Figure 2**. It includes the following components:

1) Central server: the exiting servers of the power grid company can be used; they can also be bought.

2) Master machine: its functions are shown in **Figure 3**, including fees, printing, communication and other func-

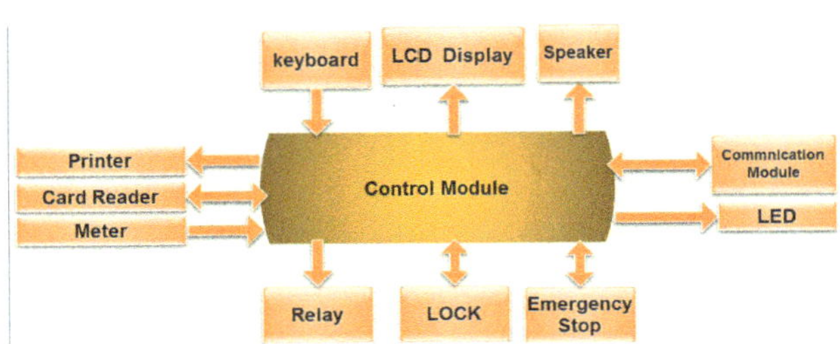

Figure 1. The structure of Chinese current AC charging piles.

tions.

3) Slave machine: its functions are shown in **Figure 4**, including charging, metering and communications.

The system has following advantages:

1) Low cost: compared with the current charging point, each slaver machine does not need trading system, display module and printer because of sharing trading system. The shell size of system will be reduced. There are

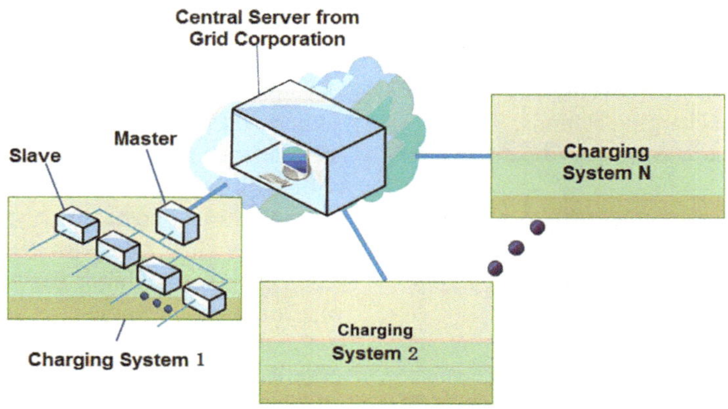

Figure 2. Low-cost smart charging system diagram.

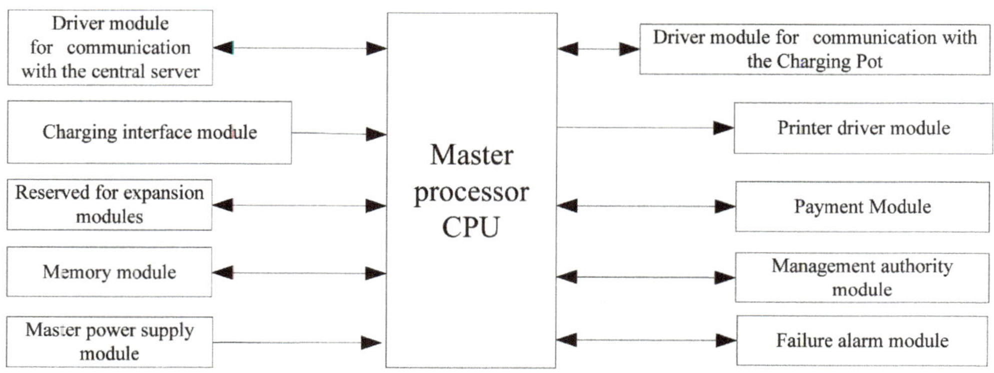

Figure 3. Master functional block diagram for low-cost smart charging system.

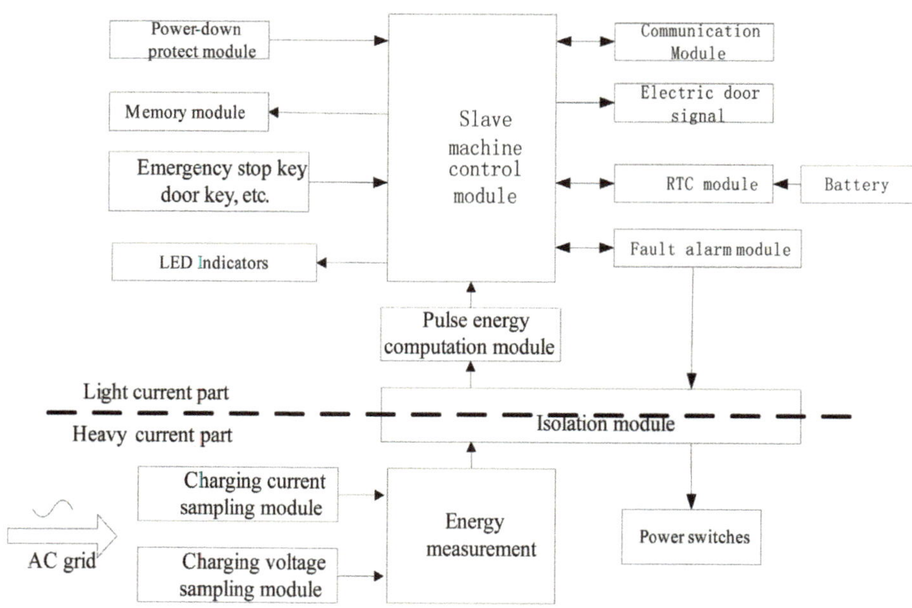

Figure 4. Slave machine functional block diagram for low-cost smart charging system.

few functions in each slaver machine, thereby the cost of slave machine will be further reduced.

2) Small size: the size is small because each slave machine has few function modules.

3) Low investment: low cost and small construction area can greatly reduce the construction cost of charging facilities.

4) Ultra low standby power consumption: if the charging point is in standby status for long time, it will consume a lot of energy. Ultra low standby power consumption is designed to reduce the power consumption of each slaver machine.

5) Expansibility: more slave machines share a trading system, and the system reserves expansion interfaces.

6) Network communications: master machine has a network communication function, the data can be regularly uploaded to the central severs, and can be backed up on the local system as well.

7) Low maintenance and update cost: when maintaining and updating the whole system, the mater machine can only be maintained and updated. The cost will be greatly down.

4.2. Payment Process of Charging

Customers can easily pay for the charging fee according to **Figure 5**. The process is described as the following:

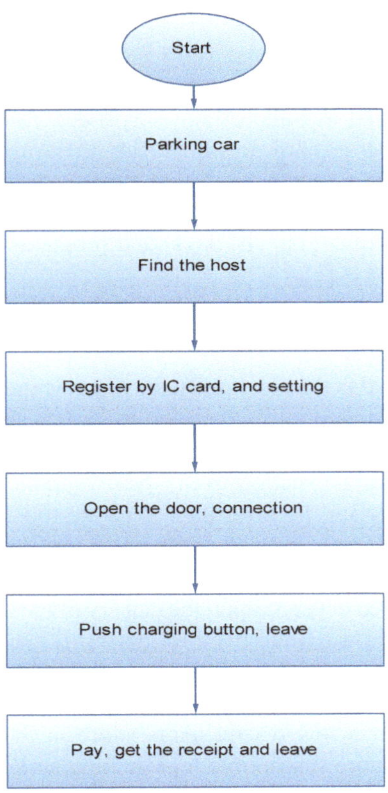

Figure 5. The payment process of charging.

1) Check the status of charging pile of parking spaces; park the car in the idle parking space. Find the host which manages the charging pile.

2) Register the charging pile by IC card and set the charging energy or the transaction amount (the setting of payment system based on the corresponding requirements of power supply bureau, the desired charging time also can be selected to ensure the full utilization of low power, and the cost of charging can be saved).

3) Open the door of the charging pile, connect the charger plug with the charging socket and shut the door of the charging pile.

4) Press the charging button if the charging indicator is lighting, and then the customer can leave.

5) Customer is needed to pay for the charging fee through the host. The transaction information is displayed and printed after the successful payment.

The above payment process of charging is only a simple of the charging system. In fact, the process can be designed according to different requirements.

5. Conclusion

Electric vehicles will be mass production in the future. Nowadays, the number of electric vehicles is small. The construction of charging facilities should be in low investment, many points. The ultimate goal of electric vehicle charging system is that all the parking places should be equipped with electric charging facilities. If these goals can be achieved, with the rapid development of key electric vehicles' parts, the electric vehicles will become the mainstream of the transport.

REFERENCES

[1] "Car Ownership." http://baike.Baidu.com/view/1379561.html

[2] BP, "BP Statistical Review of World Energy," 2009.

[3] K. X. Xu, "Development and Reform Commission: China International Petroleum Dependence of 52% Break through the Cordon," People.

[4] B. Zhao and L. Gao, "Realistic Significance of Developing Electric Vehicles in China," *Today Keyuan*, Vol. 20, 2010, p. 79.

[5] J. X. Liu, "Operability Analysis for Construction of Battery Swapping Station for Electric Vehicles," *Science and Technology Plaza*, Vol. 3, 2010, pp. 226-229.

[6] "EV Charging Pot a Better Future," *IT Manager World CEOCIO No.* 302, 2010.

[7] "AC Electric Vehicle Charging Pot Technical Specifications," *Company Standard for China Southern Power Grid Company Limited Q/CSG* 11516.4-2010.

[8] BYD QCJ7006BEV Pure Electric Car Specifications.

[9] "AC Charging Pot for Electric Vehicles," Chinese Patent No. CN201020538652.

21

Sizing and Design of the PV-WIND Energy Dryer for Medical Herbs

Emad Ahmed Sweelem, Abdel-Shafy A. Nafeh, Faten Hosny Fahmy

PV Cells Department, Electronics Research Institute, Cairo, Egypt.

ABSTRACT

Medical herbs are located in Egypt in different areas and they are very important to increase the national income of Egypt. It is necessary to dry the herbs in their growing locations. As the growing locations are located far from the natural supply grid of Egypt, renewable energy sources such as PV&WIND are preferably utilized to achieve the drying process. Moreover, since the drying process is actually used to remove water from the herbs, thus it can preserve the herbs in storage drastically by preventing the attack of microorganisms during the moist conditions of herbs and also reduce their weight and bulk (for cheaper transport). Where, this paper selects, at first, the suitable drying site, and then it suggests. The suggested drying system is completely designed, and includes the following three systems: the solar thermal system, the dryer, and the renewable energy system.

Keywords: Medical Herbs Dryer; Solar Thermal System; Renewable Energy

1. Introduction

Drying is used to remove water from herbs for two reasons: to prevent (or inhibit) microorganisms and hence preserve the product and to reduce the weight and bulk of herb for cheaper transport and storage. If drying is carried out incorrectly there is a greater loss of nutritional and qualities and more seriously, a risk of microbial spoilage. Therefore, we will describe some of the requirements for proper drying and summarize information on the various available drying techniques.

Drying can be carried out using hot air. For effective drying, air should be HOT, DRY and MOVING. These factors are inter-related and it is important that each factor is correct (for example, cold moving air or hot wet moving air are each unsatisfactory) [1].

2. Common Herb Drying Methods

Drying is the most critical operation in the production of medical herbs. Herbs can be dehydrated by various means: room drying, a conventional oven, a microwave oven or a dehydrator [2].

2.1. Room Drying

This method of drying differs from sun drying since it takes place indoors in a well-ventilated attic, room, car, camper or screened-in-porch. Herbs can be strung on a string or tied in bundles and suspended from overhead racks in the air until dry. Enclosing them in paper bags, with openings for air circulation, protects them from dust, and other pollutants. This method of drying is inefficient, takes long time (11 - 14 days) and produces low quality dried products [3].

2.2. Oven Drying

This method is the most practical way to experiment with dehydration. It protects herbs from insects and dust, and does not depend on the weather (flowers, dry well in a fan-assisted, convection oven). Non-ventilated ovens are not appropriate, because they generate too much moisture. The flowers are slotted through holes in a wire mesh rack leaving room for the stems to dangle below. The time required depends upon the density of the flowers [4].

Continual use of an oven for drying is not recommended because ovens are less energy efficient than dehydrators, and energy costs tend to be high. Also, it is difficult to maintain a low drying temperature in the oven, and herbs are more susceptible to scorching at the end of the drying period. Oven-dried herbs usually are darker, more brittle and less flavorful than herbs dried by a dehydrator.

Moreover, drying herbs in the oven causes water oils in the herb to evaporate too quickly that way. For oven drying, fresh herbs are placed in a single layer on the racks [5].

2.3. Microwave Oven

Drying herbs in a microwave is quick, simple, and gives excellent results. Two layers of paper towel are placed in the bottom of the microwave. Then a layer of herbs is added and covered with two more layers of paper towel. The microwave is run for two or more than two minuets until the herbs are dried. This method of drying can be used only in a small scale. It also requires careful attention, as the paper towels in the microwave can catch fire if hot spots occur [6].

2.4. Herbs Dryers

Herbs can be dried on trays in a dehydrator. Although the initial investment is fairly high for a dehydrator, it maintains low temperatures and uses less energy than an oven. The quality of the product is better than with any other method of drying. As with oven drying, there is no dependence on weather conditions [7].

Types of Herb Dehydrators (Dryers)

1) Firewood/fuel dryer

This technique uses fuel source to create the required heat for drying. It is quicker than open air-drying, but it requires investment in fuel. The dried product may be affected with smoke and easily degrades [8].

2) Solar thermal AC powered dryer

This type has bigger size and is usually made for heavy-duty. It is supplied by fan to circulate the hot air inside the dryer. The heating energy is provided by the sun collector. During the utilization of this type in rural areas, the air circulation fans and the pumps can be powered using DC source [9].

3) Solar thermal photovoltaic (PV) powered dryer

This dryer is called solar tunnel dryer. It had been used for drying rosemary and thyme plants produced in Thailand in 1999. The quality parameters of the dried products were selected on the basis of internationally recognized standards for dried spices and medical plants and include colour assessment, ash content, essential oil and oleoresin content and composition with regard to the important active components, cineole in rosemary and thymol in thyme. The colour preservation of dried rosemary and thyme was best achieved with the solar tray dryer [10].

3. Sizing and Design of the Suggested Renewable Energy Dryer

In this section the configuration of the suggested medicinal herb dryer system is presented. Also, a complete design and sizing for the different components of the proposed system is carried out in details.

3.1. Configuration of the Suggested Herb Dryer

The configuration of the suggested medicinal herb dryer system is shown in **Figure 1**. The diagram consists of three main systems: the solar thermal system, the dryer, and the renewable energy system. The main function of the solar thermal system is to supply the inlet air to the dryer with the required solar thermal energy that is suitable to accomplish the drying process. While, that of the renewable energy system is to feed the electrical load of the dryer with the required electrical energy, to ensure good and continuous drying operation.

3.2. Design of the Dryer and the Solar Thermal System

In this section the complete design steps of the proposed herb dryer and its associated solar thermal system is carried out. Where, the solar thermal system, which is responsible for heating the drying air, consists mainly of three components; which are the solar collector, the stor-

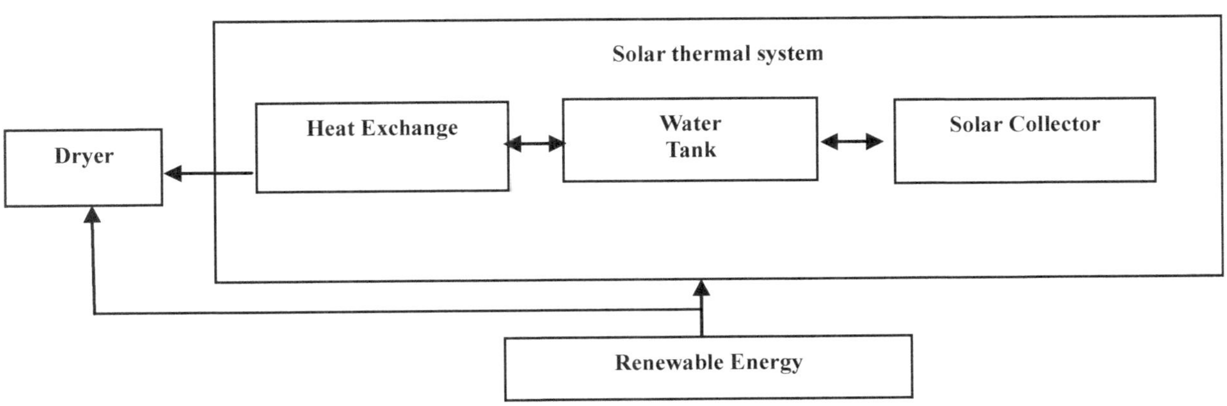

Figure 1. The medicinal herb dryer system.

age tank, and the heat exchanger.

3.2.1. The Dryer Design

The dryer design includes designing the different parameters that significantly influence the dryer performance. The dryer is designed as a cylindrical (bin) type with an inlet and outlet ducts (**Figure 2**). This type of dryer is ideally suited for drying medicinal herbs, since it is suitable for low temperature operation. The inlet duct is equipped with a blower, heat exchanger, and an electrical heater; to heat the inlet drying air. While, the outlet duct has four windows, to constitute four exits for the drying air after passing on the herb in the drying chamber. The dimensions of the drying chamber are 130 cm (diameter) × 50 cm (height). There are five wire mesh trays fixed one over the other inside the drying chamber for putting the herb to be dried. The trays are designed to distribute the air such that it follows a longer zigzag route, which increases the air/herb contact time and thus increases the drying efficiency. It is noted that a dark drying chamber is needed in order to isolate sunlight, which has a load effect on the herb. To achieve this purpose, the inner surface of the drying chamber is isolated with thin film of aluminum oxide foil.

1) The energy balance and temperature for drying

If unsaturated air is passed over wet material the air will take up water from the material. This water has to be evaporated, and the heat to do this comes from the air and the material. This air is thereby cooled. In particular, if a volume V_{air} of air is cooled from an initial temperature T_i to a final temperature T_f in the process of evaporating a mass m_w of water, then by applying the energy balance principle on the drying process, yields [11]

$$m_W L = m_{air} C_{Pair} \left(T_i - T_f\right) \quad (1)$$

where m_w could be calculated from the following equation [11]

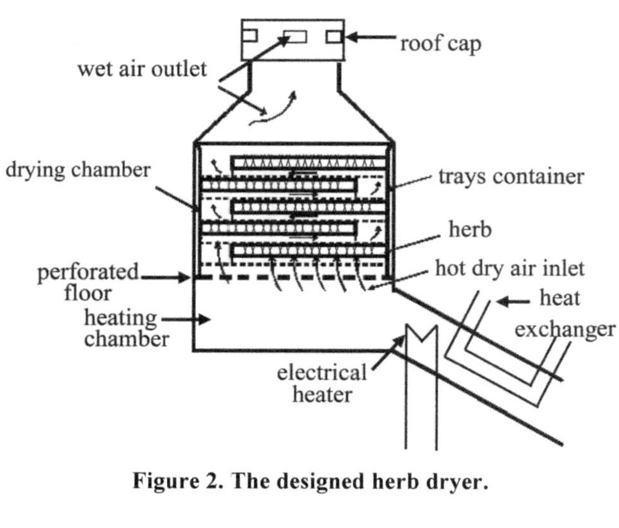

Figure 2. The designed herb dryer.

$$m_w = \frac{w_i\left(M_i - M_f\right)}{100 - M_f} \quad (2)$$

where:

m_w mass of water evaporated from the herb (kg);

m_{air} mass of the used drying air (kg), which absorbs m_w;

L the specific latent heat of vaporization of the water from the herb (=2.4 MJ/kg);

C_{Pair} specific heat capacity of air at constant pressure (=1.005 kJ/kg ˚K);

T_i initial temperature of the drying air (˚C);

T_f final temperature of the drying air (˚C);

w_i initial mass of wet herb (kg);

M_i initial moisture content of the herb (%);

M_f final moisture content of the herb (%).

In this design, it is assumed that 20 kg of Henbane, as one of the Sinai Peninsula (*i.e.*, Abu Rudies) medical herbs, wanted to be dried in 22 h. Where, the remaining two hours of the day are left for accomplishing the pre-drying process of the herbs and are chosen to be in the early morning; since gathering the herbs to be dried must be after the dew has evaporated, to minimize wilting of herbs. Then using the data of the annual average ambient temperature of Abu Rudies is 23.15˚C and the corresponding relative humidity is 57%.

The suitable initial air temperature T_i for drying of medicinal herbs may range from 35˚C to 45˚C, but it is found that the optimal drying temperature for most of the medicinal herbs is about 40˚C [2].

The final temperature of the moist drying air that leaves the dryer can be found from the Psychometric chart of **Figure 3** [12] as follows: the ambient air before entering the drying chamber corresponds to point A of **Figure 3** (*i.e.*, its temperature is 23.15˚C and its relative humidity is 57%).

If the small change in the density of the drying air is neglected under heating it from the ambient temperature of 23.15˚C to 40˚C, then point A of **Figure 3** will shift horizontally to point B (that has the same absolute humidity as point A). Note that the relative humidity of point B is lower than that of point A. After passing through the herb in the dryer, the exit air from the dryer will be cooled and becomes more moist, and this is corresponds to moving from point B to point C of **Figure 3** [12] (*i.e.*, T_f = 22.5˚C).

Since, the Henbane herb has M_i = 75% and M_f = 14%, then using Equations (4), (5) give m_w = 11.74 kg & m_{air} = 1602.047 kg. Thus, the required mass flow rate to accomplish the drying process of Henbane in the specified time is

$$\dot{m}_{air} = \frac{m_{air}}{\text{drying time}} = 0.02023\ \text{kg/sec}. \quad (3)$$

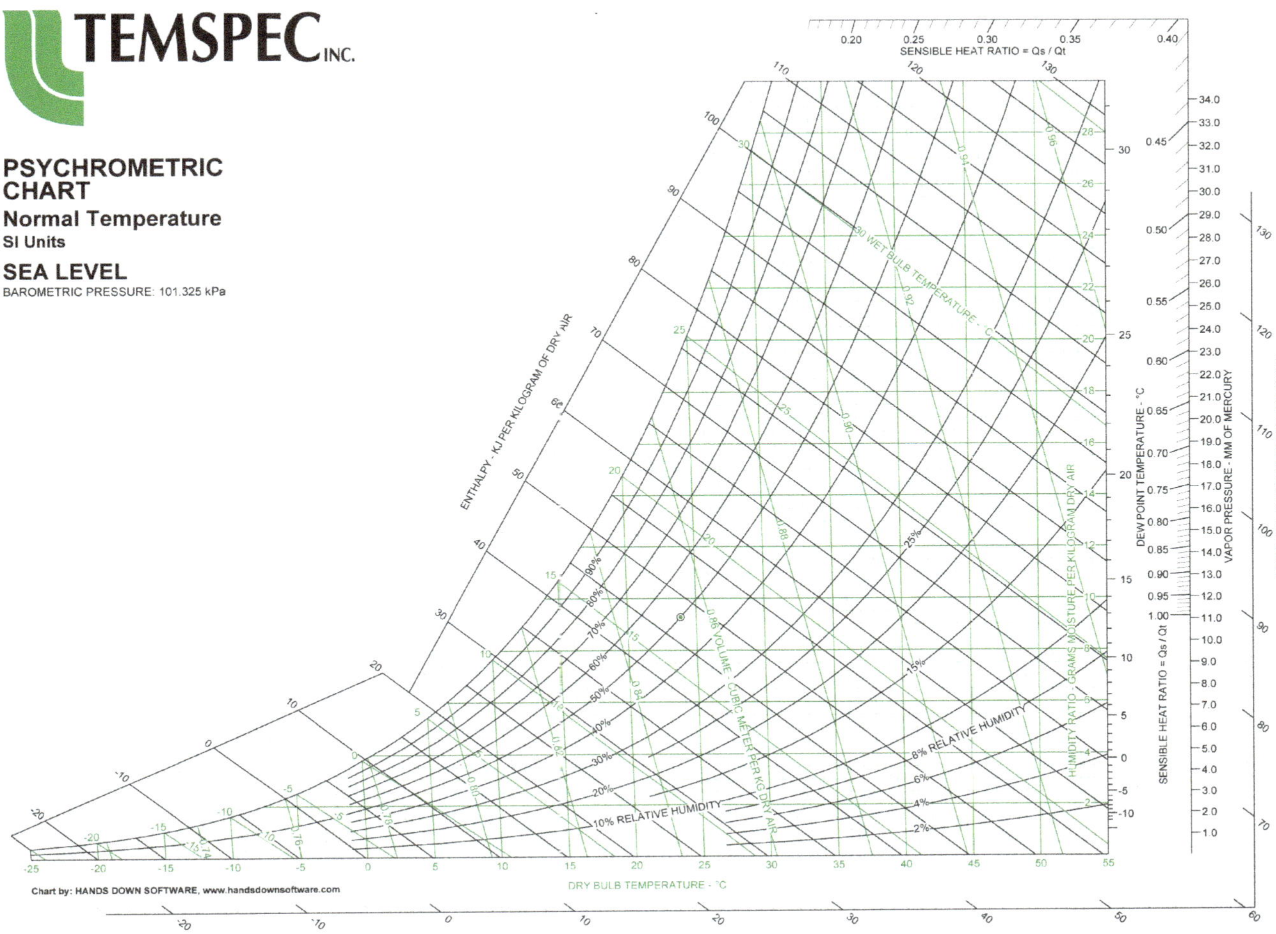

Figure 3. Psychometric chart (for standard air pressure of 101.3 kPa).

where, the volume of air needed for one drying cycle is given by

$$V_{air} = \frac{m_{air} R T_{air}}{P_{air}} \qquad (4)$$

where:

R the characteristic gas constant (=0.287 kJ/kg°K);

T_{air} ambient temperature (=23.15°C = 296.15°K);

P_{air} air pressure (=101.3 kN/m^2 or kPa);

Therefore, V_{air} = 1344.2 m^3 and the corresponding volume flow rate $\dot{V}_{air} = 0.017\ \mathrm{m^3/sec}$.

2) Coefficient of performance of the designed dryer (COP)

The performance coefficient of the dryer is defined as the ratio of the quantity of heat required evaporating the moisture content of the dried herb and the amount of heat to be supplied to the dryer, and is given by [2]

$$COP = \frac{Q_w}{Q_{dryer}} = \frac{m_w L}{m_{air} C_{Pair}\left(T_{dryer} - T_{air}\right)} = \frac{\left(T_i - T_f\right)}{\left(T_{dryer} - T_{air}\right)} = 1.0387 \qquad (5)$$

The overall thermal efficiency of the dryer is defined as the ratio of the quantity of heat required evaporating the moisture content of the dried herb and the solar energy incident on the plane of the solar collector, and is given by [13]

$$\eta_{th} = \frac{Q_w}{Q_{cin}} = \frac{Q_w}{Q_{dryer}} \cdot \frac{Q_{dryer}}{Q_{HEin}} \cdot \frac{Q_{HEin}}{Q_{cout}} \cdot \frac{Q_{cout}}{Q_{cin}} = COP \cdot \eta_{HE} \cdot \eta_{tank} \cdot \eta_C \qquad (6)$$

Assuming $\eta_{HE} = 0.9$, $\eta_{tank} \cong 1$, and $\eta_c = 0.75$, gives $\eta_{th} = 0.7$.

3.2.2. Design of the Solar Thermal System

In this work, the solar thermal system consists of a flat-plate solar collector, a vertical storage tank, and a heat exchanger.

1) The flat-plate collector

The flat-plate collector consists basically of an absorbing surface, an insulation layer, anodized aluminum section casing and a transparent cover (**Figure 4**). The collector area A_c can be computed from

$$A_c = \frac{Q_w}{60 \times 60 \times F_1 \times \eta_{th}} \approx 2\ \text{m}^2 \quad (7)$$

where, F_1 is as defined before, to be the global solar radiation in kW·h/m^2/day (of abu Radis) incident on a south facing plane, which is tilted to the horizontal by the latitude angle. Therefore, the collector is designed to have a 2 m length and 1 m width. The absorbing surface is made of aluminum. It is painted by a black color, and insulated from bottom and sides by 4 cm polyurethane insulation layer to reduce heat losses to the surroundings. Also to minimize the heat losses from the upper collector surface due to the re-radiation and convection, an absorber plate is provided with a transparent cover. This cover must be capable to transmit almost all solar radiation as possible. A glass sheet with 3 mm thickness is fixed 3 cm above the absorber.

2) The storage tank

The designed storage tank consists of a steel cylindrical vessel of 44.6 cm diameter and 132 cm height (**Figure 5**). It's insulated by a layer of 7 cm of glass wool and the other casing cover is aluminum.

3) The heat exchanger

The used heat exchanger is a parallel flow water to air type, which consists of two concentric pipes of different diameters (**Figure 6**). The hot fluid (*i.e.*, the water) passes inside the smaller diameter pipe, while the cold fluid (the air) passes through the space between the two pipes.

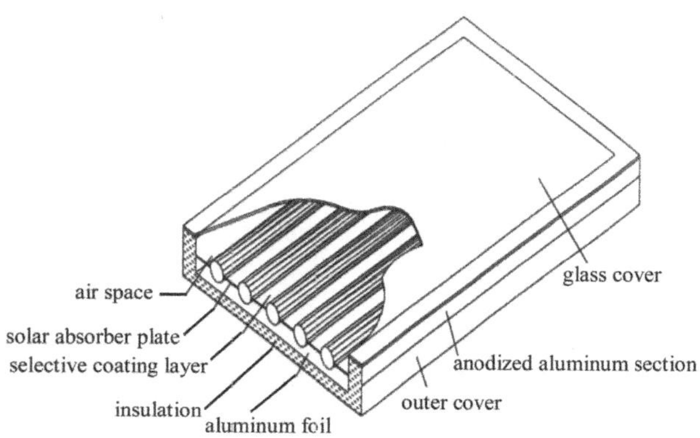

Figure 4. Description of the flat-plate collector.

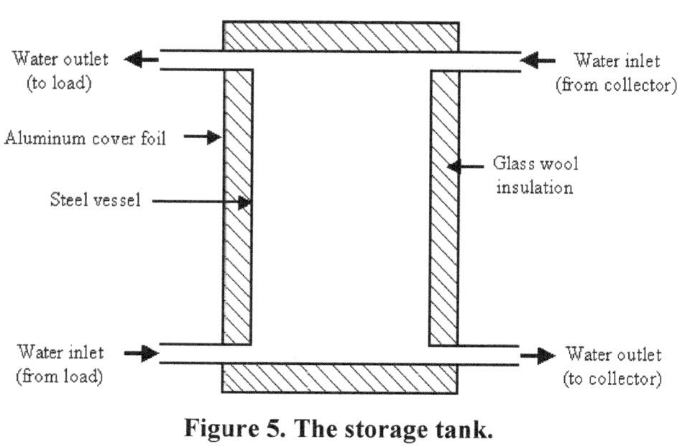

Figure 5. The storage tank.

3.3. Sizing of the Suggested Renewable Energy System

In this section the complete sizing of the proposed renewable energy system and its electrical load is carried out. Where, the renewable energy system, which is responsible for supplying the electrical load of the dryer with the required electrical power, consists mainly of three components; which are the photovoltaic (PV) array, the wind generator, and the storage battery.

3.3.1. Dryer Electrical Load

As the electrical load profile can be constructed from the suggested load data, and since the load data define all the information about the suggested electrical load (*i.e.*, number of equipments to be powered, their nominal power, system voltage, and the number of hours of operation for each equipment in a typical day). **Figure 6** shows the dryer electrical load, which consists of the following equipments: the signal lamp, the electrical heater, the blower, two valves and two pumps. These equipments constitute a typical load for the designed dryer. The data of the suggested load are illustrated in **Table 1**, which imply the dryer load profile of **Figure 7**. It is to be noted, here, that the drying process is achieved using the designed dryer in 22 h only of the day, while the remaining two hours are left for accomplishing the predrying process of the herbs and are chosen to be in the early morning (since gathering the herbs to be dried must be after the dew has evaporated, to minimize wilting of herbs).

3.3.2. Optimization Sizing of PV-Wind-Battery Hybrid Energy System

The dependence on energy resources, which are expensive and in the long run extinguishable, has led recently to an increased interest in harnessing renewable energy resources. Whether electricity generation uses conventional or renewable sources, it should have a unique function that is to continuously supply the load in the most economic way.

As can be concluded from the above, of the main concern in the design of an electric power system that utilizes renewable energy sources is the accurate selection of system components that can economically satisfy the load demand. In this section, the most economical design of an autonomous renewable energy system composed of solar, wind, and battery sets is determined using Homer Software [14].

1) Formulating the problem

The task of the optimization technique is to determine the size of each component participating in the system so that the load can be economically satisfied. For this reason, the control variables are made to represent either

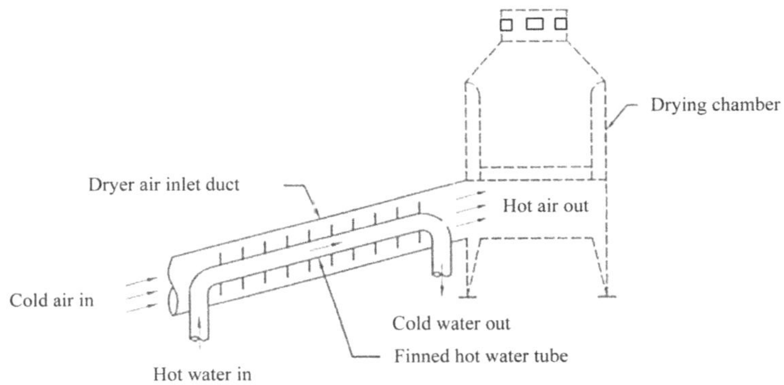

Figure 6. Schematic diagram of heat exchanger.

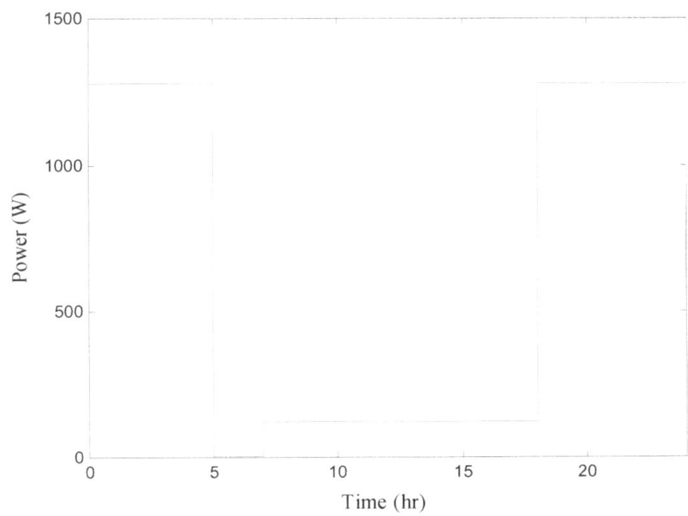

Figure 7. Typical daily load profile of the dryer.

Table 1. Electrical load data of the dryer.

Load	No.	Rated power (W)	System voltage (V)	Daily operating times (h)
Signal lamp	1	5	24	0 - 24
Electrical heater	1	1200	24	18 - 5
Blower	1	30	24	0 - 5 & 7 - 24
Water pump	2	25	24	Collector-side pump: 7 - 18 Load-side pump: 0 - 5 & 7 - 24
Water valve	2	20	24	Collector-side valve: 7 - 18 Load-side valve: 0 - 5 & 7 - 24

size or rating of system components. The objective function to be minimized is the cost of producing one kilowatt-hour of electricity, and the constraints are expressed through load requirements.

The cost function is generated by dividing the summation of the present worth of all the salvage values of the equipment, the yearly operation and maintenance costs and the initial or capital investments, by the expected energy demand which is supplied during the lifetime of the project. The cost function can be written as [15]

$$f = \left(\sum_{k=1}^{3} I_k - S_{Pk} + OM_{Pk} \right) \frac{1}{E_y \cdot N} \quad (8)$$

where: the index k is made to account for wind, solar, diesel generators and batteries; I_k is the initial investment for each component k; S_{Pk} is the present worth of the salvage value of each component k; OM_{Pk} is the present worth of the operation and maintenance costs for each component k; E_y is the yearly energy demand; and N is the lifetime of the project.

The constraints that ought to be met, while minimizing the objective function, should ensure that the load is served according to certain reliability criteria. The extent of system reliability is measured by the expected energy not served (EENS) within each sub period of operation. This value should be specified by the designer in view of the priority of the load and the adopted pricing strategy. The constraints can now be written as:

$$\sum_{k=1}^{3} E_k(i) \geq (1 - EENS) * Load(i)$$
$$\text{and} \qquad \sum_{k=1}^{3} E_k(i) \leq Load(i) \qquad (9)$$

For all the sub periods i, where $E_k(i)$ is the energy from source k during sub period i; and $Load(i)$ is the energy demand during sub period i.

2) Basic economical considerations

As Equation (8) suggests, the present worth of some annual payments as well as of salvage values are needed. Assuming a life horizon of N years, an interest rate r, and an inflation j caused by increases in price, the following can be noted.

a) If a component has a salvage value of S($) at present because it is reaching the end of its life cycle, then the salvage value is expected to be $S{\cdot}(1 + j)^N$, N years from now provided that the component is put in service at the present. The present value of $S{\cdot}(1 + j)^N$ taking the interest rate into consideration is [16]:

$$\frac{S*(1+j)^N}{(1+r)^N} \qquad (10)$$

Let $fac1 = (1+j)^N / (l+r)^N$, then $S_{Pk} = S_k * fac1$ for all components k in the system.

b) If the operating and maintenance cost of a component is OM (\$$\cdot$year^{-1}), then this tends to escalate each year at a rate not necessarily equal to the general inflation. With an escalation rate "es", the operation and maintenance costs incurred at year "y" will be $OM * (1 + es)^y$ having a present worth of

$$\frac{OM*(1+es)^y}{(1+r)^y} \qquad (11)$$

The summation of the present worth of all the annual payments is given by

$$\sum_{y=1}^{N} \frac{OM*(1+es)^y}{(1+r)^y} = OM * fac2 \qquad (12)$$

$$fac2 = \sum_{y=1}^{N} \frac{(1+es)^y}{(1+r)^y} = \frac{(1+es)}{(r-es)} * \left(1 - \left(\frac{1+es}{1+er}\right)^N\right) \quad \text{for } r \neq es \qquad (13)$$

Thus, $fac2 = N$ if $r = es$. Hence $OM_{Pk} = OM_k * fac2$, for all components, "k" in the system. At this stage it is worth mentioning that the power extracted from the wind and solar systems are calculated using the following models.

3) Wind system

A wind turbine generator (WTG) produces power when the wind speed V is higher than the cut-in speed V_{ci} and is shut-down when V is higher than the cut-out speed V_{co}, when $V_r < V < V_{co}$, (V_r is the rated wind speed), the WTG produces rated power "P_r". If $V_{ci} < V < V_{co}$ the WTG output power varies according to the cube law. The following equations are to be used in order to model the WTG in the optimization:

$$\begin{aligned} P_w &= 0 && \text{if } V < V_{ci} \\ P_w &= a \cdot V^3 - b \cdot P_r && \text{if } V_{ci} < V < V_r \\ P_w &= P_r && \text{if } V_r < V < V_{co} \\ P_w &= 0 && \text{if } V > V_{co} \end{aligned} \qquad (14)$$

where

$$a = \frac{P_r}{V_r^3 - V_{ci}^3}, \quad b = \frac{V_{ci}^3}{V_r^3 - V_{ci}^3} \qquad (15)$$

And

$$P_r = \frac{1}{2} CprAV^3 \qquad (16)$$

In the above, $Cp = 0.3$, $\rho = 1.2$ Kg$\cdot$m^{-3} at 20°C.

Efficiency of solar system ($effs$) = 0.3, Inflation rate (j) = 0.09, Interest rate (r) = 0.12,

Escalation rate (es) = 0.09, Project life span (N) = 30 years, Annual real interest rate (r) = 10%;

Battery life span (L_b) = 10 years, Experted energy not served ($EENS$) = 1%, Solar panel salvage value (s_s) = 45 \$\$/m^2, Wind turbine salvage value (s_w) = 10 \$/m^2, Diesel engine salvage value (s_d) = 80 \$/Kw, Load = see **Figure 7**.

4) Solar system

For a photovoltaic array having an efficiency "$effs$" and area "A_s", the output power "P_s" (kW) when subjected to a horizontal irradiance H (kW$\cdot$m^2) is given by

$$P_s = H * A_s * effs$$

5) Optimization coefficients

a) Wind turbine

The control variable due to the wind turbine is the total rotor area Aw in square meters. This value is constrained by both the space available and the budget of the project. Numerically, Aw is generated through linear programming, though it is the task of the designer to distribute Aw among several machines.

Given α_w, the initial price per square meter of the wind turbine, the term I_1 of Equation (16) that corresponds to the total initial price of the wind machines is equal to $\alpha_w \times Aw$. It should be noted that if the price in \$/kW is available rather than α_w then α_w can be generated by finding how many kilowatts the wind machine can generate per square meter at rated wind speed and multiplying this number by the cost per kilowatt.

With a salvage value of Sw (\$·m^{-2}), the total salvage value would be $S_1 = Sw * Aw$ and the present worth

$$S_{p1} = S_1 * fac1 \quad (16)$$

Finally, with a yearly operation and maintenance cost of α_{OMW} per square meter (\$·m^{-2}·year^{-1}), the total yearly operation and maintenance cost would be $OM_1 = \alpha_{OMW} * Aw$, and the present worth of all the yearly costs would be:

$$OM_{p1} = OM_1 * fac2 = \alpha_{omw} * Aw * fac2 \quad (17)$$

If the operation and maintenance cost is given in dollars per kilowatt per year, this value should be multiplied by the number of kilowatts generated per square meter at rated wind speed to obtain α_{OMW}. As a summary for wind turbines, we have

$$\begin{aligned} I_1 &= \alpha_w * A_w = C_1 * A_w \\ S_{p1} &= S_w * fac1 * A_w = C_2 * A_w \\ OM_{p1} &= \alpha_{OMW} * fac2 * A_w = C_3 * A_w \end{aligned} \quad (18)$$

b) Solar panels

Solar arrays can be treated following the approach used for wind rotors. The control variable is the total solar panel area as in square meters and is constrained by both the maximum available area for solar panels (for example the roof surface of buildings) and the budget preset for the panels.

With an initial cost of α_s (\$·m^{-2}), the total initial investment would be $I_2 = \alpha_s As$. The salvage value can be found by multiplying the selling price per square meter Ss by the area As and the present worth of the selling price would be

$$S_{p2} = S_2 * fac1 = S_s * As * fac1 \quad (19)$$

With a yearly operation and maintenance cost of α_{OMS} per square meter (\$·m^{-2}·year^{-1}), the total yearly operation and maintenance cost would be $OM_2 = \alpha_{OMS}$ as with a global present worth of:

$$OM_{p2} = OM_2 * fac2 = \alpha_{OMS} * As * fac2 \quad (20)$$

As a summary for solar panels, we have

$$\begin{aligned} I_2 &= \alpha_s * A_s = C_4 * A_s \\ S_{p2} &= S_s * fac1 * A_s = C_5 * A_s \\ OM_{p2} &= \alpha_{OMS} * fac2 * A_s = C_6 * A_s \end{aligned}$$

c) Storage batteries

The control variable in the case of storage batteries is their size R_b in kilowatt-hours. The lifetime of a battery (L_b) is expected to be less than that of the whole project. Hence batteries of size R_b are to be purchased at regular intervals of L_b. The total present worth of the capital investment in batteries is given by

$$I_3 = \alpha_b * R_b * \sum_{x=1}^{x_b} \left(\frac{1+es}{1+r} \right)^{(x-1)*L_b} \quad (21)$$

where L_b, is the battery lifetime, x_b is the number of times batteries should be purchased during the project lifetime: $x_b = N/L_b$ rounded to the greatest integer, and α_b is the capital cost of 1 kWh storage (\$·kW·h^{-1}).

The salvage value of the batteries is assumed negligable. With an operation and maintenance cost of α_{OMB} (\$·kW·h^{-1}·year^{-1}), the total yearly OM cost would be $OM_4 = \alpha_{OMB} * R_b$ with a global present worth of

$$OM_{p4} = OM_4 * fac2 \quad (22)$$

In summary, the battery charges can be represented by

$$\begin{aligned} I_3 &= \alpha_b * R_b * \sum_{x=1}^{x_d} \left(\frac{1+es}{1+r} \right)^{(x-1)*L_b} = c_{10} * R_b \\ OM_{p4} &= OM_4 * fac2 = C_{11} * R_b \end{aligned}$$

6) Load constraints

The constraints on the optimization problem were expressed through Equation (12). Let $e_w(i)$ and $e_s(i)$ represent the wind and solar energy, respectively, available during sub period "i" per square meter, and let the energy generated by the diesel generator be R_d times the duration of the sub period. The elements of the constraints can then be formulated as:

$$\begin{aligned} E_1(i) &= e_w(i) * A_w = C_{12}(i) * A_w \\ E_2(i) &= e_s(i) * A_s = C_{13}(i) * A_s \end{aligned} \quad (23)$$

$E_3(i) = x * R_b$ where x is the fraction of the battery capacity expected to discharge in each sub period. This implies that for each sub period i, the following should be satisfied:

$$\begin{aligned} & e_w(i) * A_w + e_s(i) * A_s + x * R_b \geq (1 - EENS) * load(i) \\ \text{and} \quad & e_w(i) * A_w + e_s(i) * A_s + x * R_b \leq load(i) \end{aligned} \quad (24)$$

Additional constraints should be imposed on the size of the system components:

$$\begin{aligned} 0 &\leq A_w \leq A_{w\max} \\ 0 &\leq A_s \leq A_{s\max} \\ 0 &\leq R_b \leq R_{b\max} \end{aligned} \quad (25)$$

7) Final form

At this stage, the optimization problem can be written in its final form as follows:

a) Minimize the cost function f:

$$(C_1 - C_2 + C_3) * A_w + (C_4 - C_5 + C_6) * A_s + (C_{10} + C_{11}) * R_b$$

b) Subject to:

$$C_{12}(i) * A_w + C_{13}(i) * A_s + x * R_b \geq (1 - EENS) * load(i)$$
$$C_{12}(i) * A_w + C_{13}(i) * A_s + x * R_b \leq load(i)$$
$$0 \leq A_w \leq A_{w\max}$$
$$0 \leq A_s \leq A_{s\max}$$
$$0 \leq R_b \leq R_{b\max}$$

8) Results of the sizing of the renewable energy system

In this part the results of applying optimization program are displayed as shown in **Table 2**.

4. Conclusions

The suggested drying system is completely designed, and includes the following three systems: the solar thermal system, the dryer, and the renewable energy system. The main function of the solar thermal system is to supply the dryer inlet air with the required thermal energy through utilizing the flat plate solar collector. While, that of the renewable energy system, which comprises hybrid photovoltaic and wind energy sources, is to feed the electrical load of the dryer with the required electrical energy.

The general principles of wind power and solar energy together with its two parts (solar thermal collectors and solar photovoltaic energy) are discussed and evaluated in details from different viewpoints. Where, Glazed flat-plate collectors are the common available and suitable collectors for using in herb drying systems. Since, these collectors are better suited for moderate temperature applications where the demand temperature is 30°C - 70°C.

Table 2. Optimization sizing results.

Item	PV-Wind-Battery system
Optimization sizing results	1.75 kW PV array 1 kW wind turbine 3.713 kWh battery
Capital cost ($)	8950
Annualized capital cost ($/y)	986
O&M cost ($/y)	132.5
Annualized cost ($/y)	1142
Net present cost ($)	10,364
Levelized cost of energy (COE) ($/Kwh)	0.211
PV production (Kwh)	4216
Wind production (Kwh)	4072
Total production (Kwh)	8288
Unserved energy (Kwh)	234
Battery throughput (Kwh)	3713
Battery life (years)	12

REFERENCES

[1] M. R. Patel, "Wind and Solar Power Systems," Library and Congress, Washington DC, 1999.

[2] M. Yahya, K. Sopian, W. R. W. Daud, M. Y. Othman and B. Yatim, "Design of Solar Assisted Dehumidification of Air Drying System for Medicinal Herbs: Pegaga Leaf," *Proceedings of the 2nd Asian-Oceania Drying Conference* (*ADC* 2001), Pulau Pinang, 20-22 August 2001, pp. 383-392.

[3] "Effects of Drying on Quality of Rosemary and Thyme Plants Produced in the Chaing Mai District, Thailand," Master Thesis of Marija Spirovska. http://www.troz.uni-hohenheim.de/research/Thesis/MScFT/Spirovska.pdf

[4] "Solar Laboratory University of the Philippines, Diliman-, Renewable Energy Technology in Asia, A Regional Research Dissemination Programme Phase II, A Summary of Activities and Achievements in Bangladesh," Regional Energy Resources Information Center (RERIC), Asian Institute of Technology, Thailand, 2002.

[5] B. K. Bala, M. R. A. Mondol, B. K. Biswas, B. L. Das Chowdury and S. Janjai, "Solar Drying of Pineapple Using Solar Tunnel Drier," *Journal of Renewable Energy*, Vol. 28, No. 2, 2003, pp. 183-190.

[6] O'Hanian and C. Hans, "Physics," W. W. Norton & Co., New York, 1985.

[7] E. W. Brown, "An Introduction to Solar Energy." http://www.ccs.neu.edu/home/feneric/solar.html.

[8] T. Markvart, "Solar Electricity," John Wiley & Sons, Inc., Hoboken, 1994.

[9] O. V. Ekechukwu and B. Norton, "Review of Solar-Energy Drying Systems II: An Overview of Solar Drying Technology," *Journal of Energy Conversion & Management*, Vol. 40, No. 6, 1999, pp. 615-655.

[10] "Commercialization of the Solar Dryer." http://www.areed.org/training/technology/solar_dryer dryer_2.pdf

[11] G. P. Sharma and S. Prasad, "Drying of Garlic Cloves by Microwave-Hot Air Combination," *Journal of Food Engineering*, Vol. 50, No. 2, 2001, pp. 99-105.

[12] www.handsdownsoftware.com

[13] H. H. El-Ghetany, "Effect of Hot Water Consumption on Temperature Distributions in Vertical and Horizontal Solar Waterv Storage Tanks," M.Sc.Thesis, Cairo University, Cairo, 1992.

[14] NREL, "Hybrid Optimization for Electric Renewable Program," NREL (National Renewable Energy Laboratory), Golden, 2009.

[15] R. Chedid and Y. Saliba, "Optimization and Control of

Autonomous Renewable Energy Systems," *International Journal of Energy Research*, Vol. 20, No. 7, 1996, pp. 609-624.

[16] W. Clark, "Energy for Survival: The Alternative to Extinction," Anchor Press, Garden City, 1974.

Feasibility Study for Self-Sustained Wastewater Treatment Plants—Using Biogas CHP Fuel Cell, Micro-Turbine, PV and Wind Turbine Systems

Ahmed Helal, Walid Ghoneim, Ahmed Halaby

Department of Electrical Engineering and Control, Faculty of Engineering and Technology, Arab Academy for Science and Technology, Alexandria, Egypt.

ABSTRACT

This paper studies the application of renewable energy sources in wastewater treatment plants to achieve self-sustainability of power. The data of wastewater treatment plant in the rural city of Toukh-EGYPT are presented as a case-study. The primary objective is to provide an entirely renewable standalone power system, which satisfies lowest possible emissions with the minimum lifecycle cost. Mass balance principle is applied on the biodegradable components in the wastewater to evaluate the volume of digester gas that is produced from sludge through anaerobic digestion process. Using digester gas as a fuel lead to study combined-heat-and-power technologies, where fuel cell is selected in order to abide by the low emissions constraint. The study assessed the electrical power obtained from fuel cell and the utilization of the exhausted heat energy for additional electrical power production using a micro-turbine. After covering the major part of load demand, the use of other renewable energy sources was studied. The strength of both solar and wind energy was determined by the case-study location. Hybrid optimization model for electric renewable (HOMER) software was used to simulate the hybrid system composed of combined-heat-and-power units, wind turbines and photovoltaic systems. Simulation results gave the best system configuration and optimum size of each component beside the detailed electrical and cost analysis of the model.

Keywords: Combined Heat Power; Economic Evaluation; Hybrid Renewable; Waste Water Treatment Plant

1. Introduction

Energy demands are increasing exponentially resulting into a rapid grow in need of conventional fossil fuels [1]. Such conventional sources are finite and fast depleting, which in turn threatens the balance of future energy demand/generation [2,3]. Renewable energy conversion devices like photovoltaic (PV), micro-turbines (MT), fuel cells (FC) and storage devices are expected to play an important role in future electricity supply and low carbon economy [4,5]. Absence of an electrical network in remote regions and the significantly high connection cost-due to large distances and irregular topography lead often the various organizations to explore alternative solutions [6]. Combining renewable energy to form standalone hybrid systems is considered as one of the most promising ways to handle the electrical requirements of these regions [7]. More and above, other environmental challenges such as greenhouse gas emissions and atmospheric temperature rise spotted the light further on the importance of renewable energy. The idea of the utilization of renewable energy sources (RES) is commonly related with sustainability. A sustainable energy system may be defined as a cost-effective, reliable, and environment friendly energy system that effectively utilizes local resources and networks [8]. For providing a sustainable energy supply, renewable energy sources appear to be one of the most efficient and effective solutions [3,9]. Beside the residential applications, the idea of studying the self-sustainability of power in industrial & service facilities looks very appealing due to their relatively high energy demand, such as wastewater treatment plants (WWTPs). This is because water pollution by nature is an energy intensive process. Also, the operating costs of waste-water treatment facilities in the recent years have increased substantially due to the increases in the unit cost of energy [10]. Using RES whether partially or totally to power WWTPs reduces the operating costs significantly.

Environmental Protection Agency (EPA) concluded that WWTPs with influent flow rate less than 19,000 m^3/day didn't produce enough biogas, through anaerobic digestion process, to make its use as a renewable source for electrical & thermal energies economically feasible [11]. This problem limits the use of biogas—for the plant in-field sustainable power generation—from the point of WWTP scale. Hence, this doesn't facilitate applying the concept of WWTP self-sustainability using renewable resources in remote and ecological areas unless other intermittent energy sources, such as wind and solar energies, have very strong potential. However, the paper demonstrates that using biogas for power and heat generation might still prove economic even in small-scale WWTP as the criterion primarily depends on the biodegradability of wastewater, not the influent flow rate. Luckily, rural areas, which commonly suffer from low or no electricity service, have strong sewage with high biodegradability which can yield more biogas [12]. The paper evaluates the economics of using biogas with other intermittent energy sources in small-scale rural WWTP to achieve self-sustainability of power. The proposed hybrid power system shall also produce the lowest emissions to match the case of rural ecological areas like oases. The paper uses the WWTP of Toukh-Egypt as a case-study. Section 2 reviews the combined-heat-power technologies working with produced biogas as a fuel. The design uses solid oxide fuel cell (SOFC) coupled with micro-turbine for maximum electrical energy production. Section 3 views the WWTP hourly electrical and thermal load demand to be covered. Section 4 views the necessary information for the software hybrid system modeling like solar & wind energy potentials and prices of the system components. Section 6 presents and discusses the simulation & optimization results.

WWTP of Toukh Center—Qalyobia in Egypt has an average flow of 8000 m^3/day. The town of Toukh lies in middle of delta region & intensively surrounded by crops and farms. Being located in rural areas, concentrations of biodegradable components in the wastewater are higher than average concentrations in urban areas. Therefore, the energy recovered from the plant through the digester would be relatively high when compared to urban treatment plants of the same size. The plant uses activated-sludge treatment process. In this process, wastewater flows continuously into an aeration tank (**Figure 1**) where air is injected to mix the activated sludge with the wastewater and to supply the oxygen needed for the organisms to oxidize the organic compounds. The mixture of activated sludge and wastewater in the aeration tank is called mixed liquor.

The mixed liquor flows from the aeration tank to a secondary clarifier where the activated sludge is settled out.

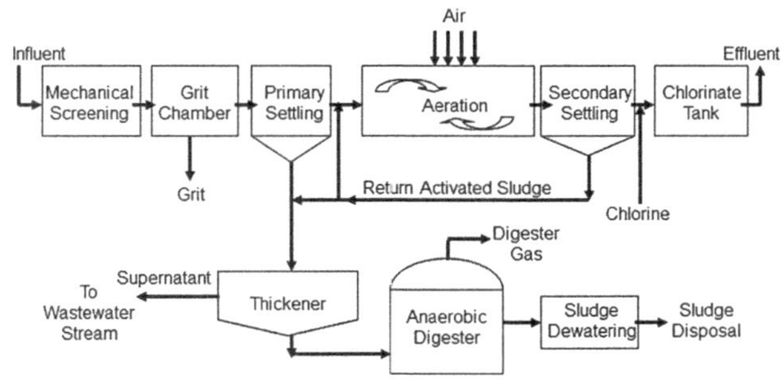

Figure 1. Conventional activated sludge treatment.

Most of the settled sludge is returned to the aeration tank, called return sludge, to maintain the high population of microbes that permits rapid breakdown of the organic compounds. The return sludge is diverted or wasted to the sludge handling system for treatment and disposal [13].

There are two main methods for sludge treatment; aerobic digestion and anaerobic digestion. Aerobic digestion is a suspended-growth biological treatment process based on biological theories similar to those of the extended aeration modification of the activated sludge process [14]. Anaerobic digestion (AD) plays an important role for its abilities to further transform organic matter into biogas (contains 65% - 70% methane by volume, 25% - 30% CO_2, and small amounts of N_2, H_2, H_2S, water vapor & other gases). The biogas, or anaerobic digester gas (ADG), can be utilized to produce energy using combined heat-power (CHP) technology [15]. The plant uses the most common scenario in sludge treatment in Egypt where the thickened sludge is directly pumped to natural dewatering without involving the digestion process [16]. Within here, the presence of anaerobic digester is assumed for sludge treatment. The production of digester gas has been estimated given the sludge daily mass, volume and concentration of biodegradable components in the raw sewage and sludge.

2. Energy Recovery in WWTP

Digester gas resulting from the anaerobic digester used in WWTP can be used to power an engine-generator SET to generate electricity, and the jacket water from the internal combustion engine can be used for digester or building heating [17]. New emerging technologies compete the combustion engines offering better emissions reduction such as fuel cell and micro-turbine. Co-generation of power and heat by these technologies is referred as combined heat-power (CHP).

Many researches have been made on the comparison between the CHP technologies over the small-scale applications. Operation, components, applications, costs, advantages & disadvantages of CHP technologies can be found in [10-26]. Selection of the technology will depend greatly on the nature of application. If cost & emissions are a major concern, reciprocating engines present lowest cost but highest emissions. FC systems have lowest emissions but highest cost. Micro-turbines offer a balance between the two factors. Startup time, part load response & efficiency also could be key performance factors according to the application type to be base load, peak shaving or emergency. Noise & size are also important factors for residential applications. Emissions reduction is always a major concern in agriculture & rural areas. Being one of the research goals, emission reduction will turn the CHP selection for the favor of fuel cells.

2.1. Selection of Fuel Cell Type

Solid Oxide Fuel Cells (SOFC) belongs to the group of high-temperature fuel cells. They are operated at temperature up to 1000°C and are of high interest because of their specific properties, especially the fuel utilization: they are able to work with a great variety of fuels (gaseous hydrogen or hydrocarbons like gasoline, diesel, kerosene, heavy oil, natural gas, or even biogas) while needing only a relatively low demand for cleaning, in particular concerning sulphur, and reforming of these fuels [27]. Amongst fuel cells for biogas conversion, the high temperature solid oxide fuel cell (SOFC) type is especially suited because of the capability of thermally integrated biogas reforming and because of manageable tolerance against fuel contaminants [28].

Due to their high operating temperature, SOFCs can convert hydrocarbons into hydrogen internally, showing global electrical efficiencies approximately of 46% even when fed by methane. SOFCs are also considered as one of the most promising candidate for a future high efficiency decentralized energy conversion model. Due to their high operating temperature and modularity, SOFCs can be easily integrated into cogeneration power plants, producing electricity, heat and cooling energy, when combined with absorption chillers [29].

2.2. Fuel Cell Model Characteristics

The fuel cell model selected for the case study is the one introduced in [30] for utilization of energy recovered from a wastewater treatment plant & using digester gas as a fuel. This study viewed the general & inside characteristic of the model fuel cell including its behavior under varying conditions. The characteristics for the fuel cell used in the study at operating point are shown in **Table 1**.

The estimated volume of digester gas produced by the anaerobic digester is 54.9 m³/hr. Only 44.6 m³/hr of the digester gas is assumed to be used. Having spare volume allows isolating the fluctuation of ADG production from the fuel cell input system. Hence, guarantees continuous availability of gas and sustainable power generation.

Table 1. Fuel cell and micro-turbine characteristics.

Fuel Cell Type	SOFC
Fuel Type	ADG
Fuel Cell Fuel Flow Rate (m^3/hr)	44.6
Fuel Cell Equivalent Input kW	290 (HHV)
Fuel Cell Electrical Efficiency	48% (HHV)
Fuel Cell Thermal Efficiency	40% (HHV)
Fuel Cell Electrical kW Output	140
Fuel Cell Thermal kW Output	117
Micro-Turbine Input Flow Rate (MJ/hr)	421 (LHV)
Micro-Turbine Electrical Efficiency (net)	22% (LHV)
Micro-Turbine Thermal Efficiency	41%
Micro-Turbine Electrical kW Output	26
Micro-Turbine Thermal kW Output	48
Overall System Electrical Output (kW)	166

As listed in **Table 1**, the net thermal power produced from the fuel cell is approximately 117 kW. The only heat load available in the plant is the digester heat load. Because of the warm climate in Egypt and the current limited purposes of thermal energy, excess thermal energy would be usually diffused to the atmosphere [31]. However, coupling a small size micro-turbine to the fuel cell can convert the excess thermal power to usable electrical power. Thus, it'll cover additional portion of plant electrical loads.

2.3. Micro-Turbines

Micro gas turbines are small gas turbines belonging to the group of turbo machines up to an electric power output of 300 kW. In order to raise the electrical output micro gas turbines are equipped with a recuperator (heat/heat exchanger). They are also equipped with a regular heat exchanger in order to use the waste heat from the exhaust gases [18]. Micro-turbine will be used for producing power from heat exhaust of the fuel cell forming what is called FC-MT hybrid model. This hybrid system offers a solution to two important problems, the low efficiency and relatively high emissions of small gas turbines, and the high cost of small fuel cell power plants [32]. MT has a smaller volume and weight but also a lower efficiency and larger emissions than a normal gas turbine. Therefore, a MT working as a stand-alone device generates not so much benefit. A fuel cell is a clean energy generator and has a considerably higher and constant efficiency even at different operating temperatures, but its volume is still extremely large [33].

The integration of FC-MT hybrid model can be done by two methods, atmospheric pressure system and pressurized system. The main advantage of atmospheric pressure system that it allows is the selection of the MGT pressure independently of the cell pressure [28]. Meanwhile, some basic studies have concluded that a pressured system may have higher system efficiency over an ambient pressure system from a thermodynamic point of view if equivalent design parameters are assumed [28]. In the pressurized system, used in the design, the fuel cell operates at an elevated pressure with the micro-turbine (**Figure 2**). Under steady state, the micro-turbine compressor is used to pressurize the air entering the fuel cell. Chemical reactions take place in the fuel cell producing both electricity & heat. With both fuel compressed by a fuel pump & the air by the compressor, the hot pressurized exhaust leaves the SOFC and goes directly to the expander section of the turbine, which drives both the compressor and the generator. The exhaust from the turbine is used through heat exchanger to heat air & fuel prior to entering the fuel cell [34].

2.4. SOFC-MT Hybrid Model

The output of the fuel cell to be directed to the micro-turbine expander is 117 kW (421 MJ/hr) which approximately equals input power required to run a 26 kW micro-turbine under full load. The characteristics of the micro-turbine used in the model are listed in **Table 1**. As a result, the SOFC-MT system will deliver 140 kW from the fuel cell & 26 kW from the micro-turbine. This yield overall electrical power output of 166 kW and 32.7 kW of useful thermal power leaving heat exchanger.

3. Plant Load Study

3.1. Electrical Load

Wastewater treatment plants are often the largest consumer of energy within local city and community government loads [35]. Energy requirements in waste-water treatment are mainly for pumping, primary treatment, secondary treatment, space heating, and sludge heating and disposal [10]. Plant loads according to its design & installation were grouped, and distributed throughout their working hours to form the load profile as in **Figure 3**. The plant has a peak load of 240 kW & minimum load of 146 kW. Diversity factor is relatively low compared to residential loads. The reason is that the aeration system in WWTP, which represents a large percentage of total plant power demand, operates 24 hours per day. Other small loads such as mechanical screen & ventilators operate occasionally; causing the respectively small deviation round the average. Maximum power deviation is 27% with 201 kW average power. Based on this load profile and the results obtained in Section 2 about the CHP output power, the load coverage percentage using the biogas energy is 69% of the WWTP peak load.

3.2. Thermal Load

Space heating is not needed in the system due to the warm Egyptian climate. Anaerobic digestion process requires heat for maintaining a stable operating tempera

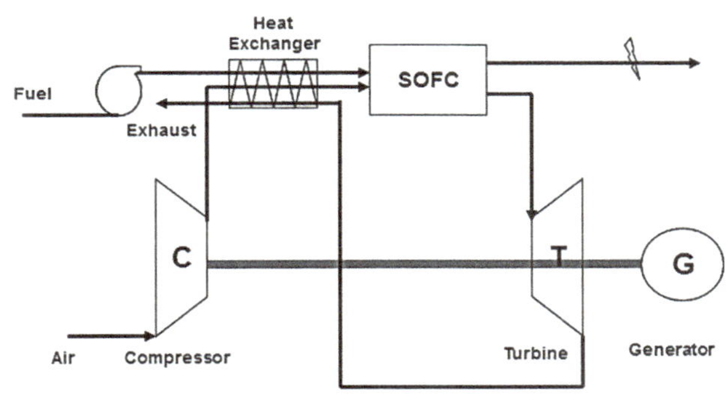

Figure 2. Pressurized hybrid FC-MT.

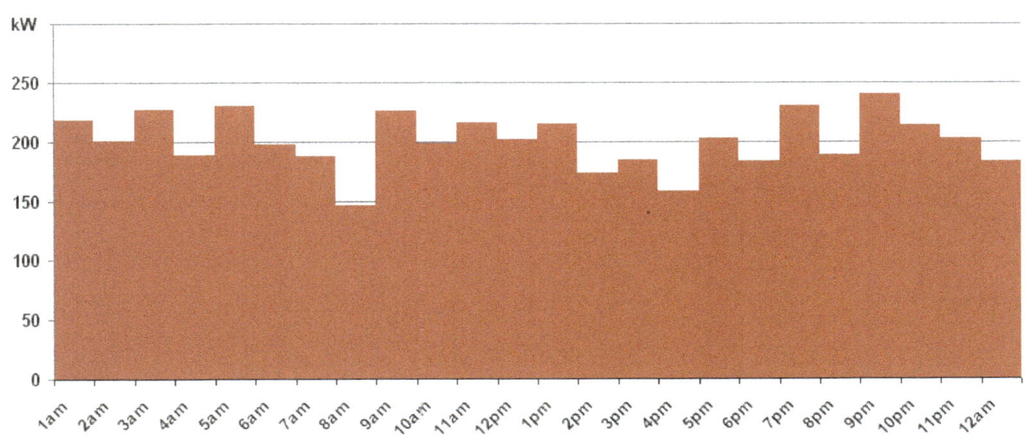

Figure 3. Plant hourly electrical load profile.

ture of the digester. This is important because the bacteria, especially the methane formers, are sensitive to temperature changes [17]. Most anaerobic digesters utilize conventional gas-fired boilers coupled to a heat exchanger in order to transfer the heat of combustion to the digested sludge [36]. The CHP system offers ready thermal power for the digester without the need for gas-fired boilers. Detailed calculations have been made to check if the digester heat load will be met by the thermal power output from the micro-turbine in FC-MT hybrid system. The operating temperature of the digester is assumed to be 35˚C which is optimal for the digestion process [10]. The heat load is sum of two quantities. The first is the heat required to raise the temperature of the input sludge stream to the operating temperature. This heat yields 9.975 kW assuming 30˚C difference between ambient and entering sludge temperatures. The second is the heat required to balance the radiation and other heat losses through the digester walls [17]. This amount of heat yields 8.6 kW. Thus, plant digester heat load is about 18.5 kW and completely covered by useful heat from micro-turbine exhaust. Ref. [17] has detailed the formula used to calculate the digester heat load.

4. System Modeling

There is still a part of the WWTP electrical load demand which needs to be covered. Other RES (Solar and wind energy) are utilized to cover this part as shown in **Figure 4**. The software tool used for the micro-power system optimization HOMER allows modeling energy resources in the site together with the energy conversion systems and hence calculates optimum configuration and size of each component.

4.1. Resources

There are three modeled renewable energy sources in the program. First is biogas from plant waste. Energy potential of the other two (solar & wind) are determined by the solar radiation & wind speed respectively. The annual average radiation of the site per unit area of horizontal surface is 5.52 kWh/m^2/day. Solar energy Data are given in **Figure 5** where clearness index, shown, is the measure of the clearness of the atmosphere. The plant site average wind speed, shown in **Figure 6** is 4.75 m/s.

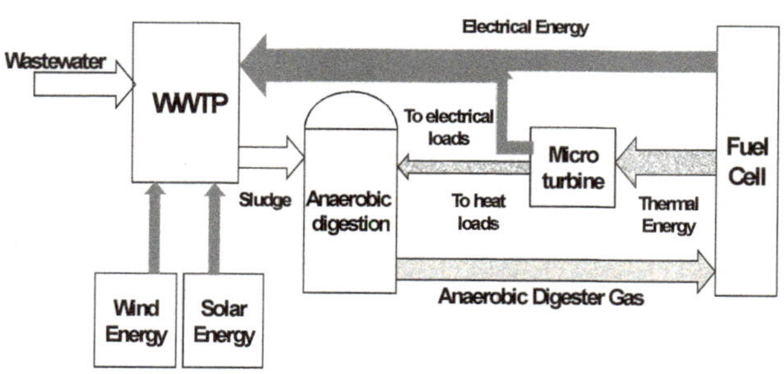

Figure 4. Energy flow diagram for the hybrid system.

4.2. Solid Oxide Fuel Cell Modeling

The capital cost for the 100 - 250 kW SOFC was estimated by EEA (energy & environmental analysis) to be 2850 $/kW as package cost & 3620 $/kW as total installed cost while O&M costs were estimated to be 0.024 $/kWh [22]. Ref. [37] indicated that capital cost of solid oxide fuel cell is about 3000 US $/kW with estimates for the reduction of the costs to 500 US $/kW in the near future, current operation and maintenance (O&M) cost of SOFC is about 0.025 US $/kWh and could be reduced to around 0.01 US $/kWh in future. In studying the fuel cell operation on digester gas [38], key assumptions were made for the fuel cell to have entry level price of 3000 $/kW and O&M cost of 0.015 $/kWh. For the lifetime, the simplicity of the SOFC design and its complete absence of liquids mean that the SOFC fuel cell should have an extremely long operating life. Units have been tested for 60,000 hours without failure and operating lives of 20 years or more can be expected [39]. Replacement cost in all system's components is assumed to be 60% of its relevant capital cost.

4.3. Micro-Turbine Modeling

EPA showed that the micro-turbine as a CHP unit has capital installed cost range of 2400 to 3000 $/kW and O&M cost of 0.012 to 0.025 $/kWh [22]. Package cost without installation was stated to range from 800 to 1650

Figure 5. Solar energy resource data.

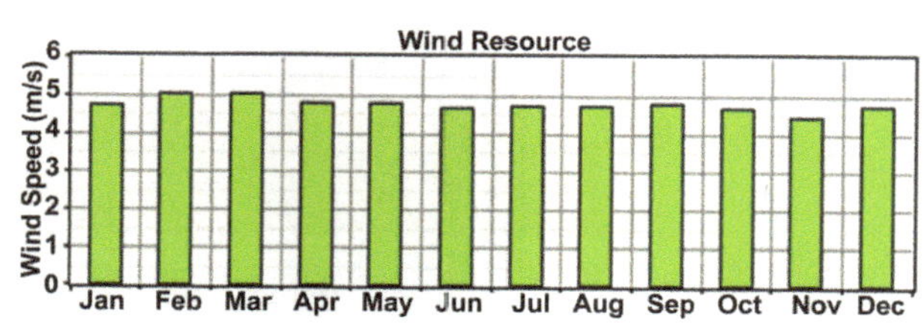

Figure 6. Wind energy resource data.

$/kW & overhaul frequency is from 30,000 to 50,000 hours [21]. Price for commercially available 30 kW micro-turbine in 2007 was 1290 $/kW as package cost & 2970 $/kW for total installed cost [34]. Ref. [23] showed a close price by energy nexus group for the same size (30 kW micro-turbine); that is 2516 $/kW. Price is given by National Renewable Energy Laboratory (NREL) as 2263 $/kW for the total installed cost [22]. FC and MT of hybrid FC-MT system are individually modeled in the software as it doesn't have a single component for FC-MT integration. However, having a constant input power from the FC-MT system makes this modeling valid.

4.4. Wind Turbine System Modeling

Whereas the rated power of new wind machines has increased year by year, the corresponding capital cost per kW dropped [40]. Ref. [39] showed that the current installation costs for an onshore wind farm at between €700/kW and €1000/kW. Smaller wind farm or residential scale turbines cost less overall, but are more expensive per kilowatt of energy producing capacity. Wind turbines of few kilowatts cost roughly $3000 to $5000 per kilowatt of capacity. Medium size wind turbines system as the one modeled would cost typically 2400 $/kW. O&M cost is entered as 30 $/kW [24]. In order to obtain most accurate results, market prices are entered in parallel with technical specs. The modeled wind turbine system is the commercial model FD10 by FSW. The specifications of the system are entered to HOMER along with its market cost.

4.5. Other Components Modeling

Cost of PV system here is based on commercial market in 2012 for medium size systems (100 - 200 kW). Collecting data considers also both technical data of the system together with its price to ensure accurate results.

The rest of components, to complete the hybrid model operation, are converters & battery bank. The inverter price is 650 $/kW in average [40,41] which is typically conformed to commercial market price. For battery system, commercial cost is according to the type of battery selected in the program. Because the output of PV & wind turbine system is variable, the software will first experience different connection topologies to determine which one of them should be connected on AC or DC side. Discussion is made for the optimum system size. The optimum system is defined as the system combination which satisfies the user defined constraints at the lowest life cycle cost or net present cost (NPC).

5. Results and Discussion

The optimization results are listed in **Table 2**. Each scenario provided four categories of optimized systems. Each

Table 2. Optimization results.

Scenario PV-Wind	*Category*	*PV (kW)*	*WTS (no.)*	*FC (kW)*	*MT (kW)*	*Battery Bank*	*Converter (kW)*	*Capital Cost ($)*	*Operating Cost ($/yr)*	*NPC ($)*
DC-AC	1	140	1	140	26	880	80	1,977,730	139,489	3,760,869
	2	230	0	140	26	1280	80	2,166,500	173,240	4,381,090
	3	0	4	140	26	1520	120	2,800,560	169,523	4,967,639
	4	225	2	140	0	1520	100	2,899,830	162,202	4,973,318
DC-DC	1	100	2	140	26	720	80	1,956,350	135,058	3,682,842
	2	230	0	140	26	1280	80	2,166,500	173,240	4,381,090
	3	0	4	140	26	1520	80	2,497,300	166,140	4,621,124
	4	230	2	140	0	1440	100	2,743,810	160,944	4,801,211
AC-AC	1	140	1	140	26	880	100	2,086,850	145,857	3,951,388
	2	230	0	140	26	1360	140	2,401,640	185,045	4,767,132
	3	0	4	140	26	1520	120	2,800,560	169,523	4,967,639
	4	240	2	140	0	1440	150	3,118,240	173,211	5,332,460
AC-DC	1	120	2	140	26	640	80	2,078,150	138,734	3,851,635
	2	0	4	140	26	1520	80	2,497,300	166,140	4,621,124
	3	230	0	140	26	1360	140	2,401,640	185,045	4,767,132
	4	160	3	140	0	1600	100	2,934,005	167,977	5,081,316

category represents unique set of subcomponents combination. The viewed result in every category is the optimum result of this category; which states the size of each component of the system that grants the lowest lifecycle cost. Optimum category in each scenario uses all the components (PV, WTS, FC and MT). The other three categories in each scenario showed possible elimination of MT or PV or WTS at higher cost. **Table 2** showed that the best connection topology is when coupling both PV & WTS systems to the DC bus (**Figure 7**) through the first category (DC-DC1) with NPC of 3,682,842$. PV system is best coupled to DC bus in sites with very poor wind speeds (scenarios DC-AC2 and DC-DC2 where WTS is not used) and same goes for WTS (scenarios DC-DC3 and AC-DC2) in case very poor solar radiation.

5.1. FC-MT Hybrid System Results

Figure 8 shows the cash flow summary. SOFC has the biggest cash flow over project lifetime with 38% share of NPC. However, it has bigger share in energy generation (61% of average yearly production). Micro-turbine generation share is 9% compared to 11% by PV system. Meanwhile, NPC of MT is only 44% of PV system NPC (**Figure 9**). This shows the economic effectiveness of coupling MT to FC and also shows the advantage of utilizing energy recovered from plant waste over the utilization of other supplementary RES.

5.2. WTS & PV System

Results for photovoltaic & wind turbine systems are summarized in **Table 3**. Although PV system everywhere has low capacity factor as it shuts down in the night, poor wind speed in the site caused the WTS to have approximately same capacity factor. In addition, the weak wind resource caused the effective NPC/kW of turbine very close to the NPC/kW for the PV system (17,764 $/kW for WTS and 16,040 $/kW for PV).

5.3. Battery Storage System

Battery storage system comes in the 2nd order as the highest NPC after the fuel cell. This is because the number of batteries installed is relatively high. This high number was required to match the variable nature of solar & wind energy. The reason for installing this high quantity of batteries (90 strings, 48 volts) as shown in **Figure 10** that the wind speed sometimes stays below the cut-in speed of the turbine resulting zero power output of the WTS system. If this occurrence takes place in the night while PV system is shut down, it falls only to the battery system to cover the electrical demand with the SOFC-MT hybrid system. Battery system can be further minimized if turbine cut-in speed is lower.

6. Conclusion

The paper presented the economic evaluation and unit sizing for a standalone micro-power system. The system serves a wastewater treatment plant using energy sources which are entirely renewable. The paper uses the data of Toukh WWTP as a case study. First, small scale CHP technologies are investigated for power production from biogas produced by anaerobic digestion process. FC has been selected as it grants the lowest emissions among other CHP technologies. The system generated power-to-heat ratio is maximized by coupling MT with FC exhaust. The FC-MT system grants 166 kW and 32.7 kW of the electrical and thermal power respectively. The Software-simulation and optimization results used all CHP capac-

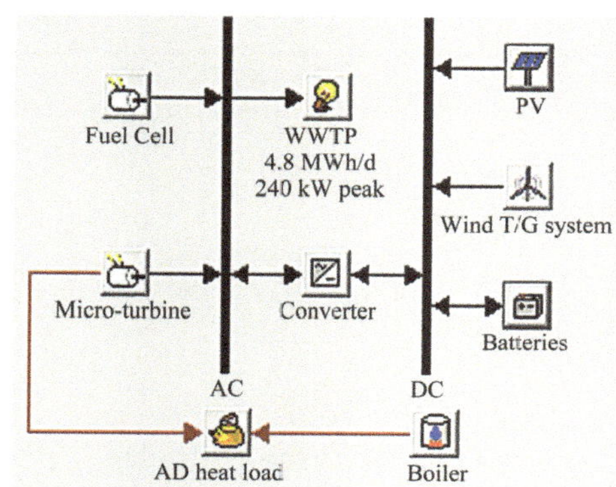

Figure 7. Optimum system configuration.

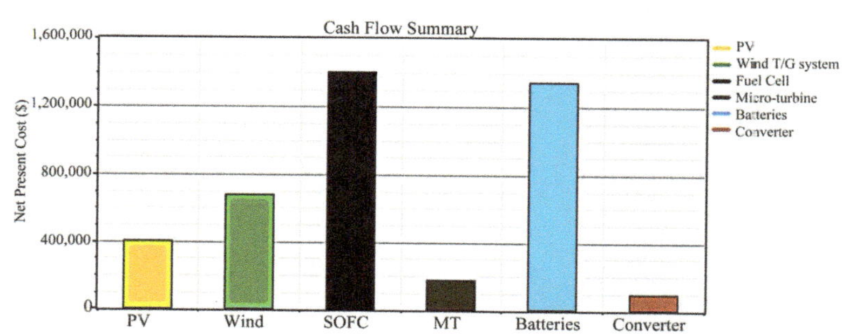

Figure 8. Cash flow summary.

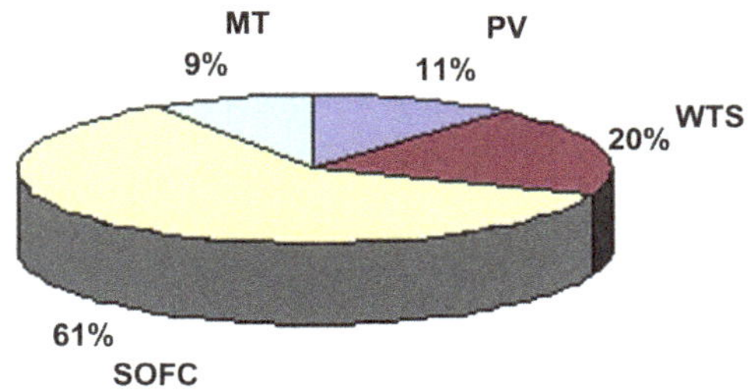

Figure 9. Average generation share.

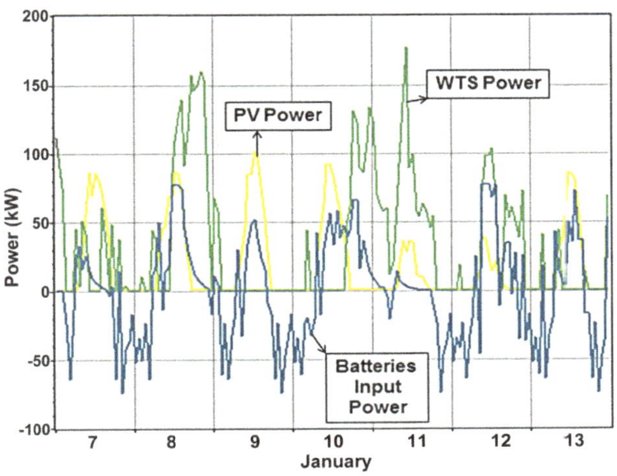

Figure 10. Relation between batteries & intermittent RES.

Table 3. PV and WTS output.

Data	*PV*	*WTS*	*Units*
Rated capacity	100	200	kW
Minimum output	0	0	kW
Maximum output	113	199	kW
Mean output	23	42	kW
Capacity factor	22.6	21	%
Total production	198,204	369,289	kWh/yr
Hours of operation	4387	4559	h/yr

ity in addition to 200 kW wind turbine system, 100 kW PV system, 720 batteries & 80 kW converter to achieve self-sustainability of power. Results showed that the best connection topology is connecting both PV & wind turbine systems to the DC bus. The use of biogas and combined-heat-power technologies proves to be more economical than solar & wind energy equipment. Battery storage system formed expensive part of the standalone system as, wind power may go lower at night while PV system is shut down; making the sustainability in these time steps totally dependent on only CHP & battery systems.

REFERENCES

[1] S. Rehman, I. M. El-Amin, F. Ahmad, S. M. Shaahid, A. M. Al-Shehri, J. M. Bakhashwain, *et al.*, "Feasibility study of Hybrid Retrofits to an Isolated Off-Grid Diesel Power Plant," *Renewable and Sustainable Energy Reviews*, Vol. 11, No. 4, 2007, pp. 635-653.

[2] M. Ball, M. Wietschel and O. Rentz, "Integration of a Hydrogen Economy into the German Energy System: An Optimizing Modeling Approach," *International Journal of Hydrogen Energy*, Vol. 32, No. 10-11, 2007, pp. 1355-1368.

[3] S. M. Shaahid and M. A. Elhadidy, "Technical and Economic Assessment of Grid-Independent Hybrid Photovoltaic-Diesel-Battery Power Systems for Commercial Loads in Desert Environments," *Renewable and Sustainable Energy Reviews*, Vol. 11, No. 8, 2007, pp. 1794-1810.

[4] R. U. Ayres, H. Turton and T. Casten, "Energy Efficiency, Sustainability and Economic Growth," *Energy*, Vol. 32, No. 5, 2007, pp. 634-648.

[5] E. R. Sanseverino, M. L. Di Silvestre, M. G. Ippolito, A. De Paola and G. L. Re, "An Execution, Monitoring and Replanning Approach for Optimal Energy Management in Micro Grids," *Energy*, Vol. 36, No. 5, 2011, pp. 3429-3436.

[6] O. Erdinc and M. Uzunoglu, "Optimum Design of Hybrid Renewable Energy Systems: Overview of Different Approaches," *Renewable and Sustainable Energy Reviews*, Vol. 16, No. 3, 2012, pp. 1412-1425.

[7] P. Yilmaz, M. H. Hocaoglu and A. E. S. Konukman, "A Pre-Feasibility Case Study on Integrated Resource Planning Including Renewables," *Energy Policy*, Vol. 36, No. 3, 2008, pp. 1223-1232.

[8] A. Hepbasli, "A Key Review on Exergetic Analysis and Assessment of Renewable Energy Resources for a Sustainable Future," *Renewable and Sustainable Energy Reviews*, Vol. 12, No. 3, 2008, pp. 593-661.

[9] A. Kornelakis, "Multiobjective Particle Swarm Optimization for the Optimal Design of Photovoltaic Grid-Connected Systems," *Solar Energy*, Vol. 84, No. 12, 2010, pp. 2022-2033.

[10] S. A. Tassou, "Energy Conservation and Resource Utilization in Waste-Water Treatment Plants," *Applied Energy*, Vol. 30, 1988, pp. 113-129.

[11] Eastern Research Group, Inc. (ERG) and Resource Dynamics Corporation, "Opportunities for Combined Heat and Power at Wastewater Treatment Facilities: Market Analysis and Lessons from the Field," US Environmental Protection Agency (EPA) Combined Heat and Power Partnership (CHPP), 2011. http://www.epa.gov/CHP/documents/wwtf_opportunities.pdf

[12] T. A. Elmitwall, A. Al-Sarawey, M. F. El-Sherbiny, G. Zeeman and G. Lettinga, "Anaerobic Biodegradability and Treatment of Egyptian Domestic Sewage," *5th IWA Conference in Small Water and Wastewater Treatment Systems*, Istanbul, 24-26 September 2002.

[13] M. L. Davis, "Water and Wastewater Engineering-Design Principles & Practice," The McGraw-Hill Companies, Inc., New York, 2010.

[14] Water Environment Federation, "Design of Municipal Wastewater Treatment Plants," 5th Edition, McGraw-Hill Professional, New York, 2010.

[15] A. Lise, J. Baeyans, J. Degreve and R. Dewil, "Principles and Potential of the Anaerobic Digestion of Waste-Activated Sludge," *Progress in Energy and Combustion Science*, Vol. 34, No. 6, 2008, pp. 755-781.

[16] M. Ghazy, T. Dockhorn and N. Dichtl, "Sewage Sludge Management in Egypt: Current Status and Perspectives towards a Sustainable Agricultural Use," *World Academy of Science, Engineering and Technology*, Vol. 57, 2009, p. 492.

[17] T. George, F. L. Burton and H. D. Stensel, "Wastewater Engineering. Treatment & Reuse," 4th Edition, Metcalf & Eddy, Inc., New York, 2003.

[18] G. R. Simader, R. Krawinkler and G. Trnka, "Micro CHP systems: State-of-the-Art," Österreichische Energieagentur, Austrian Energy Agency, 2006.

[19] Energy and Environmental Analysis, "Technology Characterization: Fuel Cell," Environmental Protection Agency, Combined Heat and Power Partnership, 2008. http://www.epa.gov/chp/documents/catalog_chptech_fuel_cells.pdf

[20] C. Soares, "Microturbines-Applications for Distributed Energy Systems," Elsevier Inc., Amsterdam, 2007.

[21] J. R. Wiser, J. W. Schettler and J. L. Willis, "Evaluation of Combined Heat and Power Technologies for Wastewater Treatment Facilities," US Environmental Protection Agency, 2010. http://www.brownandCaldwell.com.

[22] Energy and Environmental Analysis, "CHP Technologies," Environmental Protection Agency, 2008. http://www.epa.gov/chp/documents/catalog_chptech_full.pdf

[23] National Renewable Energy Laboratory, "Gas-Fired Distributed Energy Resource Technology Characterizations," Report NREL/TP-620-34783, 2003.

[24] A. Bauen, D. Hart and A. Chase, "Fuel Cells for Distributed Generation in Developing Countries—An Analysis," *International Journal of Hydrogen Energy*, Vol. 28, No. 7, 2003, pp. 695-701.

[25] S. Mekhilef, R. Saidur and A. Safari, "Comparative Study of Different Fuel Cell Technologies," *Renewable and Sustainable Energy Reviews*, Vol. 16, No. 1, 2012, pp. 981-989.

[26] K. Alanne and A. Saari, "Sustainable Small-Scale CHP Technologies for Buildings: The Basis for Multi-Perspective Decision-Making," *Renewable and Sustainable Energy Reviews*, Vol. 8, No. 5, 2004, pp. 401-431.

[27] P. Viebahn, R. Gerboni, M. Pehnt and E. Lavagno, "Final Report on Technical Data, Costs and Life Cycle Inventories of FCs," New Energy Externalities Developments for Sustainabiliy (NEED), 2008. http://www.needs-project.org/RS1a/RS1a%20D9.2%20Final%20report%20on%20fuel%20cells.pdf

[28] J. Van Herle, Y. Membrez and O. Bucheli, "Biogas as a Fuel Source for Solid Oxide Fuel Cell Cogenerators," *Journal of Power Sources*, Vol. 127, No. 1-2, 2000, pp. 300-312.

[29] F. Calise, "Design of a Hybrid Polygeneration System with Solar Collectors and a Solid Oxide Fuel Cell: Dynamic Simulation and Economic Assessment," *International Journal of Hydrogen Energy*, Vol. 36, No. 10, 2011, pp. 6128-6150.

[30] A. Liu and Y. Weng, "Performance Analysis of a Pressurized Molten Carbonate Fuel Cell/Micro-Gas Turbine Hybrid System," *Journal of Power* Sources, Vol. 195, No. 1, 2010, pp. 204-213.

[31] M. R. Ghazy, T. Dockhorn and N. Dichtl, "Economic and Environmental Assessment of Sewage Sludge Treatment Processes Application in Egypt," *International Water Technology Journal*, Vol. 1, No. 2, 2011.

[32] F. Jurado, "Study of Molten Carbonate Fuel Cell—Microturbinehybrid Power Cycles," *Journal of Power Sources*, Vol. 111, No. 1, 2002, pp. 121-129.

[33] D. Bohn, "Micro Gas Turbine and Fuel Cell—A Hybrid Energy Conversion System with High Potential," In: *Micro Gas Turbines*, 2005, pp. 13-1-13-46. http://www.rto.nato.int/abstracts.asp

[34] R. A. George and N. F. Bessette, "Reducing the Manufacturing Cost of Tubular SOFC Technology," *Journal of Power Sources*, Vol. 71, No. 1-2, 1998, pp. 131-137.

[35] "Energy Conservation in Wastewater Treatment Facilities Manual of Practice," Water Environment Federation, Alexandria, 1997, pp. 1-142.

[36] J. Nouri, *et al.*, "Energy Recovery from Wastewater Treatment Plant," *Pakistan Journal of Biological Sciences*, Vol. 9, No. 1, 2006, pp. 3-6.

[37] M. Ameri and M. J. Heidari, "The Application of Solid Oxide Fuel Cell for Power Generation in Iran: Feasibility and Cost Analysis," *The 2nd Joint International Conference on Sustainable Energy and Environment* (*SEE* 2006), Bangkok, 21-23 November 2006.

[38] R. J. Spiegel, *et al*, "Fuel Cell Operation on Anaerobic Digester Gas: Conceptual Design and Assessment," *Waste Management*, Vol. 19, No. 6, 1999, pp. 389-399.

[39] P. Breeze, "Power Generation Technologies," Newnes, New South Wales, 2005.

[40] G. M. Masters, "Renewable and Efficient Electric Power Systems," John Wiley & Sons, Inc., New York, 2004.

[41] Al-Badi, *et al*, "Economic Perspective of PV Electricity in Oman," *Journal of Energy*, Vol. 36, No. 1, 2011, pp. 226-232.

Role of Energy Storage on Distribution Transformer Loading in Low Voltage Distribution Network

Mohammad Taufiqul Arif, Amanullah M. T. Oo, A. B. M. Shawkat Ali

School of Engineering and Technology, Higher Education Division, Central Queensland University, Rockhampton, Australia.

ABSTRACT

Energy storage (ES) is a form of media that store one form of energy to be utilized at another time. Importance of ES is comprehended while intermittent nature of renewable energy (RE) generation increases and integration into the grid becomes viable in terms of economics and environment. However, technical analysis should be carried out before large scale integration into the grid. Some utilities experienced in Europe and expressed concern about issues in integrating large scale renewable energy in the areas of harmonics, voltage regulation, network protection and islanding. In Australia, distribution network (DN) is not robust compared to the European grid; moreover loads are largely distributed over large geographical areas. Installation of RE such as roof top solar photovoltaic (PV) is increasing in Australia which also boosted by the governments incentives to the individual owners. It is therefore obvious that large scale PV integration into the Australian grid is imminent. The intermittent characteristic of solar PV is expected to have greater impacts on DN in Australia compared to the DN in Europe. Therefore this paper investigated the impacts of solar PV on low voltage (LV) DN where loads connected through distribution transformer (DT) and finally further investigation was conducted with the deployment of ES into the respective load centers. It was found that storage reduced the overall peak load condition on the DT, and also reduced the energy fluctuation in the DN. It was also found that storage improved the voltage regulation on the LV side of DT and stabilized node voltage.

Keywords: Energy Storage; Distribution Network; Distribution Transformer; Impacts; Solar PV

1. Introduction

In large network, voltage drop and voltage capacity will play major role in determining various future network augmentation. Voltage regulation is an important issue of power system control as it needs to maintain steady acceptable voltage under normal operating and distribution conditions [1]. Electrical loads are connected to the DN through DT. In normal operating condition in the DN, the receiving end voltage is less than the sending end voltage due to voltage drop in cables or conductors, bus impedance and transformers throughout the system. **Figure 1** shows different voltage profiles under different load conditions considering fixed sending end voltage [2]. Under no load condition, the voltage at the receiving end is greater than the sending end as shown in **Figure 1**, this type of scenario is known as Ferranti effect. The introduction of distributed generator (DG) into the network has given great uncertainty into the network. Large scale introduction of DG such as solar PV adds uncertainty due to environmental factors such as sunshine with duration of solar radiation, temperature, cloud movement etc. This uncertainty can cause voltage rise at the receiving end while DG generates more than the required power, and also receiving end voltage may drop if DG is unable to

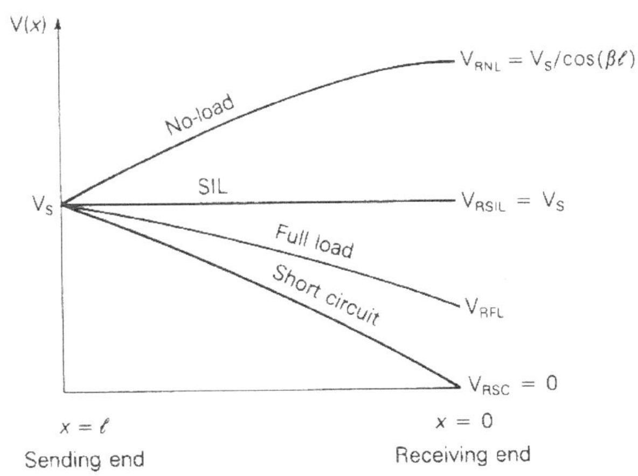

Figure 1. Voltage profile from source to end load [2].

support load especially in the evening when load demand is high. Intermittent nature of PV output could generate fluctuations and large scale such fluctuation could affect the loads connected into the network also uneven generation could exceed the capacity of connected DT. Moreover connection of DG into the network could unload one phase and overload other phase and could lead in phase unbalance condition. These unwanted characteristics can be avoided by buffering excess energy from DG and by supporting load from buffered energy when required.

Energy storage as buffer is essential with renewable energy (RE) such as solar photovoltaic (PV) for stable operation of grid by load management and by keeping voltage, frequency within allowable limit. However performance of PV module depends on load resistance, solar irradiance, cell temperature, cell shading and PV crystalline structure. Electricity generated from DG introduces bi-directional power flow at the point of connection in the DN that influences the voltage regulation, balancing between phases and loading on distribution transformer (DT). Storage can delay the bi-directional power flow by supporting load. By supporting the load demand, storage reduces load on grid and eventually increases the node voltage. Therefore investigation is required to identify the storage impacts on DT and on DN while storage and solar PV is integrated into the network.

2. Background

In LV network, loads particularly residential loads are connected in single phase line of DN through DT. Each DT has specific loading or utilization capacity at a particular time. To ensure quality in power supply to the end user, DT always operated within its safe operating limit which is set by the local distribution network service provider (DNSP). Disturbance in one DT could affect all the loads connected from that DT. Awareness about the effects of greenhouse gas (GHG) emission introduces the interest of generating energy from RE sources. Most of the developed countries set the goal to integrate electricity from RE sources and Australia set the target to introduce 20% electricity from RE by 2020 [3], moreover Australia committed national GHG emission target of 60% below 2000 level by 2050 [4]. To achieve this target roof top solar PV received the maximum boost at the LV level integration. However the output from PV clearly depends on the solar radiation, not on the load demand. Different natural factors introduce fluctuations in the solar radiation which directly adds into the connected network as inverter directly converts the available energy from DC (Direct current) to AC (Alternating current). Moreover solar radiation is not available 24 hours a day, but load demand continues for 24 hours a day. Therefore large number of such PV installation and integration into the DN could lead to some problem which was unseen before.

Austria, France, Germany, Spain, Netherlands and UK are the six leading countries representing 98% of installed PV power in European Union and expressed concern about voltage regulation, harmonics, network protection, islanding, electromagnetic compatibility (EMC) of multiple inverters, unbalanced condition due to large integration of PV into the grid [5]. Energy storage (ES) system can overcome these limitations imposed by the solar PV. ES acts as a buffer that isolates the grid from the frequent and rapid power fluctuations by high penetration of renewable resources [6]. Storage improves the grid penetration of PV energy [7]. However introduction of storage with PV into any particular phase could begin phase unbalance into the network.

Therefore investigation is necessary before large scale storage and PV integration into the distribution network for any potential impact. The following subsections describe the DN structure in Australia, distributed generation and its integration and standard guideline to integrate into the grid.

2.1. Distribution Network

Many elements constitute the DN to supply electricity to the end consumers. The high voltage (HV) transmission network transfers bulk electricity at voltages above 110 kV to the bulk supply substations (BSS). In Queensland, Australia, BSS supply power to the zone substations (ZS) and large customers via 33 kV, 66 kV, 110 kV and 132 kV networks operated as radial, parallel radial or meshed network [8,9] as shown in **Figure 2**. ZS supplies to the DN by stepping down the voltage to 11 kV, 22 kV or 33 kV. ZS provides protection and control to the local DN. In Australia, DN is a combination of, HV, SWER (Single Wire Earth Return) and LV networks. The distribution network service providers (DNSP) have to meet certain regulated voltages that supplied to the customer via DT. The standard voltages associated with DN in Queensland are:

- HV DN—11 kV, 22 kV and sometimes 33 kV;
- LV DN—415 V (Line-Line)/240 V (Line-Ground);
- SWER—11 kV, 12.7 kV and 19.1 kV.

The popular roof top solar PV mostly installed to meet residential customer needs connected at the LV DN. Although the size of this single customer is very small, however the combined total demand is significant. Similarly combined total solar PV installed in these houses have great influence on the total network performance. The combined maximum capacity of all PV installations connected without a dedicated transformer must remain <30% of the rated capacity of the DT [10]. Akatsuka Motoki *et al.* [11] mentioned fluctuation in output of megawatt class PV system may disturb stable operation of power system.

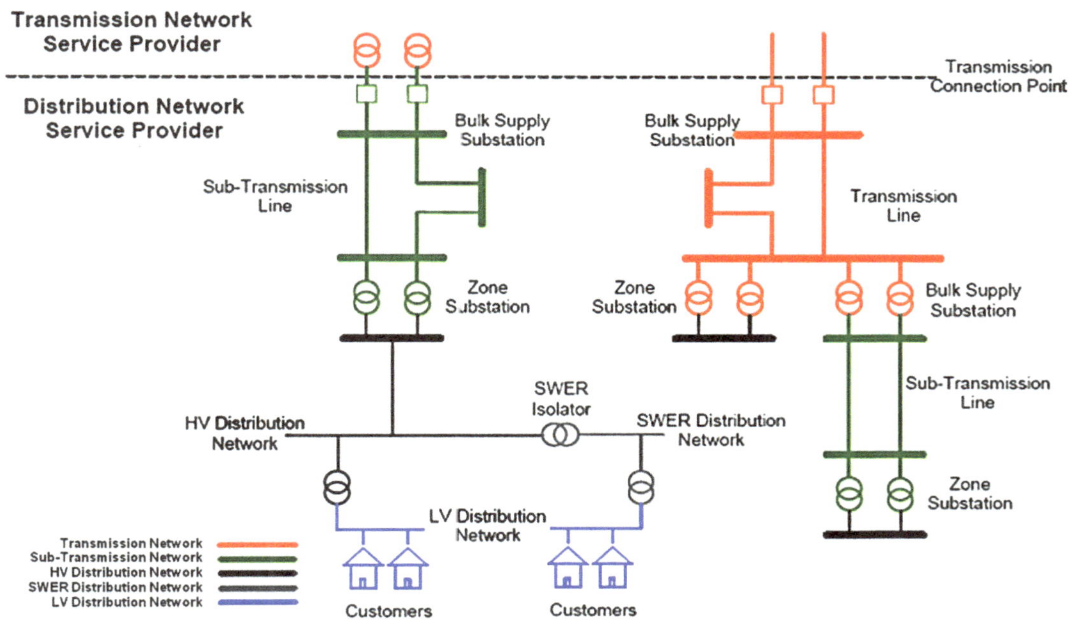

Figure 2. Power distribution network [8].

2.2. Distributed Generation

It is the production of electricity at the customer premises point-of use in the DN. This is an alternate to the conventional method. Renewable energy, co-generation plant and standby generators are considered as DG [12]. RE is defined as an inexhaustible and sustainable energy source, moreover RE reduces the CO_2 emission [13]. Two primary RE technologies integrated into the distribution level are rooftop solar PV and small scale wind turbine. However to manage generated electricity from these sources requires energy storage to store electricity when generates excess energy and to supply at a later time. Therefore storage also considered as a DG. For typical household, storage integrated grid connected PV and wind system presents a sustainable and economically viable solution [14-16].

2.3. Integration of Distributed Generators

Electricity generated by the DG is exported to the grid via grid connected inverters. Grid-Tie Inverter (GTI) converts Direct Current (DC) electricity from DGs into Alternating Current (AC) and feed into the LV DN at point of common coupling (PCC). The LV DN is connected to the HV DN through DT and the DT has defined load capacity which set by the DNSP based on the load demand. **Figure 2** shows the Network configuration of Ergon Energy where end loads are connected through DT. Maximum DT load capacity recommended by Ergon Energy is 80% however DT can sustain up to 120% of load demand for short duration. Ergon Energy in Queensland serves approximately 680,095 customers of different capacity through 90,500 DTs.

LV DN connects different types of load especially single phase residential load. Residential load profile is different than the total load profile of DN. In Australia, Queensland's electricity demand will continue to grow more than 3.5% per year [17] and average each household daily peak demand is 1.72 kW [18]. To meet the need of demand growth there should be enough supply from generator side which will eventually increase loading on DT. DNSP in Queensland [19] allows 1.3 kW of PV in each installation in rural areas for DT up to 50 kVA and in urban area 4 kW of PV in each installation for DT up to 100 kVA [19]. Therefore it is required to investigate the loading impacts on DT by adding ES with RE which is a great concern of power quality.

2.4. Power Quality and Regulatory Standards

There is no concrete standard developed yet for integration of bulk and large storage into the grid. However IEEE 1547-2003 [20] provides guideline to connect distributed resources (DR) such as solar PV, wind and energy storage to the power grid at the distribution level. AS-4777 [21] provides guideline to connect RE and storage to the distribution network (DN) via inverters up to 10 kVA for single phase and 30 kVA for three phase unit. As mentioned earlier that DT should be operated within its allowable capacity to ensure required power quality. On the other hand to upgrade DT capacity involves additional cost for DNSP. DT loaded maximally in peak demand time. As mentioned earlier residential load peak demand time is between 16:00 PM to 20:00 PM when solar PV is unable to support the load therefore overall

DT loading increases during peak demand. In Queensland Australia, approximately 10% electricity network was built to support only the extreme peak loads [17], similarly utility operators maintain costly short time generators to support peak load. By integrating proper sized ES with RE, this peak load demand can be minimized and eventually helps to reduces the cost of energy. Therefore storage can be the key enabler to support future load demand especially in peak demand period and to extend the use of RE.

When load demand is low, for example during midday residential load is very low and solar PV produces maximum energy which increases voltage at the PCC and increases load on DT. Energex and Ergon Energy indicated allowable voltage regulation as 240 V ± 6% for phase to neutral and 415 V ± 6% for phase to phase voltage [22]. Ergon Energy photovoltaic planning criteria allows voltage rise due to PV energy at PCC is 1% at 240 V line to neutral voltage [19].

If the installation of PV is not distributed among all phases in a planned way, the load demand and PV generation could lead to voltage unbalance condition among phases. Moreover it could increase the phase voltage beyond the acceptable limit. In a balanced sinusoidal system, the 3 phase-to-neutral voltages are equal in magnitude and phase difference from each other is 120 degree. Any difference in voltage magnitude and/or phase shift results in system unbalanced. According to IEEE-1195 [23], acceptable voltage imbalance in magnitude is 0.5% to 2.0% during steady state operation.

Therefore this study investigates the impacts on DT capacity, voltage rise/drop and unbalances condition due to rooftop PV and finally investigates the same impacts while storage integrated along with the PV system. In order to conduct the investigation a symmetrical scenario was developed in a software model.

3. Model Evaluation

Model was developed considering three groups of residential houses connected to the main (Grid) supply through the DT. Residential houses are connected to the single phase line of DN. **Figure 3** shows a road side view of residential electricity connection from the DN in Rockhampton in Queensland, Australia. As mentioned earlier each residential house peak load demand is 1.72 kW or 1.91 kVA considering the power factor as $\cos\varphi = 0.9$. This model considered 5 such houses are connected in each phase in each node, also considered same load for each house. Therefore 15 houses are connected in each node and total of 45 houses are connected in three nodes from the DT with a total load of 85.95 kVA. It was also considered that 5 such houses installed roof top solar PV in each node and connected to phase-1 in Node-1, phase-2 in Node-2 and phase-3 in Node-3 as shown in **Figure 4**.

All three nodes are 500 m apart from each other. Model was developed in PSS SINCAL indicating PV connected to the DN through inverter. It was considered that load was uniformly distributed among three phases in each node before installing roof top solar PV. Storage was integrated with the PV integrated loads to improve performance and maximize the use of RE. Other considerations for the model are explained below:

3.1. Load Allocation

The average Australian household electricity use is about 16 kWh/day [24]. Daily household load profile is based on the working nature of the residents and the average

Figure 3. Roadside view of residential load connection from DN.

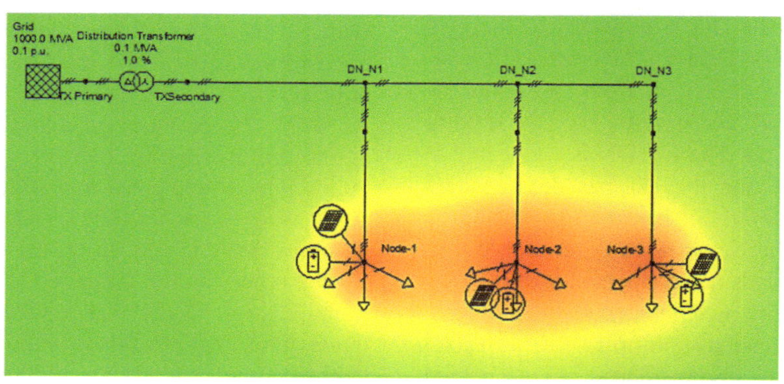

Figure 4. Model shows (load-flow) residential load connection from DN to nodes.

load pattern as shown in **Figure 5**. In Queensland, Australia, peak demand generally occurs between 4:00 PM to 8:00 PM, when most householders return home and turn on energy intensive appliances [17].

Model considered urban area load with DT (11 kV/415 V) capacity of 100 kVA. LV residential load was considered for unbalanced load flow and time varying load curve analysis. Without PV and storage, DT was 87.22% loaded in peak demand time considering total load and line impedance. Load allocation as shown in **Table 1**.

3.2. Solar PV and Storage Allocation

Average daily residential load, solar radiation in Rockhampton and allowable PV limit set by local DNSP was considered in selecting the required PV capacity. Ergon Energy, local DNSP in Rockhampton allows 4 kW capacity of PV for each urban area house. Therefore allowed PV capacity is 210% of the peak load demand of 1.72 kW. As the output from PV is not available for 24 hours period and the residential load demand is lowest when PV generates highest energy and the excess energy supplies to the grid eventually increases the voltage at the connected node. In Australia yearly average sunlight hours varies from 5 to 10 hours/day and maximum area is over 8 hours/day [25]. It was found that for the solar radiation and residential load in Rockhampton the required daily PV output should be around 2.05 times the daily load demand considering PV module efficiency of 12.5% and required storage is the same as daily total load demand considering DC system voltage as 24 V [13]. In model, inverter efficiency was considered as 97% and loss until inverter was considered 5%. For this investigation, it was considered that five houses in Node-1 installed 5 kW PV/house in phase-1, similarly 5 houses in Node-2 and 5 houses in Node-3. It was considered that storage was installed with the same peak capacity of load which is 1.72 kW in each house where solar PV was installed. **Tables 2**

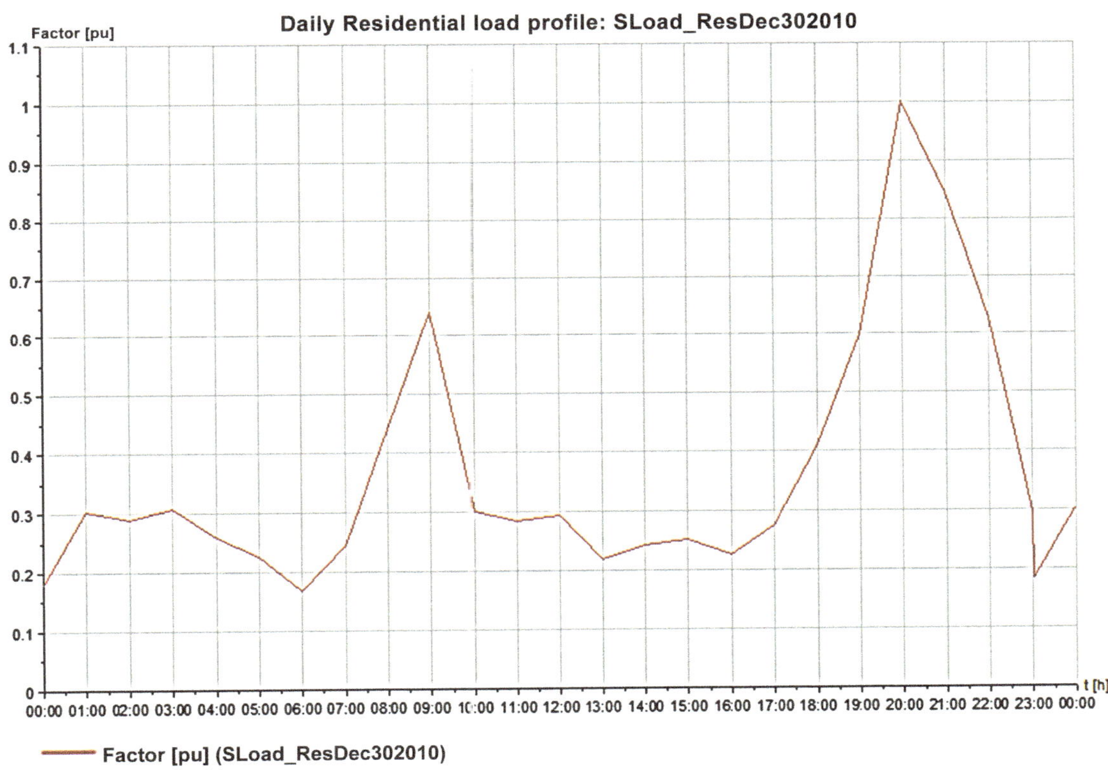

Figure 5. Daily residential summer load profile in Rockhampton [13.14].

Table 1. Load allocation in three nodes.

Node	Phase 1		Phase 2		Phase 3	
Node 1, 2, 3	*kVA*	*cosφ*	*kVA*	*cosφ*	*kVA*	*cosφ*
	9.55	0.9	9.55	0.9	9.55	0.9

Table 2. Installed solar PV in different nodes.

Node	Phase 1		Phase 2		Phase 3	
	kW	*No of house*	*kW*	*No of house*	*kW*	*No of house*
Node-1	5	5	-	-	-	-
Node-2	-	-	5	5	-	-
Node-3	-	-	-	-	5	5

and **3** shows the installed PV and storage in different nodes.

Based on allocated load, the daily utilization of DT is shown in the **Figure 6** and the peak utilization is 87.22% of DT capacity at the time of 20:00 PM.

Solar PV was added to phase-1 in Node-1 and summer time solar radiation in Rockhampton was considered. Daily solar radiation data of Rockhampton was collected from [25] and daily profile is as shown in **Figure 7**. However due to cloud movement energy level received by the PV array fluctuates. **Figure 8** shows short fluctuation due to continuously varying short interval cloud movement, **Figure 9** shows dip fluctuation for short period and **Figure 10** shows the dip fluctuation for longer

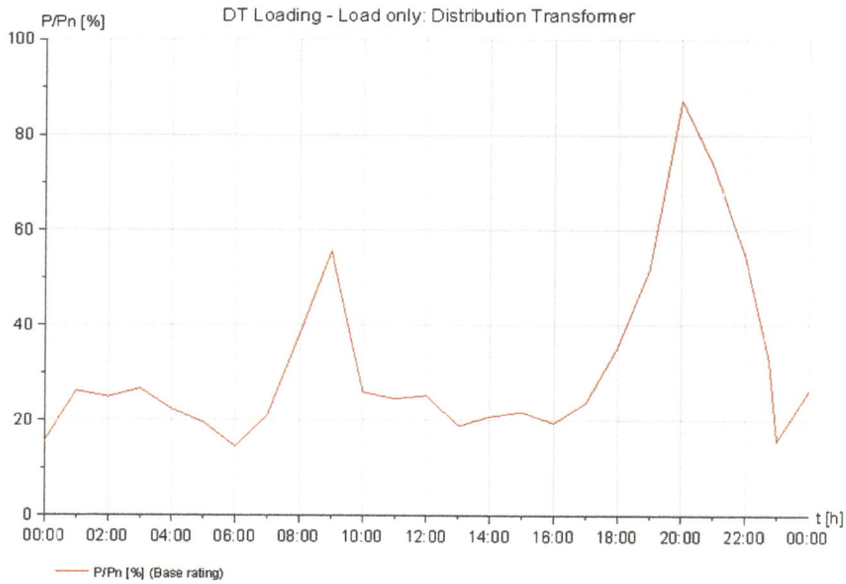

Figure 6. Distribution transformer loading due to load only.

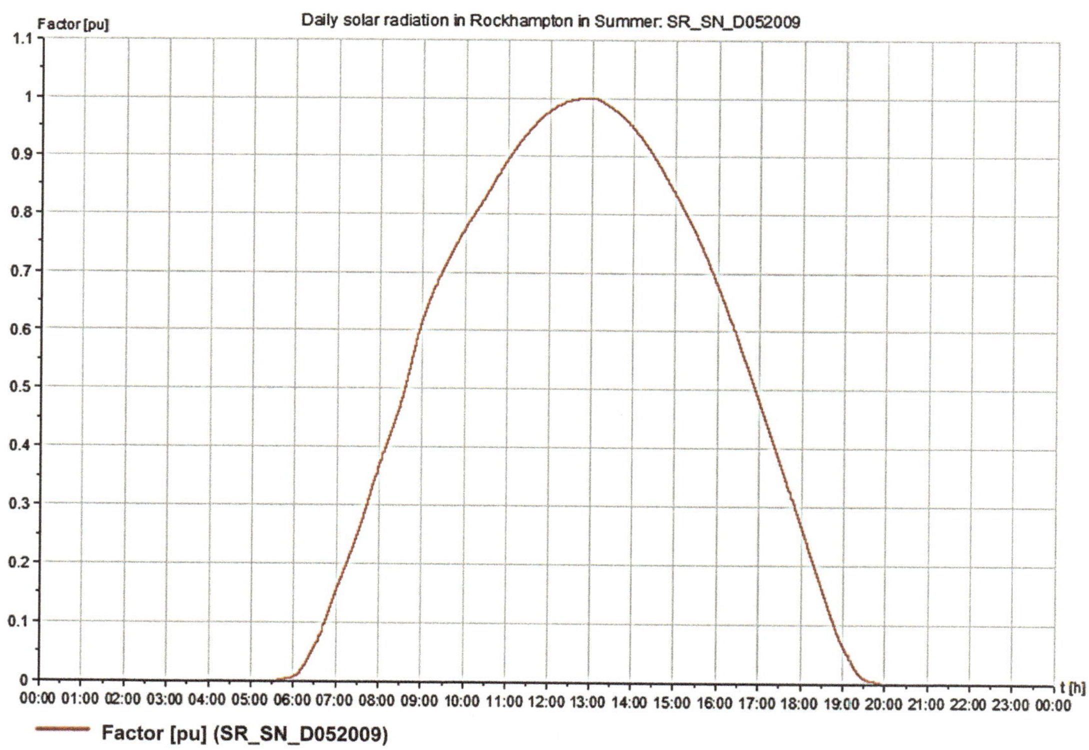

Figure 7. Daily solar radiation profile in Rockhampton, Australia.

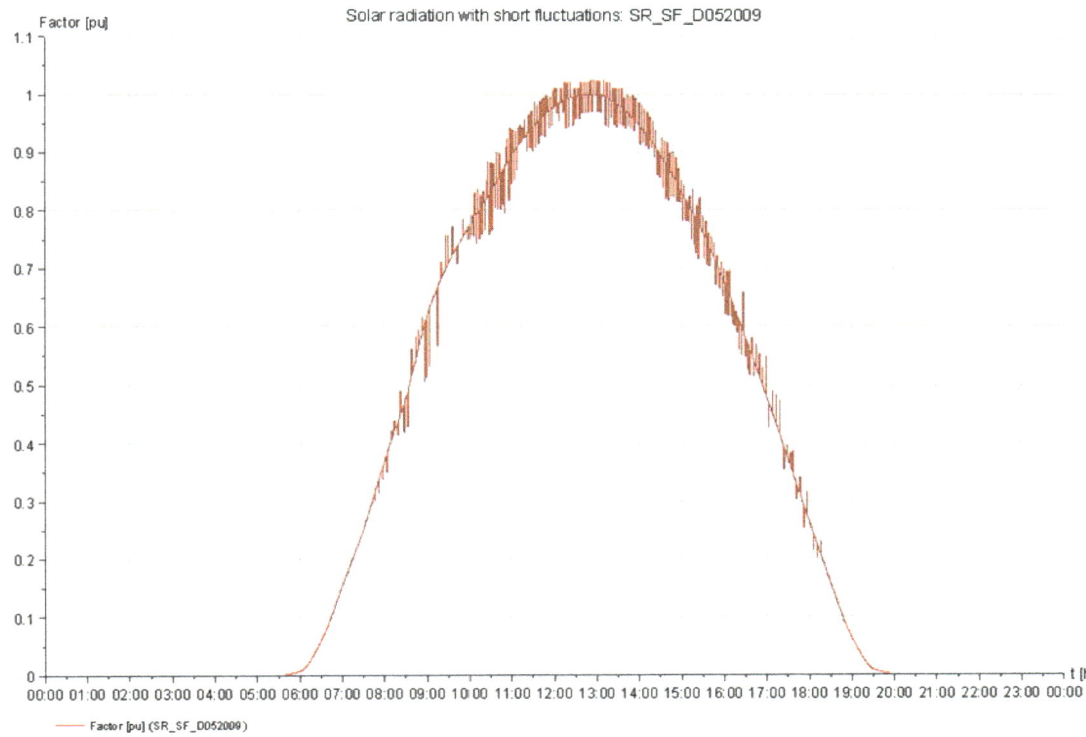

Figure 8. Solar radiation with continuous short fluctuation.

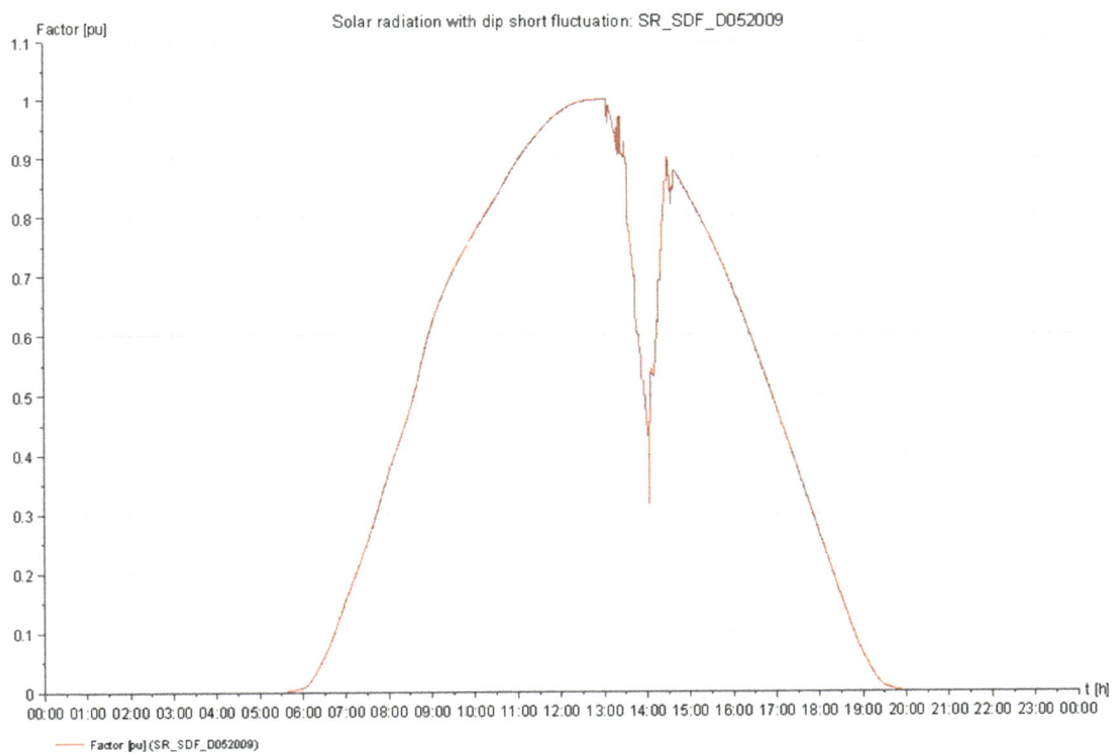

Figure 9. Solar radiation with short deep fluctuation.

period. These fluctuating solar radiations was considered in the model to investigate the response from the storage.

Model was simulated for load flow (LF) and load curve (LC) analysis. LF is an effective tool for calculat-

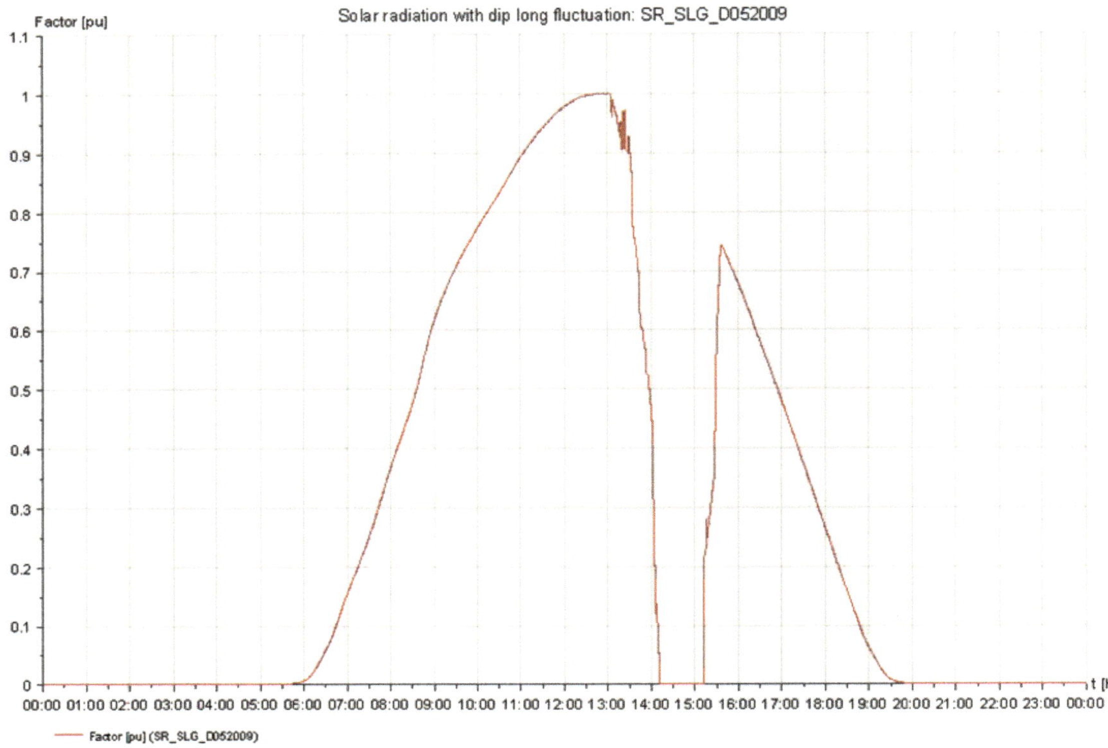

Figure 10. Solar radiation with long deep fluctuation.

Table 3. Installed storage in different nodes.

Node	Phase 1		Phase 2		Phase 3	
	kW	*No of house*	*kW*	*No of house*	*kW*	*No of house*
Node-1	1.72	5	-	-	-	-
Node-2	-	-	1.72	5	-	-
Node-3	-	-	-	-	1.72	5

Table 4. Model simulation configurations.

Configuration	Node-1	Node-2	Node-3
Case-0	Load only	Load only	Load only
Case-I	Load + PV	Load only	Load only
Case-II	Load + PV	Load + PV	Load only
Case-III	Load + PV	Load + PV	Load + PV
Case-IV	Load + PV + storage	Load + PV	Load + PV
Case-V	Load + PV + storage	Load + PV + storage	Load + PV
Case-VI	Load + PV + storage	Load + PV + storage	Load + PV + storage

ing the operational behavior of electrical transmission and distribution network. LF calculates current and voltage distribution from generation to the consumer on rated power or voltage at the node elements. LC is a load flow calculations with load values varied over time. The investigation was done in the following seven case configurations as shown in **Table 4** with PV and storage in different nodes and the findings are shown in the result section.

4. Results and Discussions

The model was simulated to evaluate the impacts of storage on DT and DN. LF simulation was done in investigating the voltage regulation on the LV side of DT while LC simulation was done in investigating the loading on DT and fluctuation handling capability of storage. The investigation identified the impacts of storage in DT loading, voltage regulation and managing energy fluctuations generated by the PV output. The findings are summarized below.

4.1. Loading on Distribution Transformer

It was found from LC simulation that PV increases the DT loading during day time (especially at 13:00 PM) when load demand was low and when storage was not integrated with the system. However maximum loading was at 20:00 PM when residential load demand is highest and when PV was not able to support the load therefore maximum load on DT remains 87.22% as shown in **Figure 11**. After integrating PV in all 3 nodes, three phases in 3 nodes becomes balanced in loading and supports the load in the morning which lowers the DT loading in the morning at 08:00 AM and afternoon at 18:00 PM as shown in **Figure 11**. However PV increases DT loading during midday (52.06% at 13:00 PM) between 10:00 AM to 16:00 PM also peak load in the evening remains same as 87.22%.

However, after adding storage only in Node-1 and connected to phase-1, storage supports the phase-1 load although which was not enough compared to the total load on phase-1, therefore it was not reflected on overall DT loading. By adding storage in Node-2 and connecting to phase-2 load, it was found that maximum/peak loading on DT now reduced to 80.51% moreover midday loading also reduced to ~42% at 13:00 PM. Gradually storage was added in phase-3 in Node-3 and found that loading on DT reduced not only during midday but also in peak demand time in the evening. **Figure 12** shows the charging and discharging period of storage connected in phase-3 in Node-3.

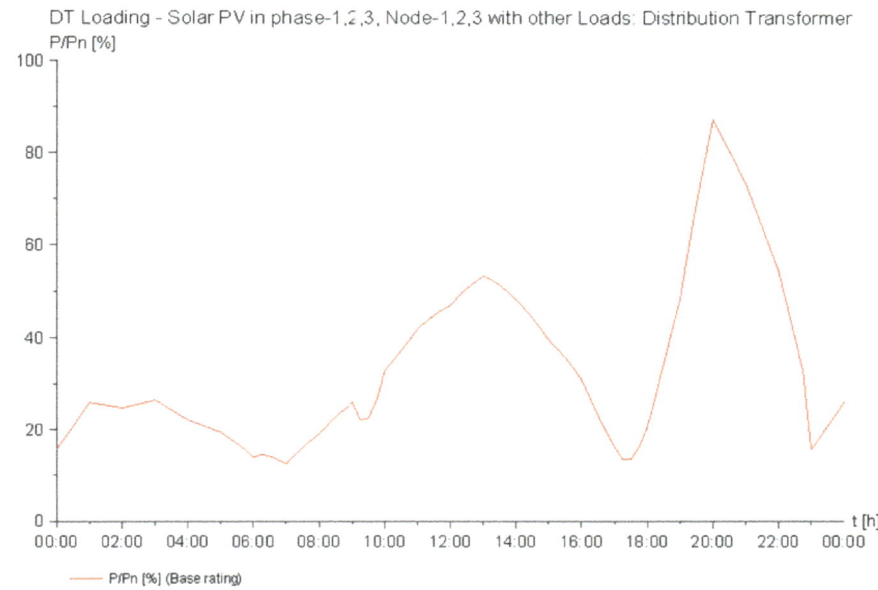

Figure 11. DT loading due to PV in all 3 phases.

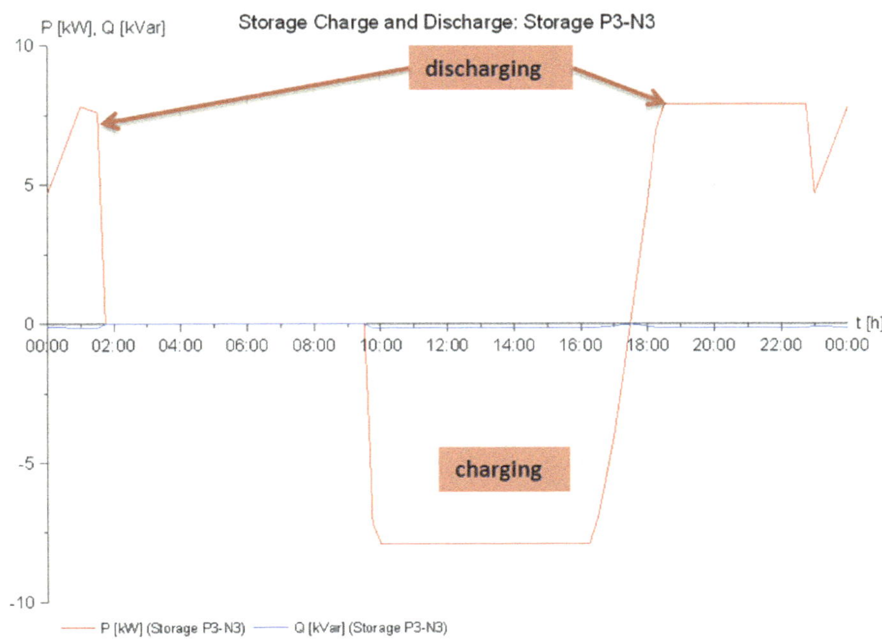

Figure 12. Charging/discharging of storage in Node-3.

After adding storage in all three phases it was found that maximum or peak time loading reduced to 66.76% as shown in **Figure 13** which is a great improvement in RE utilization and also reduces the risk of changing/upgrading the DT capacity. Moreover storage also reduces the load on DT during day time when residential load demand is low particularly during 10:00 AM to 16:00 PM and during this time maximum loading reduced to ~30% of DT capacity at 13:00 PM.

4.2. Voltage Regulation on LV Side of DT

The load flow modeling of unbalanced system shows that voltage level gradually drops from source (Grid point) to the end load. **Figure 14** is the surface plot of phase voltages with 0.5 volt interval and shows how the voltage level changes with load, PV and storage in different cases from LV side of transformer (Tx secondary) to the end node (Node-3). Case-0 shows gradual voltage drop due to load, case-I to case-III shows change in phase voltage due to PV but from case-IV to case-VI shows stable voltage in all phases after adding storage.

In six different case configurations it was observed that PV increases the phase voltages and changes phase angles which increase the risk of phase unbalance. Phase-1 voltage (V1) increases after adding PV in phase-1 in case-I, similarly PV in phase-2 and phase-3.

However, introduction of storage in case-IV, V and VI shows that storage keeps the phase voltage stable as shown in **Figures 15-17**. **Figure 15** shows the percentage change in phase-1 voltage (V1), **Figure 16** shows percentage change in phase-2 voltage (V2) and **Figure17** Shows percentage change in phase-3 voltage (V3) in different case configurations

Phase unbalance condition was observed after adding PV and/or storage in one phase and found that voltage and

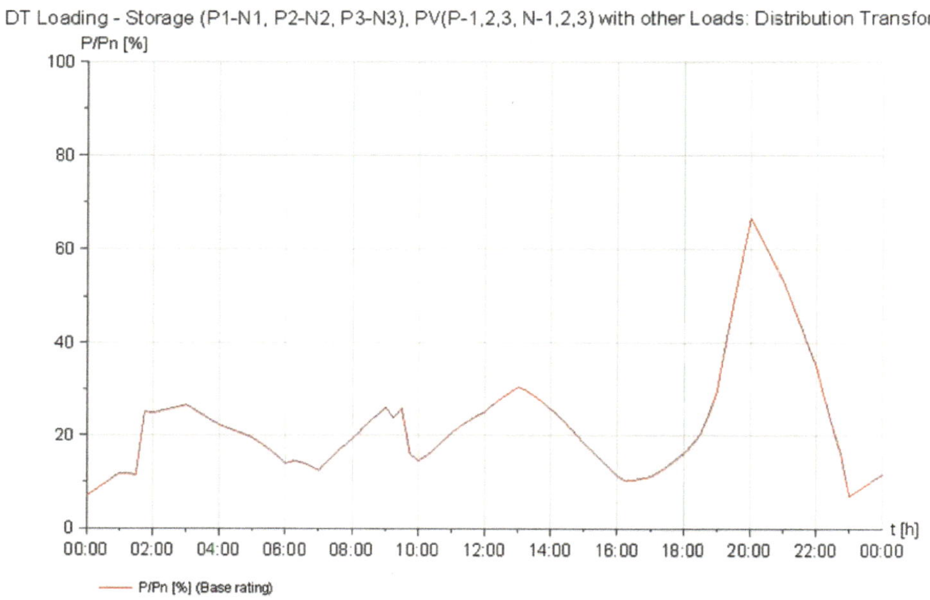

Figure 13. DT loading after adding storage in all 3 phases.

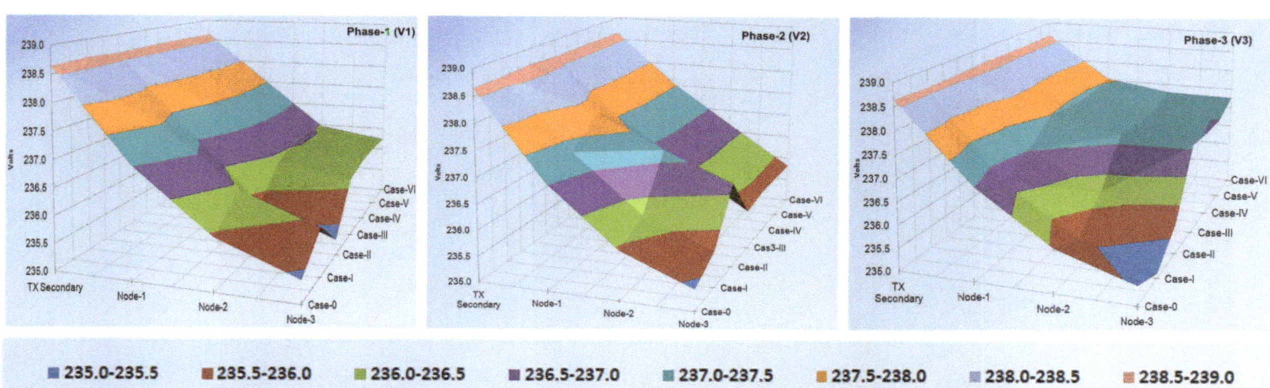

Figure 14. Surface plot of voltage change in 3 phases in different case configurations.

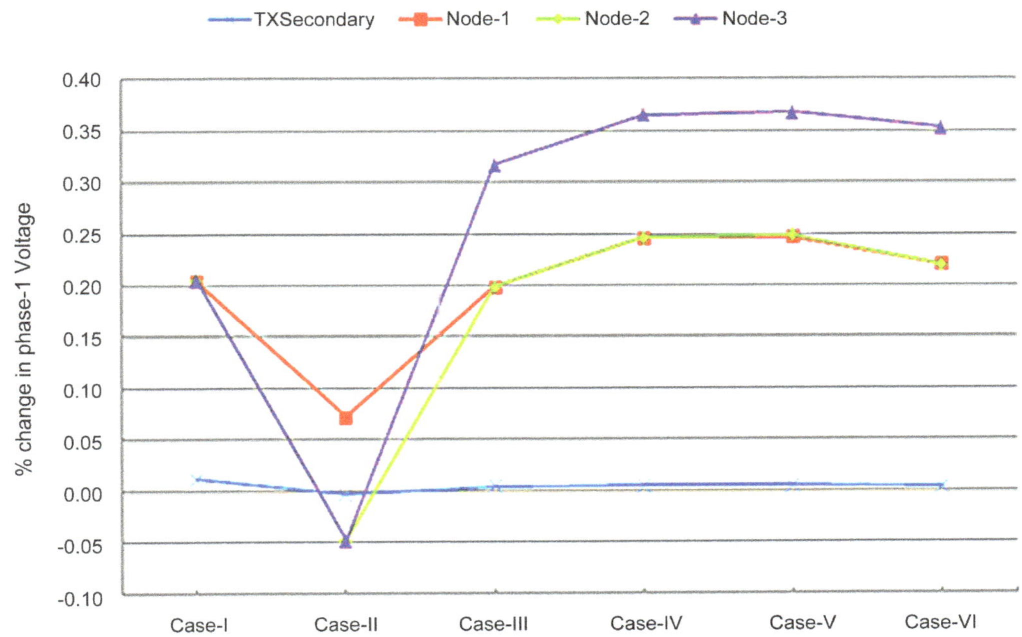

Figure 15. Percentage voltage change in phase-1.

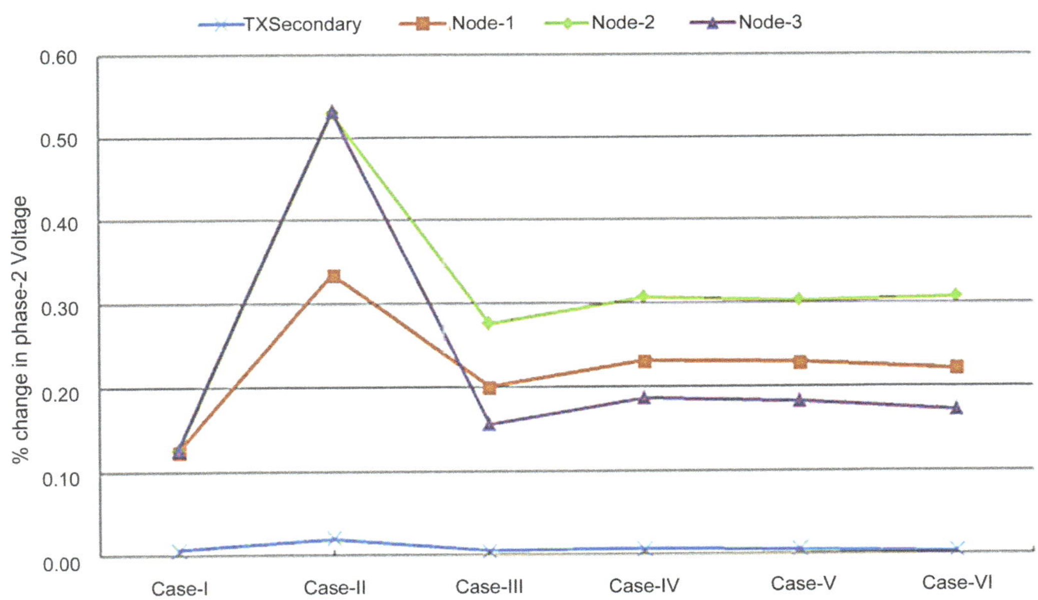

Figure 16. Percentage voltage change in phase-2.

phase angle increases with such installation. **Table 5** shows the result after adding PV and storage only in phase-3 in Node-3.

4.3. Storage Impacts during Solar Fluctuations

Cloud movement or various natural conditions could change the solar radiation profile from ideal type to solar radiation with short fluctuation, short deep fluctuation and long deep fluctuation as shown earlier in **Figures 8-10**. Due to these fluctuations in solar radiation, storage sup-

Table 5. Phase voltage and angle in Node-3.

	Load only	Load + PV	Load + PV + Storage
V1	235.406	236.266	236.550
V2	235.406	234.518	234.212
V3	235.406	236.806	237.240
Φ_{12}	120.0	120.026	120.039
Φ_{23}	120.0	118.529	118.028
Φ_{31}	120.0	121.444	121.931

port the load and DT loading with storage integrated condition (Case-VI) as shown in **Figures 18-20** respectively.

During short fluctuation in solar radiation, PV output also fluctuates and storage supports the load when the PV output reaches below the load demand. **Figures 21-23** show the storage charge by PV output and discharge to support the load while PV output fluctuates due to short, short deep and long deep fluctuations in solar radiation. Short deep fluctuation interrupts storage charging for short

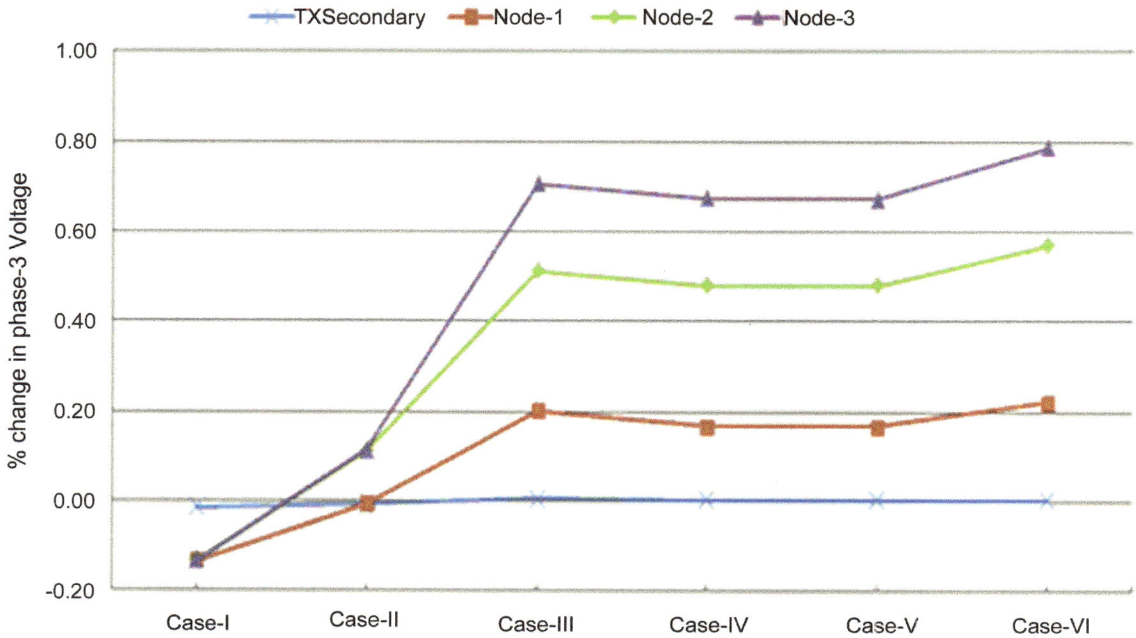

Figure 17. Percentage voltage change in phase-3.

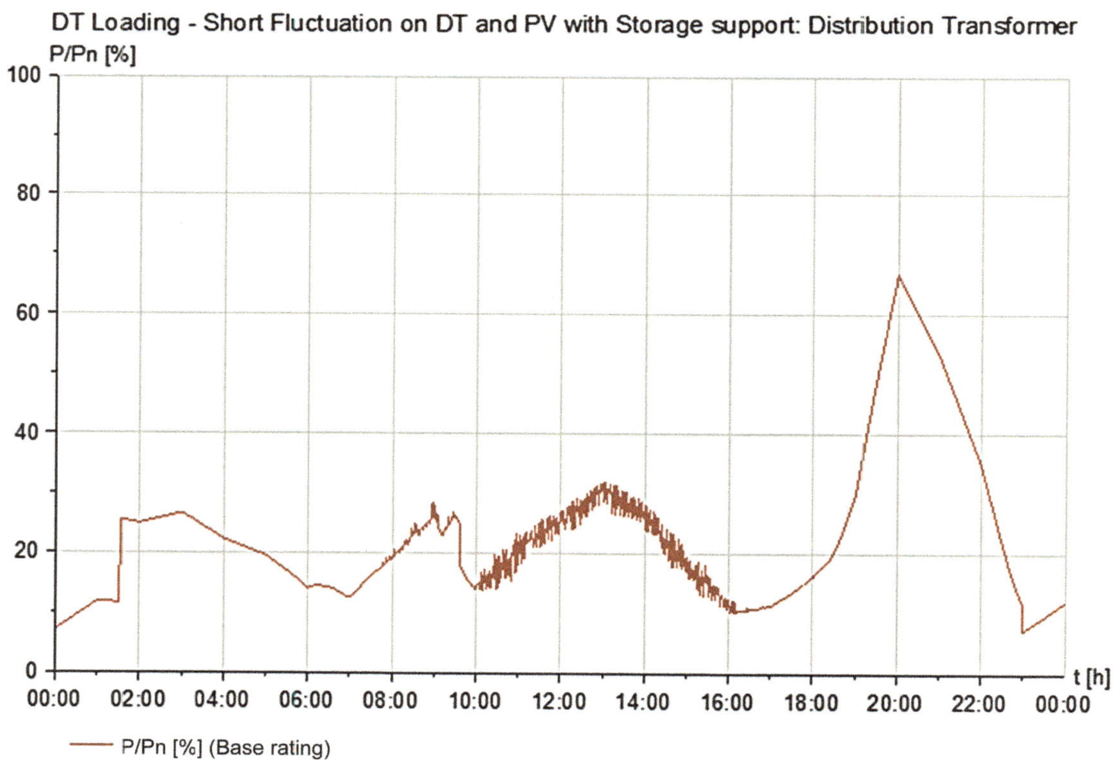

Figure 18. DT loading-PV & Storage integrated when solar radiation continuously fluctuates.

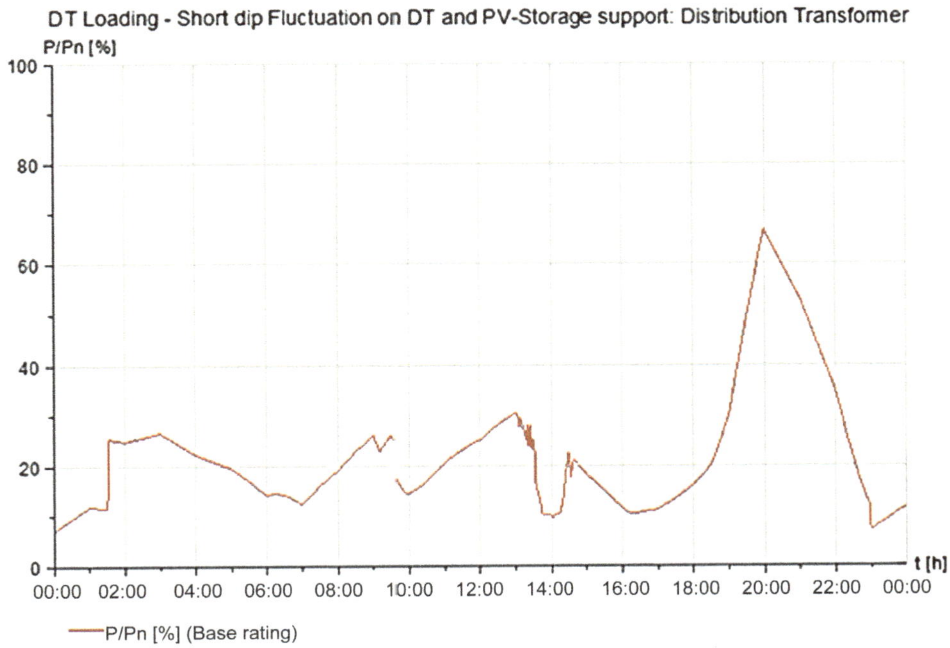

Figure 19. DT loading-PV with storage while solar radiation has short deep fluctuation.

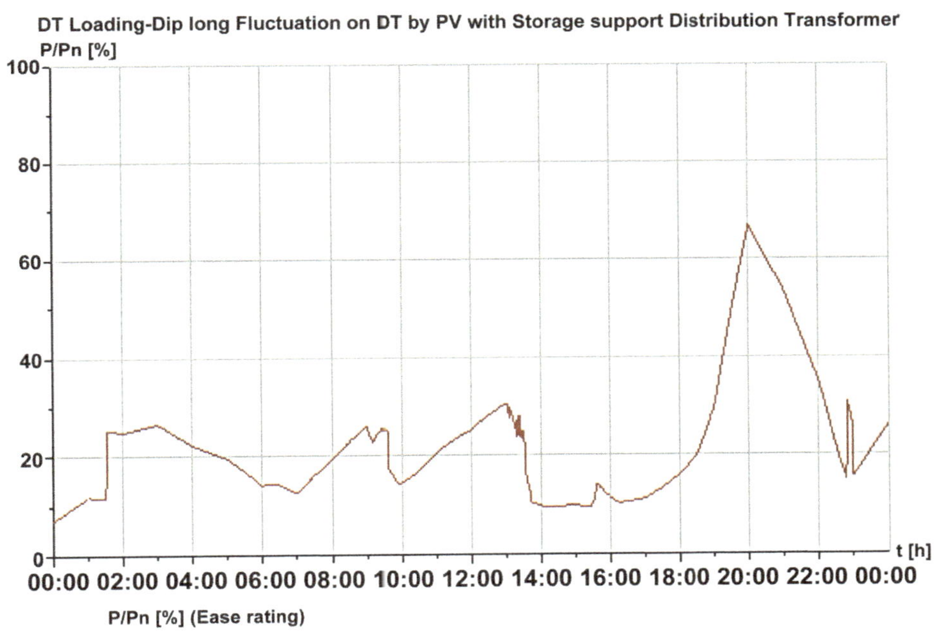

Figure 20. DT loading-PV with storage while solar radiation has short deep fluctuation.

time. **Figure 23** shows the storage charge by PV output and discharge to support the load while PV output being fluctuates due to long deep fluctuations in solar radiation. While the PV output interrupts for long time storage connected in 3 nodes in 3 phases also supports the load by discharging stored electricity as shown in **Figure 23**.

Therefore it is evident from this investigation that, storage effectively reduces the load on DT, improves voltage regulations and supports the load when RE generation fluctuates and reaches lower than the load demand.

5. Conclusions

Energy storage is the key enabler in reliable utilization of RE. This paper investigates how grid connected solar PV and storage behaves with residential load. Investigation

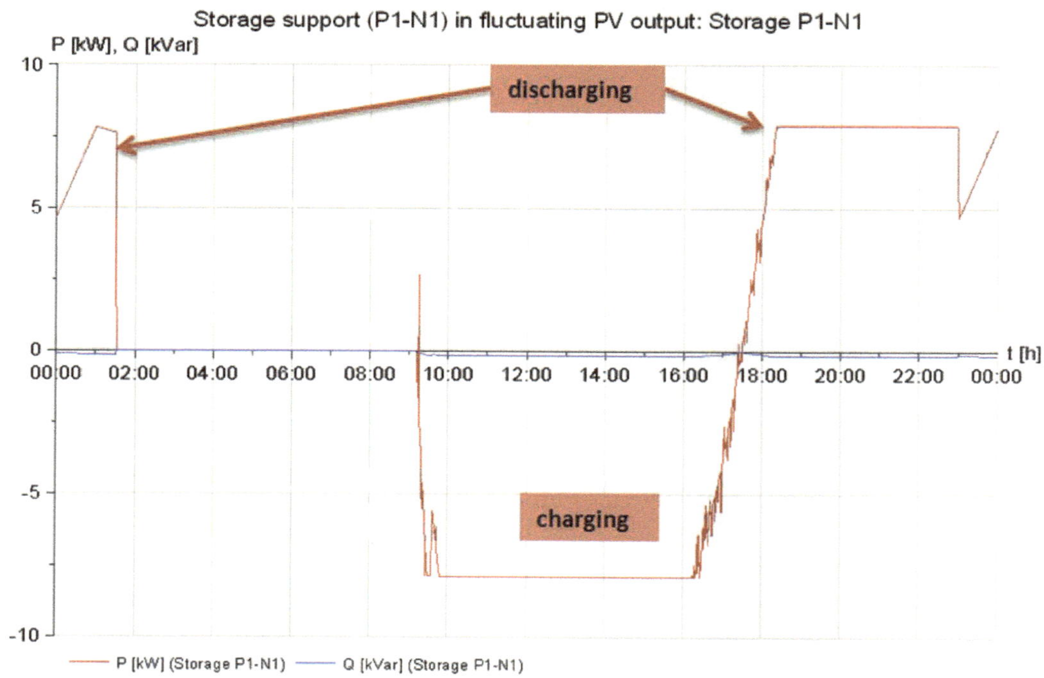

Figure 21. Storage support load when PV output fluctuates.

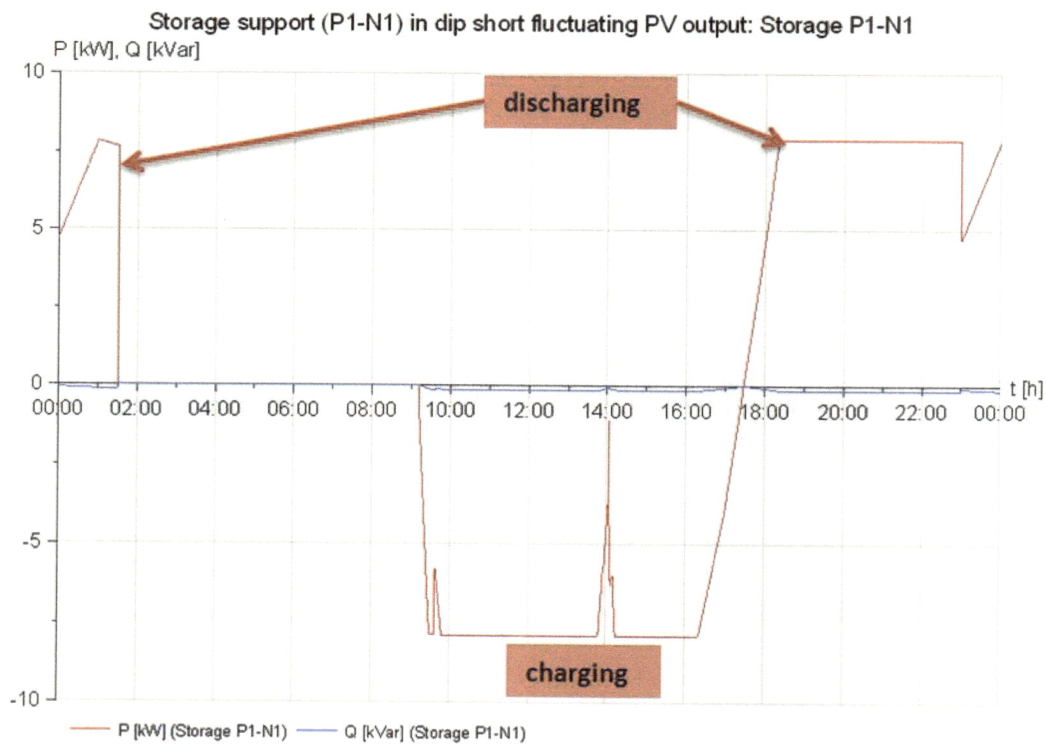

Figure 22. Storage support load when PV output has short deep fluctuation.

was done considering symmetrically distributed residential load on three different nodes from a DT. It was found after adding PV on different phases in different nodes that PV increases load on DT when it is generating maximum energy, however fails to support the load during peak demand time in the evening. PV increases phase voltages while connecting PV in different nodes and also changes the phase angle between phases which can lead

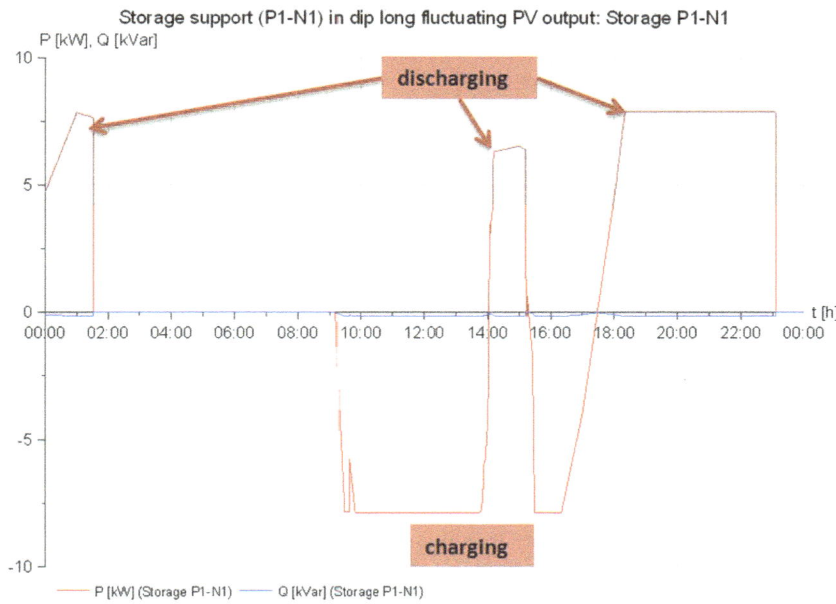

Figure 23. Storage support load when PV output has long deep fluctuation.

to the unbalanced conditions. Moreover fluctuating PV output induces fluctuations to the grid power and introduces power interruption for long deep solar fluctuation situation. After adding storage with the PV integrated loads in different nodes it was found that loading on DT reduces when storage stores excess energy during maximum generation time of PV. It was found that storage reduces more than 20% DT loading at peak demand time by releasing stored energy to support the load demand. Storage improves the voltage regulation by minimizing the unbalanced situation introduced by the PV. Moreover, storage reduces fluctuations from the PV and supports the load when fluctuation or interruptions fall into the load demand time. Therefore the investigation clearly showed the storage impacts on DT and DN by improving the loading on DT, stabilizing voltage regulation and reducing fluctuations introduced by the solar PV, accordingly storage helps in increasing PV utilization.

This study investigates considering only one DT and its associated loads in three phases. Moreover PV and storage was also not integrated with all loads therefore the impacts shown from this investigation can be a major concern for the future in large scale such integration into the grid. This study indicates the future scope of research with larger network and with various load profiles.

REFERENCES

[1] Y. Guo, D. J. Hill and Y. Y. Wang, "Global Transient Stability and Voltage Regulation for Power Systems," *IEEE Transactions on Power Systems*, Vol. 16, No. 4, 2001, pp. 678-688.

[2] J. D. Glover, M. S. Sarma and T. J. Overbye, "Power System Analysis and Design," 4th Edition, Thompson Corporation, Stamford, 2008, p. 255.

[3] ABARE, "Energy in Australia 2010," Australian Bureau of Agricultural and Resource Economics and Sciences, Department of Resources, Energy and Tourism, Australia, 2010. http://www.daff.gov.au/abares/publications_remote_content/publication_series/energy_in_australia?sq_content_src=%2BdXJsPWh0dHAlM0ElMkYlMkYxNDMuMTg4LjE3LjIwJTJGYW5yZGwlMkZEQUZGU2VydmljZSUyRmRpc3BsYXkucGhwJTNGZmlkJTNEcGVfYWJhcmVicnM5OTAxNDQ0NF8xMmEueG1sJmFsbD0x

[4] Queensland-Government, "The Queensland Renewable Energy Plan, a Clear Energy Future for Queensland," 2009. http://www.cleanenergy.qld.gov.au/queensland_renewable_energy_plan.cfm

[5] E. Caamaño, D. Suna, J. Thornycroft, S. Cobben, M. Elswijk, B. Gaiddon, T. Erge and H. Laukamp, "Utilities Experience and Perception of PV Distributed Generation," 2007. http://www.pvupscale.org/IMG/pdf/WP4_D4-2_Utilities-Experience-PVDG_v5.pdf

[6] Y. Makarov, P. Du, M. C. W. Kintner-Meyer, C. Jin and H. Illian, "Optimal Size of Energy Storage to Accommodate High Penetration of Renewable Resources in WECC System," *Innovative Smart Grid Technologies* (*ISGT*), 19-21 January 2010, pp. 1-5.

[7] A. A. Solomon, D. Faiman and G. Meron, "Properties and Uses of Storage for Enhancing the Grid Penetration of very Large Photovoltaic Systems," *Energy Policy*, Vol. 38, No. 9, 2010, pp. 5208-5222.

[8] Ergon Energy, "Distribution Feeder Database 2010," *Collected Information from Ergon Energy*, 2010.

[9] Ergon Energy, "Network Configuration Diagram," 2012. http://www.ergon.com.au/about-us/the-electricity-industr

y/electricity-distribution-pricing-methodologies/distribution-loss-factor-calculation-methodology#content-id-4876

[10] ActewAGL, "ActewAGL Guidelines for Grid-Connected Photovoltaic Installations via Inverter," 2009.

[11] R. H. M. Akatsuka, H. Kita, T. Ito, Y. Ueda and Y. Saito, "Estimation of Battery Capacity for Suppression of PV Power Plant Output Fluctuation," *35th IEEE Photovoltaic Specialists Conference* (*PVSC*), 20-25 June 2010, pp. 000540-000543.

[12] Queensland-Government, "Distributed Generation," Department of Energy and Water Supply, 2012. http://www.cleanenergy.qld.gov.au/distributed-generation/default.htm

[13] M. T. Arif, A. M. T. Oo and A. B. M. S. Ali, "Estimation of Required Storage System for Solar PV and Wind Energy and Feasibility Analysis of Storage Systems," A. Zobaa, Ed., 2013. http://www.intechopen.com/books/energy-storage-technologies-and-applications/estimation-of-energy-storage-and-its-feasibility-analysis

[14] M. T. Arif, A. M. T. Oo and A. B. M. S. Ali, "Investigation of Energy Storage Required for Various Location in Australia," *Central Queensland Regional Engineering Conference* 2012, Queensland, 10-11 August 2012.

[15] H. Nfaoui, J. Buret, A. A. M. Sayigh and P. D. Dunn, "Modelling of a Wind/Diesel System with Battery Storage for Tangiers, Morocco," *Renew Energy*, Vol. 4, No. 2, 1994, pp. 155-167.

[16] A. Zahedi, "Technical Analysis of Electric Power System Consisting of Solar PV Energy, Wind Power and Hydrogen Fuel Cell," *Australasian Universities Power Engineering Conference* (*AUPEC* 2007), Perth, 9-12 December 2007, pp. 1-5.

[17] "Queensland Energy Management Plan," Department of Employment, Economic Development and Innovation, Queensland Government, 2011. http://rti.cabinet.qld.gov.au/documents/2011/may/qld%20energy%20management%20plan/Attachments/Qld%20Energy%20Mgt%20Plan.pdf

[18] Origin Energy, "Household Peak Demand," 2012. http://www.originenergy.com.au/3404/Mt-Stuart-Power-Station

[19] A. Thomas, "Photovoltaic Planning Criteria, Network Planning and Development Distribution Planning and Capability, Ergon Energy, Australia," *Technical Note*, 2011.

[20] IEEE, "IEEE STD 1547-2003, IEEE Standard for Interconnecting Distributed Resources with Electric Power Systems," *Standard*, 2003, pp. 1-16.

[21] Standard Australia, "AS 4777.1-2005, Grid Connection of Energy Systems via Inverters, Part 1: Installation Requirements," 2005. http://www.saiglobal.com

[22] "Network Standard, Standard for Network Performance, Joint standard between ENERGEX and Ergon Energy," *Standard.*

[23] IEEE and Standards, "IEEE STD 1159-1995, IEEE Recommended practice for Monitoring Electric Power Quality," *Standard*, 1995.

[24] Energy Matters, "Average Household Electricity Consumption per Day in Australia," 2012. http://www.energymatters.com.au/climate-data/grid-electricity-usage.php

[25] Bureau of Meteorology, "Weather Data," Bureau of Meteorology, Australian Government, 2011. http://reg.bom.gov.au/

Permissions

The contributors of this book come from diverse backgrounds, making this book a truly international effort. This book will bring forth new frontiers with its revolutionizing research information and detailed analysis of the nascent developments around the world.

We would like to thank all the contributing authors for lending their expertise to make the book truly unique. They have played a crucial role in the development of this book. Without their invaluable contributions this book wouldn't have been possible. They have made vital efforts to compile up to date information on the varied aspects of this subject to make this book a valuable addition to the collection of many professionals and students.

This book was conceptualized with the vision of imparting up-to-date information and advanced data in this field. To ensure the same, a matchless editorial board was set up. Every individual on the board went through rigorous rounds of assessment to prove their worth. After which they invested a large part of their time researching and compiling the most relevant data for our readers. Conferences and sessions were held from time to time between the editorial board and the contributing authors to present the data in the most comprehensible form. The editorial team has worked tirelessly to provide valuable and valid information to help people across the globe.

Every chapter published in this book has been scrutinized by our experts. Their significance has been extensively debated. The topics covered herein carry significant findings which will fuel the growth of the discipline. They may even be implemented as practical applications or may be referred to as a beginning point for another development. Chapters in this book were first published by Scientific Research Publishing Inc.; hereby published with permission under the Creative Commons Attribution License or equivalent.

The editorial board has been involved in producing this book since its inception. They have spent rigorous hours researching and exploring the diverse topics which have resulted in the successful publishing of this book. They have passed on their knowledge of decades through this book. To expedite this challenging task, the publisher supported the team at every step. A small team of assistant editors was also appointed to further simplify the editing procedure and attain best results for the readers.

Our editorial team has been hand-picked from every corner of the world. Their multi-ethnicity adds dynamic inputs to the discussions which result in innovative outcomes. These outcomes are then further discussed with the researchers and contributors who give their valuable feedback and opinion regarding the same. The feedback is then collaborated with the researches and they are edited in a comprehensive manner to aid the understanding of the subject.

Apart from the editorial board, the designing team has also invested a significant amount of their time in understanding the subject and creating the most relevant covers. They scrutinized every image to scout for the most suitable representation of the subject and create an appropriate cover for the book.

The publishing team has been involved in this book since its early stages. They were actively engaged in every process, be it collecting the data, connecting with the contributors or procuring relevant information. The team has been an ardent support to the editorial, designing and production team. Their endless efforts to recruit the best for this project, has resulted in the accomplishment of this book. They are a veteran in the field of academics and their pool of knowledge is as vast as their experience in printing. Their expertise and guidance has proved useful at every step. Their uncompromising quality standards have made this book an exceptional effort. Their encouragement from time to time has been an inspiration for everyone.

The publisher and the editorial board hope that this book will prove to be a valuable piece of knowledge for researchers, students, practitioners and scholars across the globe.

List of Contributors

Murali Bottu and Mariesa L. Crow
Electrical & Computer Engineering, Missouri University of Science & Technology, Rolla, USA

Andrew C. Elmore
Geological Engineering, Missouri University of Science & Technology, Rolla, USA

Stan Atcitty
Wind Energy Technologies, Sandia National Laboratories, Albuquerque, USA

Mustapha Elamouri
Department of Technology, Preparatory Institute for Engineering Studies of Sfax (IPEIS), Sfax, Tunisia

Fathi Ben Amar
Department of Technology, Preparatory Institute for Engineering Studies of Sfax (IPEIS), Sfax, Tunisia
Electric Networks and Machines Research Unit, National Institute of Applied Sciences and Technology (INSAT), Tunisia

Sampath Naveen Kumar, Natarajan Rajalakshmi and Kaveripatnam Samban Dhathathreyan
Centre for Fuel Cell Technology, ARCI, IIT Madras Research Park, Chennai, India

Jian Hu
College of Information and Engineering-Technology, Sichuan Agricultural University, Ya'an, China

Md Rahat Hossain, Amanullah Maung Than Oo and A. B. M. Shawkat Ali
Power Engineering Research Group (PERG), Central Queensland University, Rockhampton, Australia

Moses Koilraj Gnanadason
Mechanical Engineering, PRIST University, Thanjavur, India

Palanisamy Senthil Kumar
Mechanical Engineering Department, KSR College of Engineering, Tiruchengode, India

Vincent H. Wilson
Engineering & Technology, PRIST University, Thanjavur, India

Gajendiran Hariharan
Faculty of Mechanical Engineering, VOC Anna University College of Engineering, Tuticorin, India

Navaneethakrishnan Shenbaga Vinayagamoorthi
Department of Mechanical Engineering, Anna University of Technology, Tirunelveli, India

Mehdi Moradian
Department of Electrical Engineering, Sahand University of Technology, Tabriz, Iran

Faramarz Mahdavi Tabatabaei
Saman Gostar Company (Distributor of SANTERNO, Italy), Tehran, Iran

Sajad Moradian
Department of Electrical Engineering, Iran University of Science and Technology, Tehran, Iran

Nur Mohammad, Muhammad Quamruzzaman and Mohammad Rafiqul Alam
Department of Electrical & Electronic Engineering, Chittagong University of Engineering & Technology, Chittagong, Bangladesh

Mohammad Rubaiyat Tanvir Hossain
Department of Electrical & Electronic Engineering, Bangladesh University of Engineering & Technology, Dhaka, Bangladesh

Md Multan Biswas
Department of Electrical and Electronic Engineering, Bangladesh University of Engineering and Technology, Dhaka, Bangladesh

Md Shafiul Azim and Tonmoy Kumar Saha
Department of Electrical and Electronic Engineering, Khulna University of Engineering and Technology, Khulna, Bangladesh

Umama Zobayer and Monalisa Chowdhury Urmi
Department of Electrical and Electronic Engineering, Stamford University Bangladesh, Dhaka, Bangladesh

Muhammad Qamar Raza, Muhammad Usman Haider and Farooq Sharif
Department of Electrical Engineering, The University of Faisalabad, Faisalabad, Pakistan

S. Muhammad Ali
Department of Electrical Engineering, COMSATS Institute of Information Technology, Abbottabad, Pakistan

Muhammad Zeeshan Rashid
Department of Electrical Engineering, Islamia University of Bahawalpur, Bahawalpur, Pakistan

Yuan Huang, Junyong Liu, Xiaodong Shen and Tingting Dai
School of Electrical Engineering and Information, Sichuan University, Chengdu, China

Irfan Ullah and Seoyong Shin
Department of Information and Communication Engineering, Myongji University, Yongin 449-728, South Korea

Atsushi Yona and Tomonobu Senjyu
University of the Ryukyus, Okinawa, Japan

Mohammad T. Arif and G. M. Shafiullah
Central Queensland University, Power and Energy Centre, Institute for Resources Industries and Sustainability, Rockhampton, Queensland 4702, Australia

Funabashi Toshihisa
Meidensha Corporation, Tokyo, Japan

Chul-Hwan Kim
Sungkyunkwan University, Suwon City, South Korea

Emad M. Natsheh and Alhussein Albarbar
Advanced Industrial Diagnostics Centre, School of Engineering, Manchester Metropolitan University (MMU), Manchester, UK

Xiangyang Zhao and Shiyang Liu
School of Automation Science and Electrical Engineering, Beihang University, Beijing, China

Ebisa Negeri and Nico Baken
Network Architectures and Services, Delft University of Technology, Delft, The Netherlands

Marjan Popov
Electrical Sustainable Energy, Delft University of Technology, Delft, The Netherlands

Bo Wang
Shenzhen Power Supply Bureau Co., LTD, Shenzhen, China

Emad Ahmed Sweelem, Abdel-Shafy A. Nafeh and Faten Hosny Fahmy
PV Cells Department, Electronics Research Institute, Cairo, Egypt

Ahmed Helal, Walid Ghoneim and Ahmed Halaby
Department of Electrical Engineering and Control, Faculty of Engineering and Technology, Arab Academy for Science and Tech-nology, Alexandria, Egypt